TURBOJET
HISTORY AND DEVELOPMENT 1930–1960

This book
is dedicated to the Memory
of our Son, David

TURBOJET
HISTORY AND DEVELOPMENT 1930–1960
VOLUME 1
GREAT BRITAIN AND GERMANY

Antony L. Kay MRAeS

THE CROWOOD PRESS

First published in 2007 by
The Crowood Press Ltd
Ramsbury, Marlborough
Wiltshire SN8 2HR

www.crowood.com

© Antony L. Kay MRAeS 2007

All rights reserved. No part of this publication may be reproduced or transmitted in any form or by any means, electronic or mechanical, including photocopy, recording, or any information storage and retrieval system, without permission in writing from the publishers.

British Library Cataloguing-in-Publication Data
A catalogue record for this book is available from the British Library.

ISBN 978 1 86126 912 6

Typeset by Florence Production Ltd, Stoodleigh, Devon
Printed and bound in Great Britain by TJ International Ltd, Padstow

Contents

Foreword by Captain Eric M. Brown   7
Acknowledgements   8

Introduction   9

PART I   GREAT BRITAIN   11

1. Early Work at the RAE   12
2. Frank Whittle and Power Jets   20
3. Towards Britain's First Jet Aircraft   28
4. W.2 Variants and the Spread of Turbojet Technology   34
5. Bypass and Ducted Fan Projects from Whittle and Power Jets   41
6. Metropolitan-Vickers Electrical Co. Ltd   45
7. The de Havilland Engine Company   50
8. Rolls-Royce   54
9. Meteor: The World's First Operational Jet Aircraft   61
10. Vampire and Venom   68
11. Britain's First Jet Aircraft at War   75
12. Further de Havilland Developments   77
13. Rolls-Royce: Nene and Avon   85
14. Rolls-Royce: Turboprops, Bypass and Lift Engines   105
15. Armstrong Siddeley Motors Ltd   118
16. Bristol   137
17. D. Napier & Son Ltd   158

PART II GERMANY — 167

18	Ernst Heinkel A.G.	168
19	Towards an Official Programme	175
20	Heinkel and the Struggle for a Production Turbojet	178
21	Junkers Flugzeug- und Motorenwerke A.G.	186
22	Bayerische Motoren Werke A.G. (BMW)	198
23	Flight-Testing and Production	209
24	German Turbojets Go to War	211
25	Heinkel-Hirth's 109-011	220
26	Further Junkers Developments	227
27	Further BMW Developments	230
28	Daimler-Benz A.G.	239
29	Focke-Wulf Flugzeugbau GmbH	244
30	Dr-Ing. h.c. F. Porsche KG	245
31	Dispersal and Revival	247

Appendix One
Report on the Whittle System of Aircraft Propulsion (Theoretical Stage) – October 1935 – and Related Documents — 257

Appendix Two
Engine Data Tables — 263

Abbreviations and Acronyms — 267

Bibliography and Further Reading — 268

Index — 269

Foreword

Captain Eric M. Brown, CBE, DSC, AFC, MA, Hon.FRAeS, RN

As a test pilot at the Royal Aircraft Establishment I was fortunate to be heavily involved with the early days of jet flight and to have worked closely with the great jet pioneer, Sir Frank Whittle. I witnessed the first flight of Britain's first jet aircraft and was privileged later to take part in the flight testing of that historic Gloster E.28/39, as well as the wartime German jets, the Arado 234B, Messerschmitt 262 and Heinkel 162.

With this background I felt I knew a lot about early jet history, but this book has opened my eyes. The extent of its geographic coverage of jet developments is comprehensive, and its in-depth commentary on both the technical aspects and political aspects of the history of the turbojets is fascinating. A huge amount of research has obviously gone into producing such a satisfying end result.

The simple, evocative title of *Turbojet* conceals a wealth of information on little-known facts of this momentous invention, which has impacted on the life of everyone on this planet. The search for the Holy Grail of automotive propulsion has made a great story which just had to be written, and how well this has been done.

Captain Eric Brown.

Author's note

Captain Eric Brown is the Fleet Air Arm's most decorated pilot. In his test flying career he has flown a world record 487 basic types of aircraft. He is a past President of the Royal Aeronautical Society, an Honorary Fellow of the Society of Experimental Test Pilots and a Master Pilot of Russia. In 1995 he was inducted into the US Navy's Carrier Aviation Test Pilot Hall of Honor, the only non-American to have received this accolade.

Acknowledgements

When writing a book of this size and complexity, one needs all the help one can get and so I would like to thank the following: Ted Oliver, Ian Whittle, Nigel Eastaway (Russian Aviation Research Trust), Tim Kershaw (Jet Age Museum), Eddie Creek, Ian McKenzie, Ken Dyer, Ken Quimby, Richard Haig (Rolls-Royce Heritage Trust), Alan Vessey (Napier Power Heritage Trust), Cecilie Kay, Dr András Katona (Közlekedési Múzeum, Budapest) and Kimble D. McCutcheon (President, Aircraft Engine Historical Society, Inc.).

Particular encouragement and enthusiasm was received from Ian Whittle, son of Sir Frank Whittle, in the course of many conversations. Without in any way depreciating the help of others, I want to mention a particular debt that I owe to Ted Oliver in Frankfurt, who freely and willingly gave me so much help with research and translation, and read my material whenever he was asked. His help is all the more remarkable when one considers that he was contemplating writing a similar book himself. Enthusiastic and skilful assistance of this order is a rare thing.

Finally, I do hope the reader will find the end result worthwhile and I am always interested to hear from anyone who wishes to add any interesting fact to the story. Personal anecdotes and photographs are particularly valued. Hopefully, new material can be incorporated in a revised edition.

Antony L. Kay,
Selsey, West Sussex, England

Introduction

This history covers the worldwide development and utilization of the aeronautical gas turbine in its turbojet, turboprop and turboshaft forms up to the end of the 1950s. The reason for stopping there is that, by then, designing such engines had moved from the very difficult times of gestation to times when such work became more of a certain science. This was due not only to the vast reservoir of knowledge and experience that been built up but also the computer was evolving into many useful forms to help the designer, whereas before he only had a slide rule and drawing board. The old adage that 'Necessity is the Mother of Invention' was never more true than in the case of the turbojet. Such an engine was the only way to go, once the piston engine and airscrew had reached their limit of performance. Following the end of the Second World War, during which everyone had come to recognize that the aeronautical gas turbine was the future for aircraft, aviation entered into what became known as the Golden Age. There were a great many developments of engines and aircraft going on, of a bewildering variety, often accompanied by ample funding. It was a time of wonderment and excitement, unlike today when superb machines are taken for granted.

What began as an apparently simple idea of jet propulsion became increasingly complex and the brains of physicists and engineers were greatly taxed all over the world in solving many problems. As Lord Hives of Rolls-Royce had promised when told that the turbojet was really quite simple, they would soon design the simplicity out of it! Development of the turbojet, turboprop and turboshaft has been so successful and rapid that engines of great power, efficiency, reliability and ruggedness now make air travel possible for all and not, as before, only for the rich. Military aircraft too have evolved out of all recognition and powerful gas turbines for land, sea and industrial use have been developed from their aero-engine counterparts.

This history aims to give the broadest possible survey of the early development of gas turbine engines and the aircraft that used them. Besides the hardware, the people involved and their exploits are included. In the beginning there was little notion of the vast amount of time, effort and money that needed to be expended on development. Consequently, the work of many engineers was stifled before it came to fruition, and governments and politicians were often unwilling to sponsor development beyond a certain point. There are many instances of such unhappy outcomes in these pages.

However, the pace of development of the turbojet was such that it was soon ahead of airframe design, a circumstance that rarely pertained with piston engines with their longer development times. The power of the turbojet and turboprop was enough to propel an aircraft to higher speeds than its aerodynamics would allow and to higher altitudes than non-pressurized cockpits would permit. Indeed, the power of the early turbojets was often disbelieved: a simple but convincing demonstration at Power Jets involved tossing a house brick into the exhaust of a turbojet to see it fly away like a piece of paper. Respect for the new engines was also needed and there were a number of instances, some fatal, of people and equipment being sucked into the air intakes of turbojets.

In the beginning, there was usually resistance from piston aero-engine makers all over the world to the new forms of powerplant. They were quite happy to go on filling the largely military orders for piston engines. Therefore, it often fell to companies without engine experience to build or help to build the first turbojets and turboprops, such companies being in the fields of heavy engineering, steam turbines and airframes. Examples of these companies are Metropolitan-Vickers, General Electric, AEG and Ganz in heavy engineering and Heinkel and Lockheed in airframes.

The main components of the gas turbine, whether used as a turbojet, turboprop or turboshaft, are the air compressor, combustion chamber and turbine in that order. In general, help in designing compressors came from aerodynamicists and research institutions, while that for designing turbines came from the steam turbine world, the aerodynamicists and the metallurgists. With these components, the main problems to solve were those concerning vibration, lubrication and cooling. As for combustion chambers, possibly the most troublesome element of the gas turbine, the unprecedented combustion intensity required had to be researched from first principles since industry had little experience to offer.

Early pioneers failed to produce practical engines because the efficiencies required of the various components were not recognized and then, when they were, suitable materials (such as heat-resisting metals) and technologies were not to hand. Neither was the large funding required. Finally the pioneers Frank Whittle in Great Britain and Hans von Ohain in Germany kick-started the development of the turbojet. Whittle was a capable engineer initially starved of funds, while von Ohain was a physicist with the backing of a wealthy aircraft company. These visionary men had different aims: Whittle was looking for a means of propelling an aircraft faster and higher (and initially wasted seven years trying to convince unbelievers), while von Ohain was looking for an engine that would give an aircraft a much smoother flight than piston engines had hitherto. In the end, it was derivatives of Whittle's engines that spread all over the world, whereas none of von Ohain's engines had reached production status by the end of the Second World War. The two men later became friends, neither wishing to detract from the other's achievements. Of course, many other pioneers also played their part, such as Herbert Wagner and Max Adolf Mueller in Germany and A.A. Griffith and Stanley Hooker in Great Britain.

This history is arranged as far as possible chronologically, first by companies and then by events. In a book casting as wide a net as this, there will inevitably be mistakes and the author hopes the reader will forgive these. While this first volume deals with the work of Great Britain and Germany, the main countries pioneering aeronautical gas turbine development, this is far from the whole story and the work of the rest of the world is dealt with in the second volume.

Regarding data, figures are given in the metric form followed by the Imperial form in brackets. The movement afoot in Europe to oust Imperial for metric is not approved of by the author. Thrust is given in kp (kiloponds), which is simply thrust in kilograms. One figure is given for specific fuel consumption and denotes kg/kp.hr (or lb/lb.hr). Quite often it has not been possible to specify a figure exactly, for example whether a weight is for a dry engine or not. As for turboprops, shp stands for shaft horsepower and eshp stands for equivalent shaft horsepower (including an allowance for residual thrust).

There is virtually no limit to the power possible with the gas turbine. The turbojet, for instance, is already some hundred times more powerful than the first examples, and further development of its efficiency and reliability will continue. Finally, in reading about the endeavours of the pioneers, let us remember that they all rely to some extent or other on the work of those who went before, a debt acknowledged even by the great Sir Isaac Newton.

PART ONE

Great Britain

In considering the history of the turbojet, nowhere are the roots found more firmly embedded than in Great Britain. Of course, the greatest of the country's pioneers was Frank Whittle but, as we shall see, many other people played their part in the development and took up the baton passed on by Whittle. Long before the most serious, practical work on the turbojet began in the late 1920s, others were putting forward ideas in Britain, largely concerned with the turbine, that may have inspired future engineers.

The embryonic forerunner of the modern continuous combustion turbine can be attributed to John Barber of Nuneaton, Worcestershire, who patented his invention in 1791. His ideas were covered by Patent Specification No. 1833. He proposed that coal, wood or oil be heated externally in a metal retort to produce a gas that was mixed with air in a receiver and then piped to an 'exploder' or combustion chamber. The resulting burnt gases were impinged through a nozzle upon the vanes of a wheel (or turbine) that was geared to other wheels. Some work was mechanically extracted to work pumps, one of which provided compressed air for the combustion process. Barber had the forethought to inject water into the combustion chamber, by means of the other pump, to prevent the pipes and turbine nozzle melting. The resulting steam also augmented the velocity and density of the jet impinging on the turbine.

In 1850 the English inventor W.F. Fernibough patented a type of gas turbine that was also to operate on the products of combustion from wood or coal. Again, steam was to be mixed with the gases for cooling purposes and the air for combustion was to be supplied under pressure. The resultant steam and hot gas was to be impinged upon a turbine wheel from a tapered nozzle.

Naturally, great inefficiencies in these types of machines would have meant all the work would have been used for compressing air and little of use, if anything, would be left over. Although only two inventors have been mentioned above, a great many patents were awarded in Britain alone to inventors of turbines, there being more than 200 during the eighteenth century for example. Thus, although broad concepts were well known, nobody had produced any useful work from reaction turbines, the Devil, as they say, being in the detail.

In 1884, the thirty-year-old genius Charles A. Parsons, an engineer highly skilled in theory and practice, produced the first turbine, albeit steam driven, with a really useful power output. His pioneering work on the steam turbine also influenced later gas turbines and thence the turbojet. Also in 1884, Parsons patented a machine having the essential features of the modern gas turbine. Extracts from his patent, No. 6735, are illuminating. He observes that his (multi-stage) turbines, if driven, can become pumps:

> and can be used for producing a pressure in a fluid ... Such a fluid pressure can be combined with a multiple motor according to my invention so that the necessary motive power to drive the motor for any required purpose may be obtained from fuel or combustible gases of any kind. For this purpose I employ the pressure producer to force air or combustible gases into a closed furnace of any suitable kind ... into which furnace there may be introduced other fuel (liquid or solid). From the furnace the products of combustion may be led, in a heated state, to the multiple motor, which they will actuate. Conveniently, the pressure producer and multiple motor can be mounted on the same shaft, the former driven by the latter.

Of course, the 'pressure producer' and 'multiple motor' were what we now call an axial compressor and multi-stage turbine respectively. Parsons realized that the gas turbine cycle was a long way off, awaiting the development of suitable heat-resisting materials, but he did consider 'water or other fluid to cool the blades'.

In 1901 Parsons patented an axial-flow compressor and constructed an experimental model of it, a measure of how skilful he was with his hands. Several other axial compressors followed, for industrial applications, but, owing to their low efficiencies, further development was dropped in favour of the centrifugal compressor. Other pioneers are discussed in the sections devoted to their own countries.

While it is by no means the sole arbiter of advancing technology, Great Britain can claim many firsts in the annals of the turbojet and turboprop: the first operational jet aircraft (the Meteor); the first turboprop-powered aircraft; the first true jet airliner (the Comet); and the first turboprop passenger aircraft (the Viscount). There are also many firsts in the engines themselves, which will be discussed later. As in most other countries, however, where government funding was necessary to support development the politicians often misunderstood the technicalities and advantages to be gained, and let the engineers and the country down with unhelpful policies. Probably the most poignant example of this was the lack of adequate support given to the country's foremost turbojet pioneer, Frank Whittle. There are many other examples connected with both engine and aircraft programmes. Nevertheless, Britain did kick-start turbojet technology for most of the world and its engines and jet aircraft were sold widely.

Pioneering work on British gas turbines for aeronautical use began with the work of Dr A.A. Griffith at the RAE using axial compressors and with the work of Frank Whittle using centrifugal compressors. These separate lines of development are discussed next.

CHAPTER ONE

Early Work at the RAE

In 1920 Dr Stern at the Air Ministry Laboratory in South Kensington, London, was asked by the Aeronautical Research Committee (ARC) to write a report on the expected problems to be met in developing a 1,000hp gas turbine to drive an airscrew. Stern concluded that the idea was impractical on the grounds that such an engine would be too large and heavy and that its fuel consumption would be uneconomic. Furthermore, he doubted that suitable heat-resisting metal could be developed.

A.A. Griffith

Alan Arnold Griffith, senior scientific officer of the Physics and Instruments Department of the Royal Aircraft Establishment (RAE), Farnborough, studied Stern's report and recommended that the National Physical Laboratory (NPL) should research heat-resisting steels while he studied the problems of compressors and turbines. Griffith, son of a journalist, had studied mechanical engineering at Liverpool University and had, by 1921, received a doctorate. He had begun work at the RAE on the workshop floor in 1914 when the establishment was known as the Royal Aircraft Factory. (Confusion between its initials and those of the Royal Air Force, established in 1918, prompted the change to RAE.)

In July 1926 Griffith wrote his classic report 'An Aerodynamic Theory of Turbine Design'. Leading on from the comprehensive theories of Prandtl on lift and drag, Griffith proposed that changes in velocity and pressure generated by axial compressor blades should be reactions to the blade (or aerofoil) shapes. At the end of his report, Griffith suggested the use of a single-shaft turbine engine with a multi-stage axial compressor to drive an airscrew via reduction gearing. At a conference held at the RAE in October 1926 Griffith put his proposals to a small committee from the Air Ministry and the RAE. Those present were unanimously in favour of carrying out preliminary investigations to verify the theories. The Air Ministry and the ARC then agreed to some initial research and work began in 1927.

The first experiments at the RAE were on compressor and turbine blade designs using cascades of aerofoils. During 1929 a small test rig was built, under Griffith's supervision, consisting of a single-stage axial compressor and a single-stage turbine. Despite the fact that the casing for these had an internal diameter of only 102mm (4in) and blade heights of a mere 13mm (½in), a highly encouraging overall efficiency of 91 per cent was achieved during tests in 1929. To develop suitable aerofoil sections for his blades, Griffith later made use of his earlier work on airscrews at the RAE, for which he had achieved efficiencies of about 82 per cent. However, the results of these tests were not available in time to incorporate into the compressor/turbine test rig.

The cascade wind tunnel developed at the RAE was extremely important to the development of axial compressors and turbines. It enabled rows, or cascades, of blades to be readily set up and tested, in complete contrast to the considerable time and work involved in making up rotors and stators of blades for revolving tests. The experimental research workers Harris and Fairthorne carried out the cascade tests.

In November 1929 another report from Griffith to the ARC concluded that a gas turbine driving an airscrew would be lighter, smaller and more efficient than a contemporary piston engine setup. By then he had been promoted to principal scientific officer at the Air Ministry Laboratory in South Kensington, where he continued to develop his axial-flow compressor ideas. The turboprop he proposed was of about 500hp and very complex. This curious form of contra-rotating, contra-flow gas turbine was actually a hot gas producer and could not directly deliver any mechanical power. It consisted of a row of fourteen discs, free to spin on a fixed, central shaft,

Dr A.A. Griffith of the RAE carried out pioneering work on the axial compressor and, as early as 1926, advocated the axial-flow turbojet. His work was later continued at Rolls-Royce, where Ernest Hives urged him to 'Carry on thinking'.

each disc carrying outer turbine blades and inner compressor blades. There were no stator blades. The turbine blades drove the inner compressor blades that fed air to a combustion chamber, having a rotating burner, at one end. The gas flow, thus heated, was turned through 180 degrees to enter the outer, turbine section and then the exhaust outlet. (Presumably, from there it would have entered a free turbine geared to an airscrew.) Each disc of turbine and compressor blades rotated in the opposite direction to its neighbour and labyrinth seals were used, but a major problem with the scheme was that of preventing hot gases leaking from the turbine to the compressor section. If this happened it would lower

12

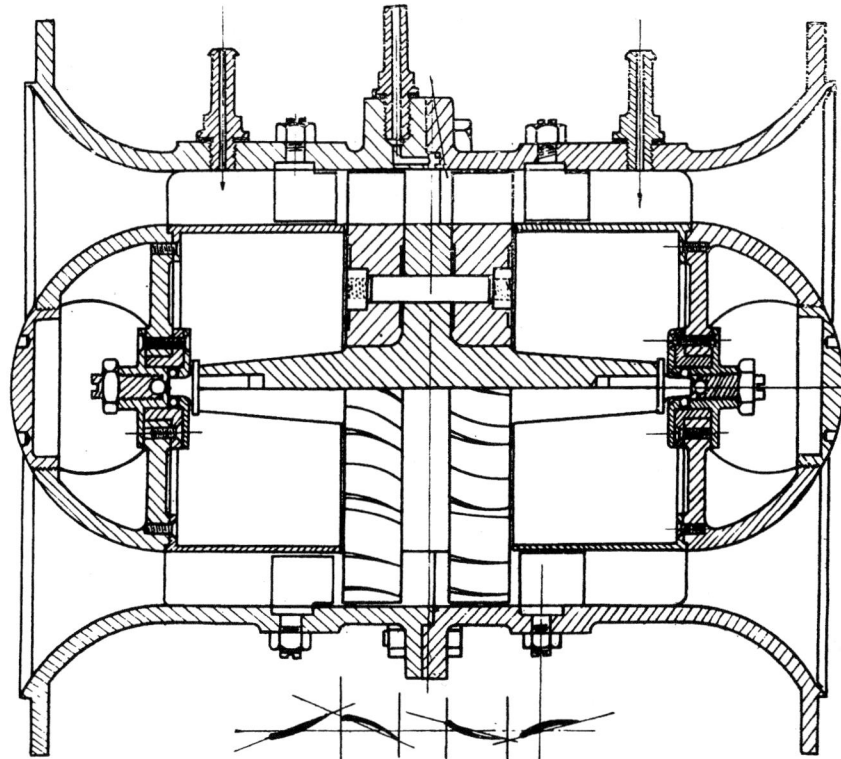

Drawing and photograph of the RAE 1929 model turbine.

RAE to look again into the business of adopting the gas turbine for aeronautical use.

Hayne Constant

Construction began on an eight-stage axial-flow test compressor, designed by Griffith and Hayne Constant and finally completed in 1938. Tizard had persuaded Constant to return to the RAE to assist Griffith and Baxter with gas turbine research, after a period spent lecturing at Imperial College, but some nine years had been wasted in which little work towards a gas turbine was done. According to Constant, 'what was believed possible in 1936 could equally well have been done earlier. All that had changed was man's outlook.'

Constant produced a report by March 1937, a month before Whittle's first engine ran, entitled 'The internal combustion turbine as a prime mover for aircraft' (RAE Note E.3546). This report was highly persuasive and the Engine Sub-Committee of the ARC vigorously recommended developing the gas turbine to drive an airscrew. Constant had surmised that a turboprop could be built that would equal or exceed the performance of then-current piston engines, except near ground level. He concluded that developments in heat-resisting metals and better compression ratios would increase the superiority of the

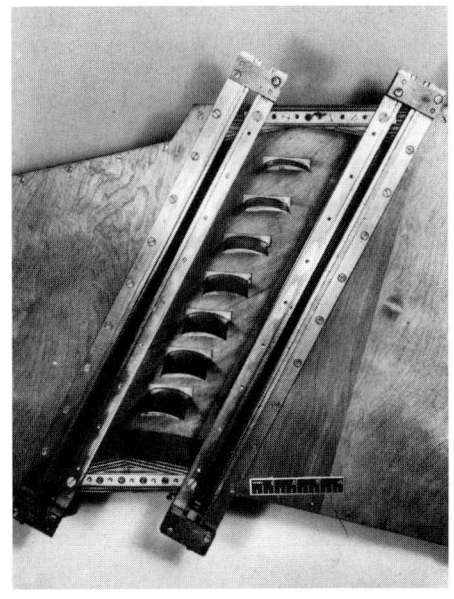

Compressor blade aerofoils arranged for RAE cascade tunnel.

the efficiencies of the components, which would, in turn, aggravate the leakage problem. There were also aerodynamic and mechanical difficulties to be overcome.

Griffith proposed building an experimental unit on the above lines but, in April 1930, the ARC concluded that the existing state of knowledge did not predict that the gas turbine would be superior to the piston engine. In 1931 Griffith moved back from South Kensington to the RAE at Farnborough, but it was not until July 1936 that the ARC authorized the RAE to proceed with Griffith's turboprop scheme. This rekindling of interest in gas turbine research was due to the awareness of Sir Henry Tizzard, the new chairman of the ARC, of the formation of Frank Whittle's Power Jets company and his perception that it would be as well for the

gas turbine and, additionally, he stressed its simplicity over the piston engine.

Early axial compressors and gas turbines

By 1937 there were alloys available with quite good high-temperature properties, for example Hadfield's ERA/ATV, which had a useful creep strength up to 700°C. Thus, the RAE was confident in designing the turbine end of an engine and therefore considered the first priority was to develop an axial compressor of good efficiency. There was little previous work to draw upon. Charles Parsons had designed and operated a number of axial compressors from 1904 onwards, but they were based on 'reversing' turbines and had stalled blading of low efficiency. Some data could be drawn from previous RAE work on turbo-compressors for piston engines.

The eight-stage axial test compressor mentioned above, designated 'Anne', was of 152mm (6in) diameter. Unfortunately, due to a faulty oil seal, it overheated on its very first run and the blades were stripped off within 30 seconds. This compressor was then rebuilt and was running by October 1938. This time the blades had greater clearances and were designed according to the ideas of Jacob Ackeret after Griffith had visited the Brown Boveri company in Switzerland in 1937. From then on, RAE compressors used 50 per cent reaction blading, whereby the pressure rise was shared equally between the rotor and stator blades. This made for the most efficient compressor. Experiments with 'Anne' continued until it was destroyed during a German air raid by Ju 88 bombers of the Luftwaffe's KG 54 on 13 August 1940, otherwise known as Adler Tag ('Eagle Day'), when the Luftwaffe's all-out air assault on Britain began. Although the optimum pressure ratio of 'Anne' was only 1.5:1, this was at an efficiency of 82 per cent. The team of brilliant aerodynamicists working under Griffith included A.R. 'Taffy' Howell and D. Carter, who 'wrote the bible' on axial compressor design.

The next step was to build a turbojet engine but, although the RAE had some workshop facilities, it was hampered by not having the capacity to carry out any large-scale manufacturing in its research work. It was therefore arranged that the Metropolitan-Vickers Electrical Company (MV) should carry out detail design and manufacture of test engines. Many engine arrangements were studied and those with a centrifugal compressor, under Scheme A, were soon abandoned. Engines with an axial compressor, under Scheme B, were selected for development as providing less frontal area and bulk and promising a higher efficiency.

The first requirement was for an engine to develop power on the test rig in order to demonstrate the practicability of the gas turbine. Accordingly, no attempt was to be made initially to minimize weight or bulk, in order to avoid unnecessarily prolonged development. Compactness, lightness and efficiency could be striven for later.

There was also a fear of trying to develop too high a pressure ratio in a single compressor, not least because of difficult starting. The idea was therefore mooted of a compound engine having two mechanically independent compressors, one low-pressure followed by a high-pressure one, each driven by its own turbines. Study of a large number of arrangements included a double compound engine with a power turbine in parallel to the compressor turbine. (These ideas presaged the modern twin-spool, bypass turbojets with their low-pressure and high-pressure sections.) However, the mechanical complexity of the concentric shafts could not be faced at that time at the RAE and this scheme was abandoned at an early stage. It is a matter of debate whether or not the RAE should have been influenced by mechanical complexities in its decision.

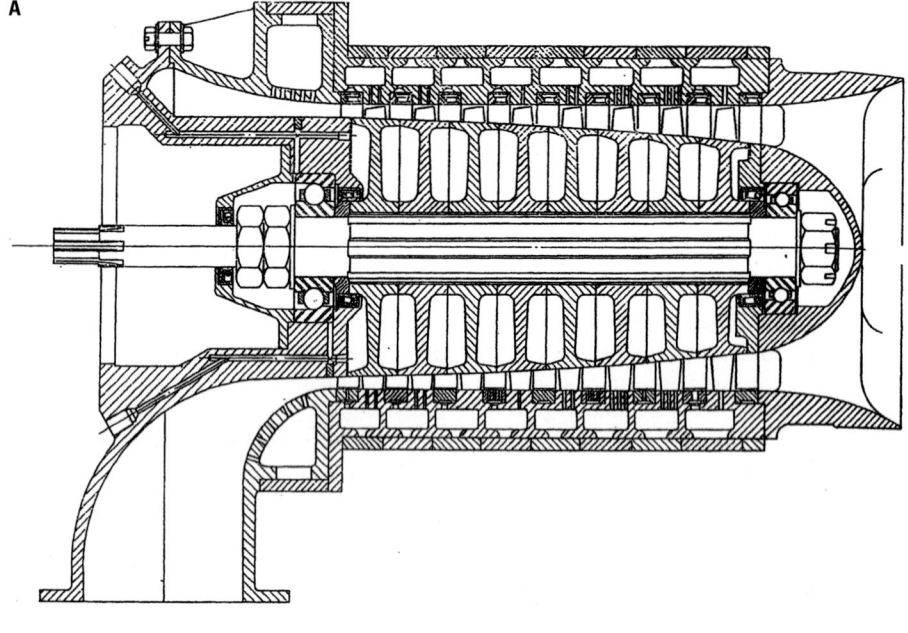

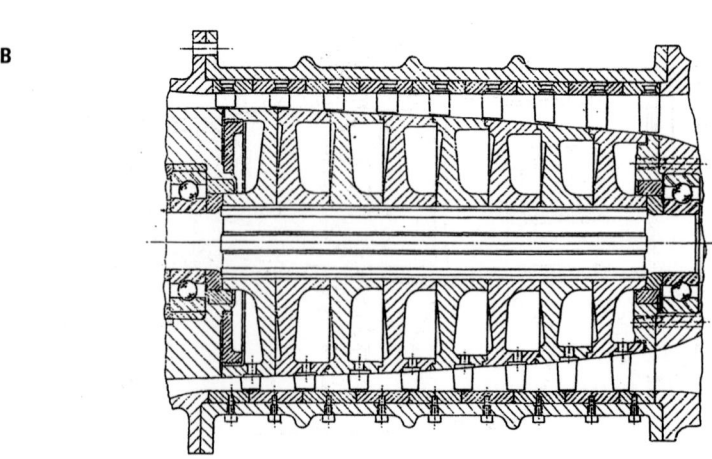

RAE 'Anne' 8-stage axial compressor: (A) as first constructed; (B) as tested.

RAE 'Anne' 8-stage experimental axial compressor.

There then arose the idea of dispersing the low-pressure and high-pressure elements of the engine, with a separate power turbine, as shown in the drawing. This made the mechanical side much simpler but at the expense of the air and gas flow requiring twelve right-angled bends. It was then decided not to fix the final arrangement of the elements until tests had been made on a single turbo-compressor unit, which would confirm the soundness or otherwise of the mechanical details.

The unit, known as B.10 'Betty', was therefore constructed and this represented the high-pressure section of the complete dispersed compound engine. 'Betty' was designed to give sustained running but no useful surplus power. It featured a nine-stage axial compressor of 0.47m (1ft 6½in) diameter and a four-stage turbine, both built onto drum rotors. The idea of a drum rotor (instead of disc construction) was to match its expansion and contraction more closely with that of the casing of the engine. The turbine rotor also had double cone end pieces to relieve thermal expansion stresses. In those days, there was considerable worry over blade fouling and other problems expected from thermal expansion and contraction of the engine's components. At one time even a very complex scheme of a water-cooled turbine and rotor was considered, but air cooling was later seen to be very satisfactory.

Betty's compressor and turbine were tested separately, the latter with steam. A combustion chamber was developed and the complete unit was finally assembled and tested in October 1940. The only problems encountered were bearing failures and, although their water cooling was successful, simpler air cooling was later used instead. The turbine rotor, running red hot at high speed with an inlet temperature of 675°C, remained free from distortion.

By now it was clear that aerodynamic losses due to elbows and volutes could not be tolerated and that, for aircraft use, mechanical complication was unavoidable in order to obtain compactness. Therefore, coaxial machinery was necessary in order to obtain a suitable compound engine. Next it was questioned if, in fact, a compound engine was necessary just to get easy starting, since pressure ratios of only 5:1 were thought to be adequate for the immediate future. On the other hand, the RAE was aware that higher pressure ratios would be needed in the more distant future. Such were some of the complexities of design decisions at that time.

The steam turbine company of C.A. Parsons Ltd was also drawn into this research work. In 1939 it completed and tested an eight-stage axial compressor designated 'Alice', which had a similar

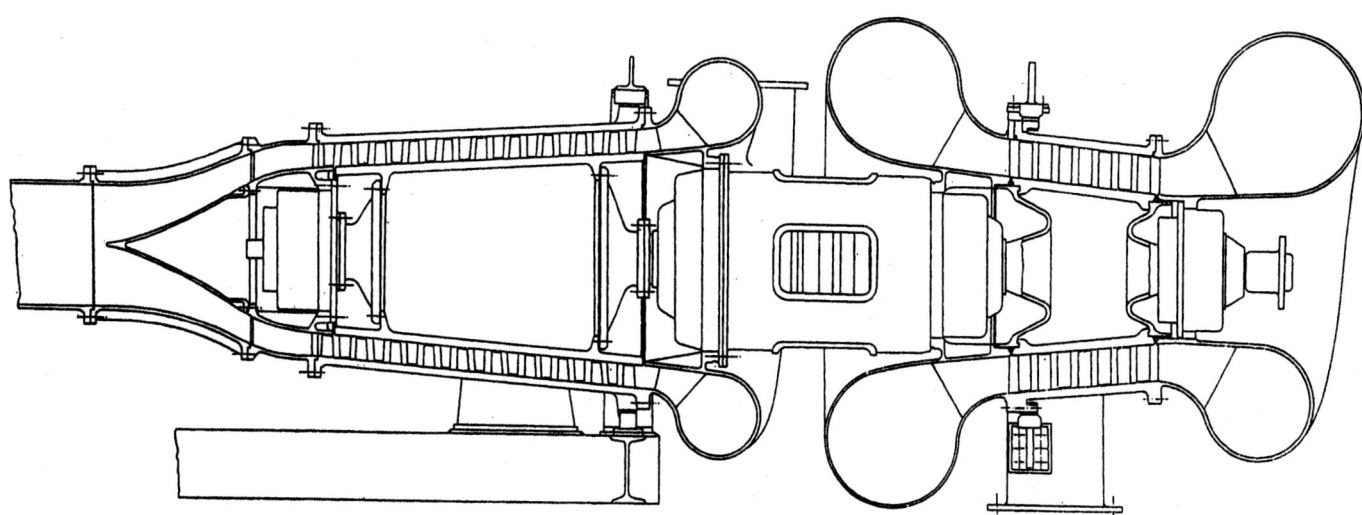

RAE B.10 'Betty' turbo-compressor. Note double-cone end pieces of turbine rotor to relieve thermal expansion stresses.

EARLY WORK AT THE RAE

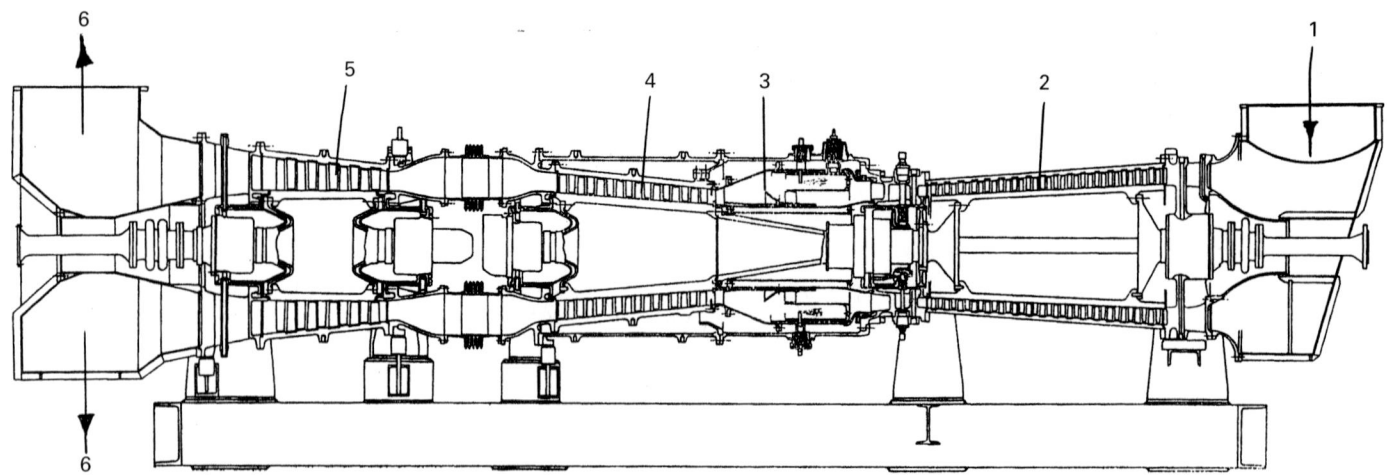

RAE D.11 'Doris' turbine engine: (1) air intake; (2) 17-stage axial compressor; (3) annular combustion chamber; (4) 8-stage HP turbine; (5) 5-stage LP turbine; (6) exhaust ducts.

ABOVE: RAE D.11 compressor with rig-test ducts; LEFT: D.11 HP turbine rotor.

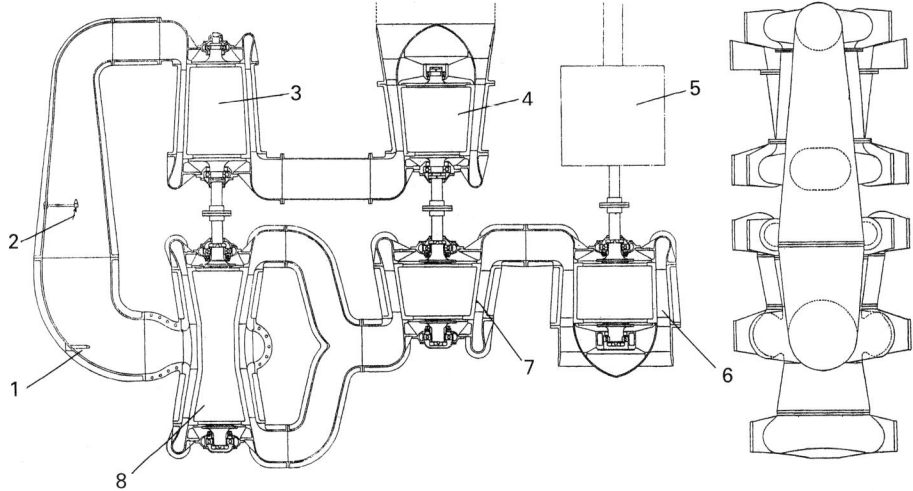

Arrangement of RAE Dispersed Compound turbine engine: (1) starting air nozzle; (2) fuel nozzle; (3) HP compressor; (4) LP compressor; (5) gearbox; (6) power turbine; (7) LP turbine; (8) HP turbine.

pressure ratio to 'Anne' and 'Betty' but an efficiency of 85 per cent.

By spring 1939 it was concluded that the desirable pressure ratio of 5:1 could be obtained and so several designs of a turboprop were made. They projected an axial compressor, annular combustion chamber and two turbines, one to drive the compressor and the other an airscrew. A design designated D.11 'Doris', aimed at 2,000shp, was selected for development. It was a very long, complex engine and comprised a seventeen-stage axial compressor, an annular combustion chamber, an eight-stage high-pressure turbine to drive the compressor and a five-stage low-pressure free turbine to drive an airscrew. This five-stage turbine was fed by gas from the exhaust of the high-pressure turbine. Air was drawn into the compressor through a duct at right angles to the axis and a shaft for starting and driving the auxiliaries

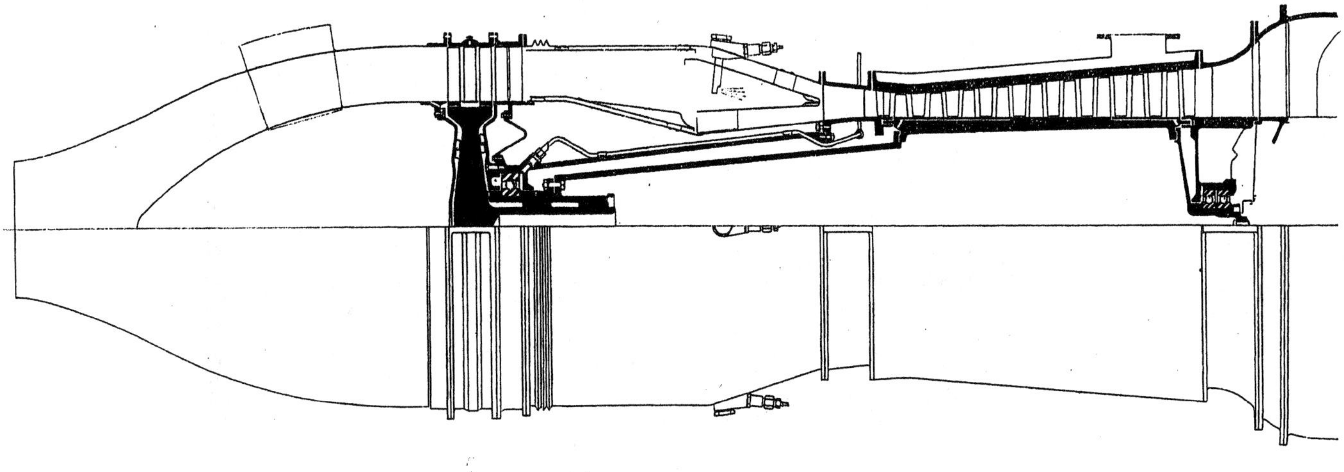

Arrangement of RAE turbojets: (A) F.1; (B) F.1A.

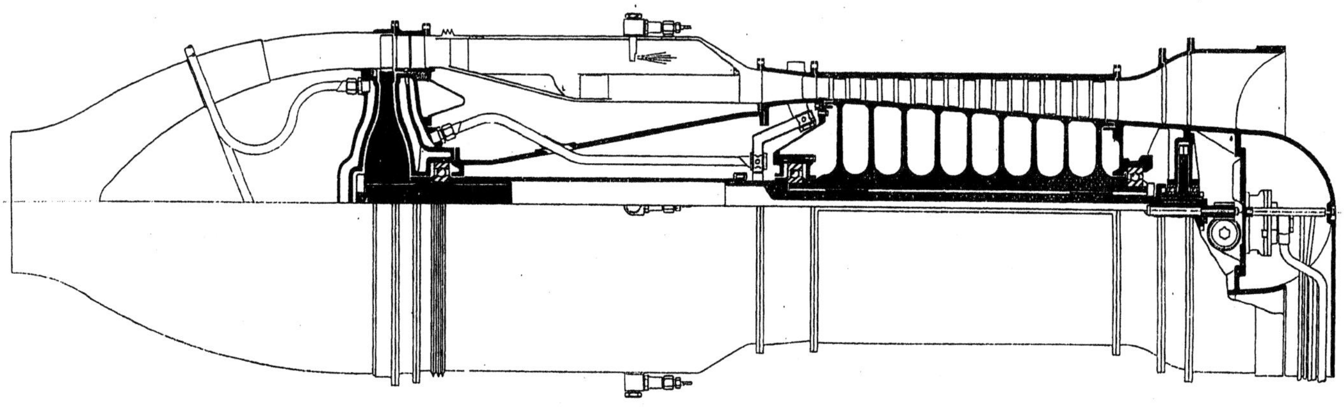

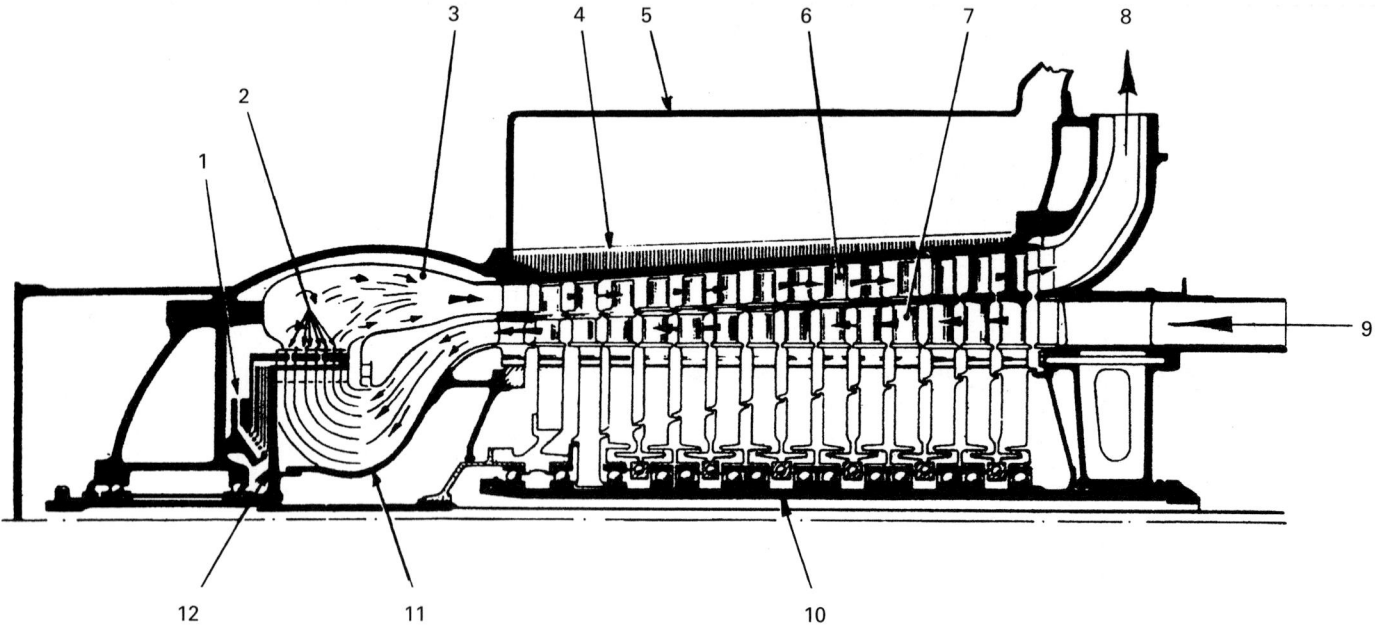

Rolls-Royce Griffith C.R.1 contra-flow engine: (1) fuel inlet; (2) fuel spray; (3) combustion chamber; (4) finned aluminium alloy turbine casing; (5) cooling air casing; (6) turbine section of disc; (7) compressor section of disc; (8) exhaust outlet; (9) air inlet; (10) stationary shaft; (11) stationary member; (12) rotating burner.

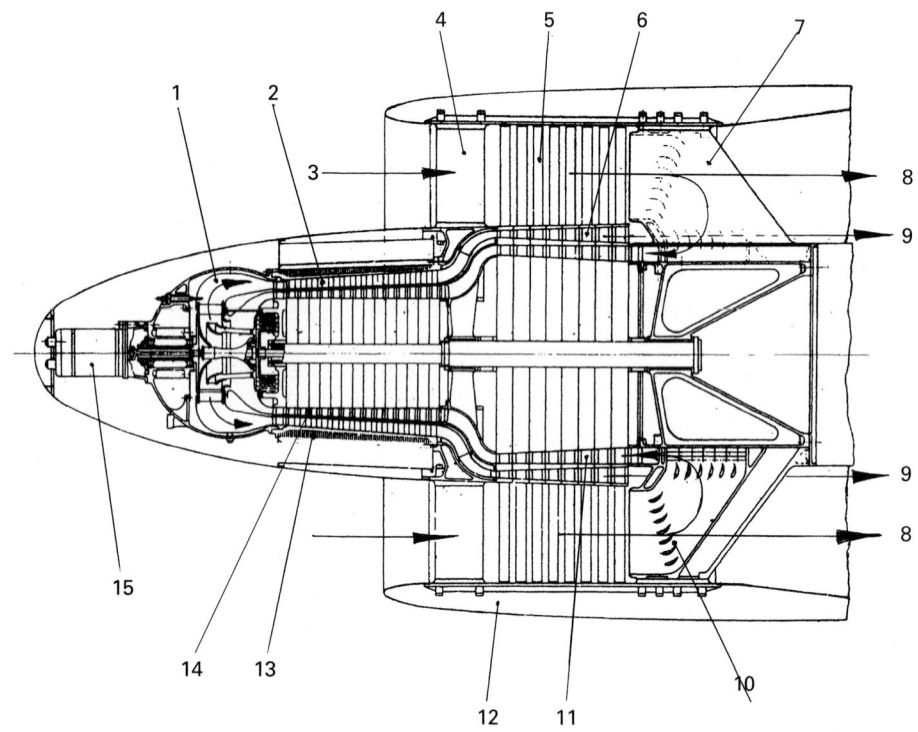

Griffith's contra-rotating, contra-flow, ducted-fan turbojet project: (1) combustion chamber; (2) HP turbine (contra-rotating); (3) annular air intake; (4) support vane; (5) 8-stage ducted fan; (6) LP turbine; (7) hollow vane air channel and support; (8) cold air discharge; (9) hot exhaust gas; (10) cascade vanes; (11) LP axial compressor; (12) fan duct; (13) cooling fins; (14) HP axial compressor (contra-rotating); (15) starter motor.

passed through this duct. Similarly, at the opposite end, another right-angle duct carried away the exhaust gases and the shaft for the airscrew passed through this duct.

Construction of the D.11 'Doris' began in 1940 but at a slow pace, partly due to the limited manufacturing facilities at the RAE and partly due to doubts about the all-axial engine in the light of Frank Whittle's success with the centrifugal engine (see below). As a result, only the compressor of the D.11 was tested (in 1941) and was found to suffer from Mach number effects and needed redesigned blading. This compressor had a pressure ratio of 4:1 for an efficiency of only 78 per cent (it was not until 1944 that a high-speed wind tunnel for cascade tests was available at the RAE, as opposed to the earlier low-speed one, enabling compressibility and Mach number effects to be studied). Later in 1941 it was decided that the D.11 engine design was out of date and it was abandoned.

Metropolitan-Vickers involvement

In any event, with the advent of war in September 1939 a jet propulsion project, which had been proceeding in parallel with D.11, came to the fore and then

Components of the RAE 'Ruth' 6-stage axial compressor, built by Fraser and Chalmers.

took priority. This followed a visit by Dr D.R. Pye, Deputy Director of Scientific Research, to Whittle's Power Jets company when he became convinced that it was better to pursue the turbojet and not the turboprop. The RAE then suggested to Power Jets (*see below*) that such a turbojet engine be constructed along the lines of the D.11 but with the airscrew turbine omitted; that company, however, was far too busy with its own projects to enter into such a collaboration. Therefore, by July 1940 Metropolitan-Vickers had become involved with the RAE's turbojet project instead.

For this turbojet the RAE designed a new nine-stage axial compressor designated 'Freda'. It was designed for a mass flow of 22.70kg/sec (50lb/sec), a pressure ratio of 4:1, a speed of 7,390rpm and to have a tip diameter of 563mm (22.2in). Its blading was of free vortex design using RAF 27 profiles. The blades were made in RR.56 by a new pressing process perfected by High Duty Alloys Ltd, which gave inexpensive blades with a good finish and very accurate and consistent profiles. In tests the 'Freda' compressor gave excellent results and the best so far obtained with an axial type.

'Freda' was followed by 'Sarah', which had a high-pressure stage identical to 'Freda' but with an additional five stages of low-pressure blades, making fourteen stages in all. Considerations of vortices and Mach number effects were taken into account during its design and it gave good results in tests during 1943. Its pressure ratio was 5:1, efficiency 83 per cent and speed 8,000rpm. 'Sarah' was built by the Armstrong Siddeley company as part of its turbojet designated ASX (*see* Chapter 15).

F.1 and F.2 turbojets

The 'Freda' compressor was incorporated into the RAE-designed turbojet designated F.2. This had been preceded by the F.1 and F.1A designs evolved by Hayne Constant. Produced in December 1939, the F.1 turbojet was planned with a nine-stage axial compressor with a rotor of disc construction, an annular combustion chamber and a single-stage, water-cooled turbine. Its compressor boundary layers were to be controlled by air bleeds and the bearings were to have oil bath lubrication, without circulation. This was the design that Power Jets had been invited to collaborate on. The F.1 had a static thrust of 975 kp (2,150lb), its maximum temperature being 800°C. Its compressor had an air mass flow of 17.24kg/sec (38.0lb/sec) at a pressure ratio of 4:1 and a speed of 9,450rpm. Its diameter was 0.685m (2ft 3in) and length 2.37m (7ft 9½in).

During 1940 the F.1 turbojet design evolved into the F.1A, which was to have the greater thrust of 1,220kp (2,690lb), mass flow of 21.55kg/sec (47.5lb/sec) and a speed of 7,470rpm. The changes were considerable. The compressor was now to be built on a drum and the turbine was to be of the two-stage type and with air cooling instead of water. The change to a two-stage turbine was made to reduce the overall diameter but this, and the removal of the centre bearing, increased the weight. The F.1A design was handed over to Metropolitan-Vickers in July 1940 and led to the design of their F.2 turbojet (*see below*).

In June 1939 Griffith left the RAE to work for Rolls-Royce. There he continued to develop his ideas, designing an extremely complicated contra-flow ducted-fan turbojet that had fourteen high-pressure and six low-pressure compressor stages; only the high-pressure section was built and tested, from 1942 to 1944. The scheme was abandoned in 1944 owing to insuperable problems, but it formed the foundation for later very successful engines.

Work continued at the RAE on axial compressors, usually with the collaboration of outside companies. For example, a six-stage compressor known as 'Ruth' was engineered by (British) GEC to the RAE's aerodynamic design in 1940. Its pressure ratio was only about 2.50:1, but at an efficiency of 83 per cent. Work was also carried out in 1941 on six- and eight-stage axial superchargers, designated E5, their pressure ratio being 3.7:1.

CHAPTER TWO

Frank Whittle and Power Jets

The effects of Frank Whittle's pioneering work on the turbojet were to spread all over the world. Born on 1 June 1907, Whittle began with a modest education but much curiosity and showed an early interest in engineering – his father had a small engineering works – and aeronautics. By 1928 he had graduated from the RAF College, Cranwell, been awarded the top prize for engineering studies, assessed as above average to exceptional as a pilot and commissioned to the rank of Pilot Officer. That year he also began thinking about the possible future developments for aircraft and in his science thesis wrote about the possibilities of the gas turbine becoming a prime mover in aeronautics. Eighteen months later, after a tour with a fighter squadron, he was on a Flight Instructor's Course at the Central Flying School, Wittering. It was here that he conceived the notion of using the gas turbine to supply a propulsive jet directly and thereby dispense with the airscrew. The turbojet, he foresaw, would overcome the altitude limitations of the piston engine and the speed limitations (about 644km/h or 400mph) of the airscrew.

He submitted his ideas to the Air Ministry in October 1929 but they were turned down on the grounds that a gas turbine was involved. Also, the Air Ministry was advised by none other than A.A. Griffith that Whittle's ideas did not merit further investigation! In Whittle's words, 'Griffith's report damned it with faint praise'. This is curious when, as we have seen, the very next month Griffith was proposing a gas turbine to drive an airscrew. At that time the general consensus in engineering circles was that the gas turbine had no future in aeronautics. The main objections were that the available metals would not permit high enough working temperatures and that efficiencies of compressors and turbines were too low for turbo-shaft applications. Whittle, however, considered quite rightly that all these components could, in time, be suitably developed.

Once he had conceived the idea of using the gas turbine to provide a propulsive jet, Whittle took into account in his calculations factors such as the beneficial lower temperatures at higher altitudes, the ram effect of forward motion on the compression process and the fact that the turbine had to drive the compressor so that exhaust expansion losses in the turbine were only a fraction of the whole exhaust process. In short, Whittle discovered that, for a jet propulsion unit, lower component efficiencies than those required for a successful turbo-shaft unit were acceptable. This was crucial in his perception that turbojet power would be far easier to accomplish than any form of turboprop.

For the compressor, a centrifugal type, perhaps in two stages, was proposed and a compression ratio of at least 4:1 would be needed. (It was only in Whittle's patent application that two stages of axial compressor were mentioned. This was probably at the suggestion of Pat Johnson, who knew it would be wise to include the axial word in the patent, even though Whittle had no intention of using anything but a centrifugal compressor.) Several combustion chambers were proposed and a thrust of about 450kp (993lb) was considered.

Jet propulsion patent

Despite the Air Ministry's lack of interest, Whittle applied for his first patent on 16 January 1930. The drawing for this was simplified for patent purposes and was not as shown to the Air Ministry. This patent, No. 347,206, was granted in mid-1931 and its details were published in 1932. In the meantime, during 1930 Whittle tried to interest a few companies in his ideas but without success. Not least against him was the worldwide Depression, which had begun with the 1929 Wall Street crash and left precious few funds about for developing new ideas.

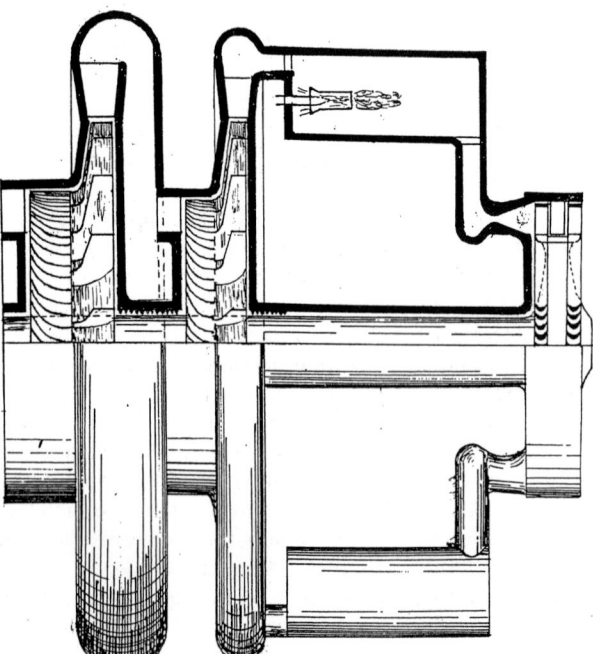

Frank Whittle's concept of the turbojet, as presented to the British Air Ministry in October 1929.

Frank Whittle, the turbojet pioneer, using his slide rule, one of the chief means of calculating in those early, pre-computer days. Ian Whittle

Following this lack of interest, Whittle did not give up on developing his turbojet ideas and in his spare time he conceived many new schemes. Improving on his original concept, he thought of using the double-sided centrifugal compressor instead of two stages. He also thought of the turbofan and of thrust augmentation and afterburning. The idea of the turbofan came to him because he perceived that the jet engine would be totally uneconomic for conventional aircraft with airframes unsuited to high speed and high altitude. Thus, the larger, slower-moving mass of air from the turbofan would be more suitable for lower speeds and lower altitudes. It was not until Whittle's patent expired in January 1934 (he was unable to pay the renewal fee) that he gave up hope of seeing his ideas put into practice and he began focusing on an electrohydraulic automatic transmission device that he was developing.

Formation of Power Jets Ltd

In any case, he was busy with his RAF career, first as an instructor and then as a test pilot. As one of the RAF's best pilots, Whittle took part in the Hendon airshows. Admirably supported by the RAF, he was posted to the Officer's Engineering School at Henlow and then, in 1934, he was sent to Cambridge University to study for a degree in Mechanical Science. Whittle served in the RAF for 21½ years, and spent ten years of these under training, mostly in engineering. At the end of the so-called Tripos Course, he obtained a First Degree. In the middle of the course, however, he was approached by two ex-RAF officers, R.D. Williams and J.C.B. Tinling, who were interested in doing something about Whittle's turbojet ideas. Although not very optimistic, he agreed to try. In order to form a company to develop the turbojet, an investment banker was needed and eventually the interest of O.T. Falk & Partners was aroused. This bank then sought the opinion of M.L. Bramson, a consulting engineer, who, crucially, reported very favourably on Whittle's scheme (*see* Appendix 1 for this report). Meanwhile Whittle was channelling all his new-found engineering and scientific knowledge into his turbojet project and was working on the design of a simple, experimental engine that comprised a single-stage, double-sided centrifugal compressor, a single combustion chamber and a single-stage turbine. O.T. Falk then gave this design to the British Thomson-Houston (BTH) steam turbine works at Rugby to carry out engineering and design work. Finally, the bank helped in the formation of Power Jets Ltd on 15 March 1936, with the President of the Air Council as a party to the agreement. The initial sum subscribed was £2,000 and the Air Ministry was also a shareholder.

Whittle would have liked to build and test the compressor, combustion chamber and turbine as separate elements and then move on to a complete engine, but he was forced to select the cheaper but riskier option of building the complete experimental test engine straight away. Accordingly, Power Jets ordered from BTH (whose Chief Turbine Engineer was F. Samuelson, later succeeded by R.W. Colingham) on 17 June 1936 the manufacture of this engine, complete except for the combustion chamber, instruments and some accessories.

The problems that had to be solved were formidable and outside the experience of regular engineering companies. The pressure ratio, air consumption and combustion intensity required were all far beyond anything previously achieved: over 3,000 shp at high efficiency was required from the turbine, which had a mean diameter of 355mm (14in) and a blade length of 61mm (2.4in). Whittle soon found the main problems to lie in the development of the combustion chamber and sought assistance from the company of Laidlaw, Drew & Co. in experiments, beginning on 2 October 1936, that led to its design and manufacture. A combustion intensity

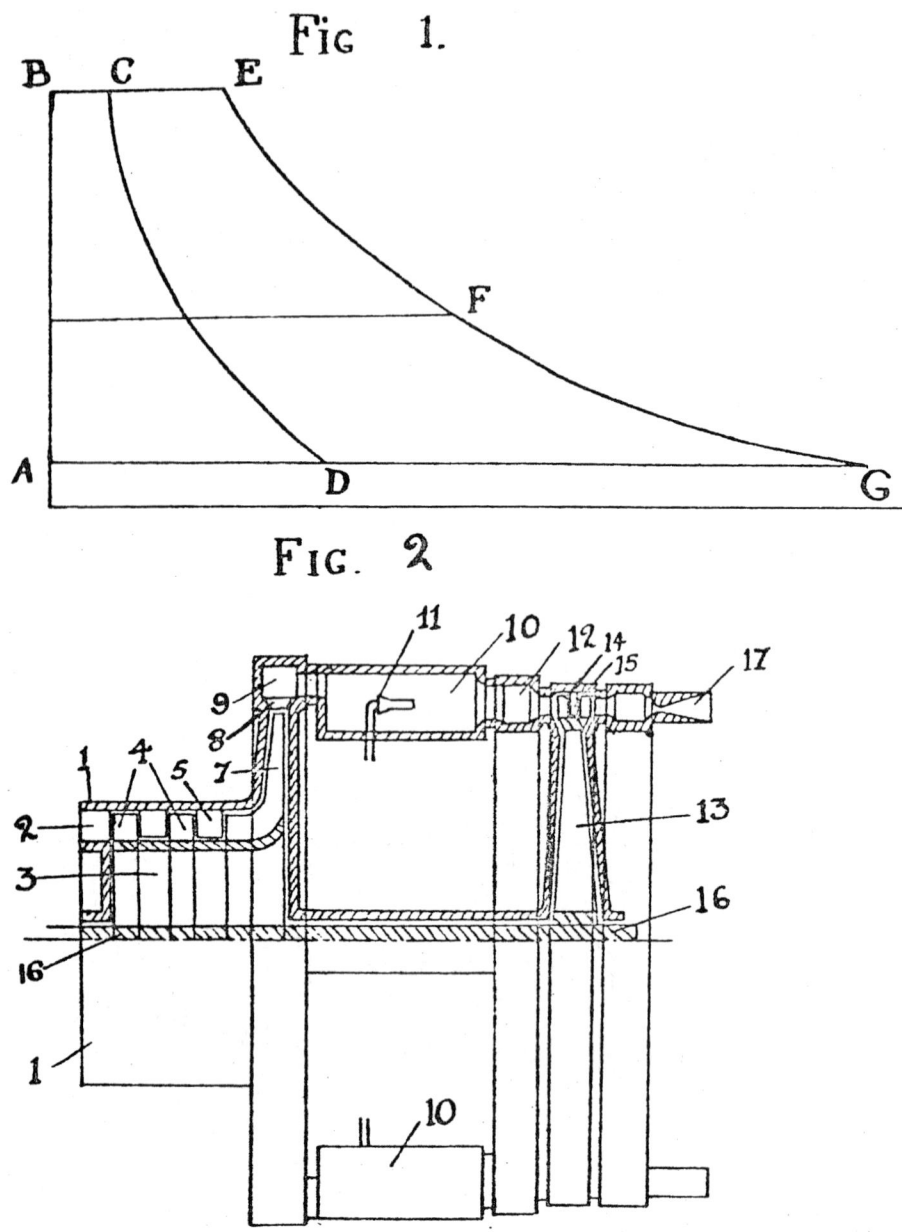

The sketch in Frank Whittle's original turbojet patent No. 347,206, applied for on 16 January 1930 and granted on 16 April 1931. This is reproduced on a reduced scale.

pressor had a diameter of 483mm (19in) and had thirty blades on each side. The single-stage turbine had sixty-six blades; these and the disc were made of Firth-Vickers Stayblade heat-resisting, austenitic steel, the disc being surrounded by a water jacket for cooling. Leading from the compressor outlet was a diffuser, followed by a large duct that curled round to form the combustion chamber. There were no inlet nozzles for the turbine but, instead, Whittle elected to try out a 'nozzle scroll' that encased the turbine.

Testing of the first build of the WU began at Rugby on 12 April 1937 and continued until 23 August. It was mounted on a trailer complete with instruments and controls but with the fuel and cooling water separate. The intention was to measure the thrust with a spring balance, but the weight of the electric starter motor meant that the wheels of the trailer had to be removed. Diesel fuel was used and there was no difficulty in starting the engine. About four times the engine accelerated out of control, usually due to puddling of fuel in the combustion chamber or a sticking burner needle valve. On these alarming occasions, everyone in the vicinity of the engine would make a run for safety, leaving a frightened Whittle at the engine controls: he wittily remarked that people did more running than the engines at that time! No reliable readings were taken but the concept had been proved. Clearly there was a great deal of work to be done, not least on the combustion system and compressor. At the end of testing, this engine was reconstructed by BTH and it was decided to move testing to Lutterworth, where Power Jets rented the old Ladywood Works foundry from BTH, since BTH considered it too dangerous to allow speeds above 12,000rpm in the open works.

First 'U' type turbojet and first experimental engines

Certain assumptions were made in designing the first experimental WU (Whittle Unit) type engine, including a performance of a small 900kg (1,985lb) aircraft flying at a speed of 800km/h (497mph) at 21,335m (70,000ft). This speed and height were considered to give the least drag and most economy and a net thrust of 50kp (111lb) would be required. Other assumptions included a compressor efficiency of 80 per cent, a turbine efficiency of 70 per cent and a static thrust of 630kp (1,389lb). This was all on the optimistic side but formed a starting point for the design of an engine of the smallest size in which the necessary accuracy of machining could be obtained without excessive manufacturing costs. The WU engine was not intended for flight. Its double-sided centrifugal com-

of twenty-four times contemporary combustion was required.

Air Ministry interest

The Air Ministry now began to take more interest in the work but funds remained very tight. This, of course, hampered development, especially as worn-out parts often had to be reworked and reused. Although there was no official security requirement for Whittle's work, Air Ministry contracts automatically had a secrecy condition and this made the raising of funds in the commercial field difficult. In 1939 it was found that about six of Whittle's patent drawings had been reproduced in the

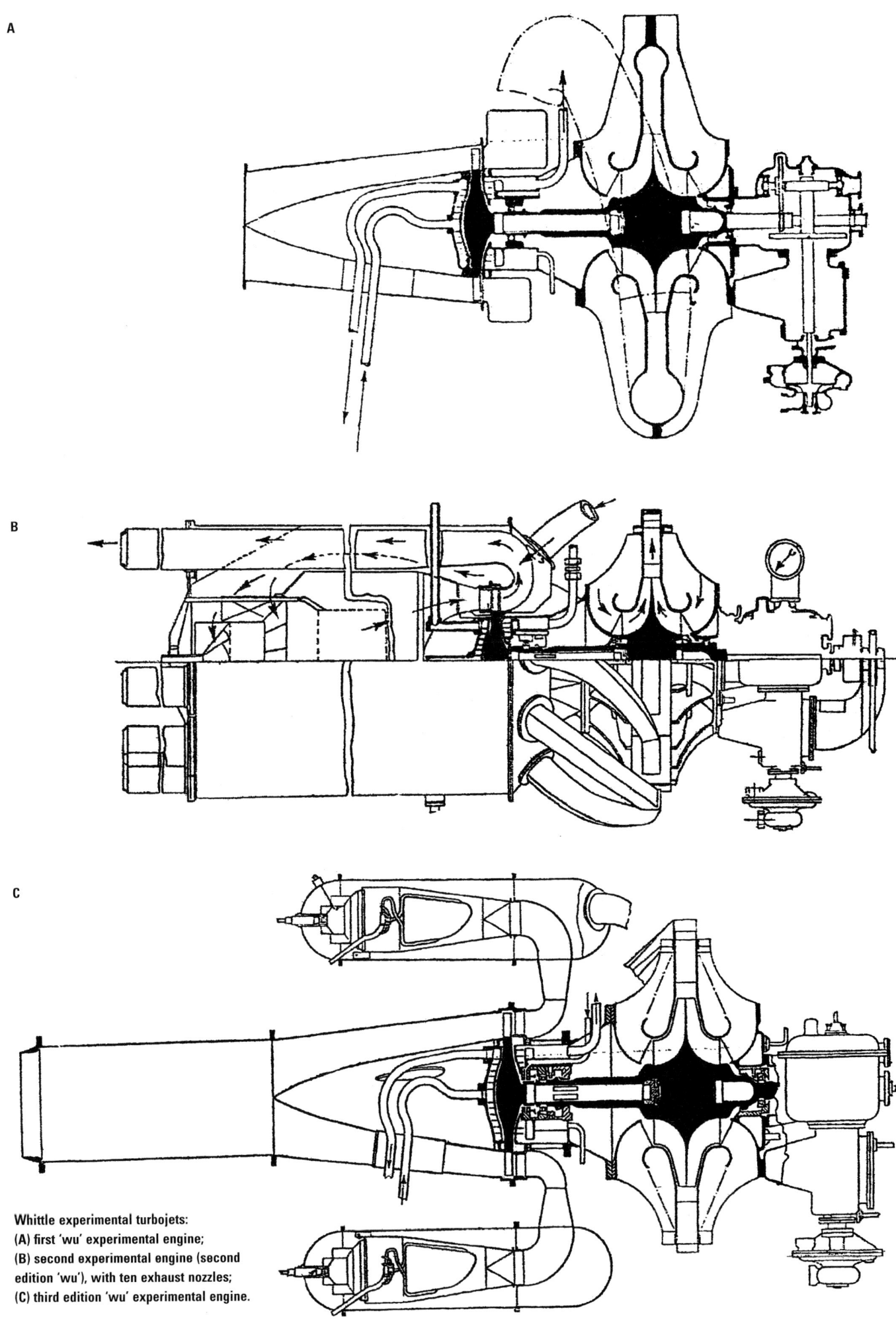

Whittle experimental turbojets:
(A) first 'wu' experimental engine;
(B) second experimental engine (second edition 'wu'), with ten exhaust nozzles;
(C) third edition 'wu' experimental engine.

The first experimental Whittle 'WU' turbojet with its starting panel. Ian Whittle

German magazine *Flugsport*. After the Second World War it was discovered that the German Embassy in London had bought copies of the patent from HMSO in 1932 and distributed them to research institutes and aeronautical companies in Germany. However, the patents were published for all to see and, as Whittle himself was only too aware, the difficulty was not in thinking of the turbojet but in bringing it into being. During 1936 Whittle had also patented his ideas on turbofan designs, bypass designs and afterburning.

When the WU had been rebuilt, as the 'second edition', considerable redesign was incorporated while using as much as possible of the original, in the interests of economy. Among the design aims were the obtaining of as much symmetry as possible and an improved diffuser system. The large, single air duct curving around the engine was replaced by ten fabricated pipes leading from the compressor casing. There was still a single combustion chamber, although longer, and with ten fuel vapour nozzles and also swirl vanes for the primary air. The turbine nozzle and blades were considerably modified but the single exhaust tube was replaced by ten smaller ones.

Testing of the 'second edition' WU began on 16 April 1938 at Lutterworth, but on 6 May it was severely damaged by a turbine failure. There were also cracks in the buttress ribs of the compressor. By then, only nine test runs had been made, totalling five hours, the maximum speed reached being 13,000rpm and the thrust 218kp (480lb). The 'test cell' at Lutterworth was an existing room with a circular hole in one of the outer walls, through which the jet exhaust pipes projected. The engine exhausted across the road to the company's offices and sometimes staff could not leave until the engine was shut down. The engine was on a wheeled trolley and the control panel was in a room separated from the test cell by a brick wall. Observations could be made only through two small openings in this wall. Unfortunately it was necessary for an observer to go into the test cell, with the engine running at full speed, in order to take a reading from the spring balance just below the exhaust pipes. These test facilities, which had no heating or sound insulation, were later replaced by much more sophisticated facilities.

After another redesign, BTH reconstructed the WU as the 'third edition', again using as much material as possible from the previous engine. This constant use of old parts was a false economy brought about by the government's scandalous parsimony and only increased the development time needed. In redesigning this engine, one exhaust pipe was returned to, but now ten combustion chambers were used, each approximately one-tenth the size of the previous single one, but otherwise with similar features. There were interconnecting pipes between the chambers. The single exhaust pipe contained, at first, a fuel vaporizer inside for the combustion system. The previously cracked buttress ribs on the compressor were machined away and the blades were thinned down.

Testing of the 'third edition' WU commenced at the end of October 1938 and continued until it was wrecked beyond repair on 22 February 1941 by another turbine failure. Water cooling of the turbine was still used at this stage but the De Laval bulb roots of the turbine blades proved unsatisfactory (after 169 hours) and, subsequently, fir-tree roots (invented at Power Jets) were used. Also, in March 1939, compressor blades cracked near the tips and a new compressor used twenty-nine instead of thirty blades, thus altering the harmonic value of the compressor and reducing the stresses.

Despite the short life of the third edition of the 'WU' engine, a great deal was learned that assisted subsequent developments. Towards the end of its life, it ran several times non-stop for up to 10 hours at a

The third experimental Whittle 'WU' turbojet, with ten combustion chambers, at the BTH Ladywood works in 1938.
Ian Whittle

The wrecked turbine of Whittle 'WU' test engine in 1941, the blade bulb roots having torn through the wheel rim. By this time the fir-tree root for the blades had been developed and these proved to be entirely satisfactory. Ian Whittle

cruising speed of 14,000rpm. It was found that the water cooling of the turbine was unnecessary and so this was not used again. When main bearings failed, this was attributed not to faulty design but to faulty fitting brought about by the fatigue of the overworked engineers. The biggest problem, however, that of combustion, proved the hardest to solve.

By August 1940 Power Jets had four combustion test rigs in operation at Lutterworth. In addition the Ministry of Aircraft Production (MAP), which had been created earlier that year to boost aircraft production, incorporating the RAE and the Technical Departments of the Air Ministry, arranged for the company to use a large compressor, which had been used in the construction of a pilot tunnel under the Thames at Dartford, to carry out high-pressure tests. Most of the tests were carried out on fuel vaporization systems but, frustratingly, what worked on the test rigs rarely worked in the engine. Eventually, in autumn 1940 the Asiatic Petroleum Company produced a shell-type combustion chamber using atomized fuel spray injection. This was developed by a team led by I. Lubbock at the company's Fulham (London) laboratory and it was eventually turned into a good combustion system at Lutterworth.

A combustion chamber test arrangement at the BTH Ladywood works, using a barrel of white spirit fuel, an electric motor with pump and temperature and pressure gauges. Note the 'welding mask' type of periscopic viewer used to observe the combustion. Ian Whittle

Special alloys for the turbojet

There has already been mention of special metals for the gas turbine, and the subject will return increasingly as this book progresses. Nowhere were these more important or instrumental to success than in the story as it relates to Britain, where the production of special steels and alloys was very advanced and usually manufactured by specialist companies, and so this is an appropriate point to make a few notes on what is actually a big subject.

The most severely tested metals in the turbojet or turboprop engine are, of course, those used for the turbine blades and, to a lesser extent, the combustion system and turbine rotor disc. At the time of this history, the maximum engine temperatures dealt with were in the order of 2,000°C in the combustion chamber, soon diluted with the secondary air flow to between 650°C and 850°C at the turbine inlet and down to about 550°C at the exhaust outlet.

The alloys and steels used in high-temperature components needed to deal with many injurious effects at these high temperatures, including corrosion, oxidation and scaling from the hot gases, fatigue stresses and insufficient rigidity under all conditions. More problematical was that turbine rotor blades had to resist high centrifugal forces under high temperatures and therefore had to have good creep characteristics, creep being a permanent elongation or deformation.

While nickel-chromium steel has good heat-resisting properties, its creep characteristics are not so good. On the fabrication side, its forging temperature approaches its melting temperature, while at room temperature its hardness makes machining difficult. As early as 1937 there were heat-resisting alloys being made by companies such as Hadfield Steel that had acceptable creep strength, but only up to about 700°C. The H. Wiggin company (an affiliate of the Mond Nickel company) therefore put much effort into developing its so-called Nimonic alloys, which contained extra alloying metals to give the desired characteristics up to higher temperatures. The first simple alloys, containing 80 per cent nickel and 20 per cent chromium, were used for high electrical resistance elements for fires and furnaces and these formed the basis of the subsequent Nimonic alloys.

The first such alloy produced for the turbojet, around 1940, was Nimonic 75, which had the addition of titanium as a stiffening element. It was easy to work in sheet form, had a low coefficient of expansion and was resistant to oxidation. It proved ideal for combustion chambers. Nimonic 75 was also used for forged and cast components such as turbine inlet guide vanes and nozzles and exhaust nozzles. Temperatures up to 800°C could be resisted.

In 1941 the alloy Nimonic 80 was developed, which had good creep resistance and became the alloy of choice for turbine blades. It was then replaced by Nimonic 80A, in which better properties were the result of closer composition and manufacturing controls. To cope with higher turbine temperatures, Nimonic 90 was produced for temperatures up to 900°C. This alloy consisted of approximately 60 per cent nickel, 20 per cent chromium and 20 per cent cobalt, plus traces of titanium and aluminium to give stiffness.

Nimonic alloy development was continuous but, because of its high cost, less expensive steel was employed where cooling air was possible. For early turbine discs, such as those used in Whittle's engines, austenitic steel (carbon steel) known as Stayblade and made by Firth-Vickers was used but disc failures demanded the use of stronger alloys. Improved austenitic steels from the Jessop company, such as G.18B and H.27, were later used. Typical non-magnetic, austenitic steel used between 12 and 18 per cent chromium and 8 to 12 per cent nickel, with or without traces of tungsten, molybdenum, titanium or other elements. These steels could not be hardened by heat treatment and were difficult to machine.

The austenitic steels were then largely replaced with ferritic steels (or iron alloys) for such uses as turbine discs, since they had high creep strength, good scale-resisting properties and were readily forged and machined. An example was Jessop's H.40, which contained small quantities of chromium, manganese, silicon, molybdenum, carbon and vanadium. Another, which could be used for turbine blades as well as discs (up to 650°C), was Jessop's H.46. which had a high content (11.5 per cent) of chromium and small amounts of niobium, molybdenum and vanadium.

CHAPTER THREE

Towards Britain's First Jet Aircraft

Until the summer of 1939 the Air Ministry regarded Whittle's work, and that at the RAE, as long-term research, but once Whittle's WU had run with some conviction it was realized that there was some possibility of a viable aero-engine. Dr D.R. Pye, the Director of Scientific Research, became a complete convert to Whittle's work after he had viewed the WU engine running for 20 minutes on 30 June 1939. Pye's enthusiasm was such that he felt he needed to convert Whittle! The official response was dramatic and rapid: on 12 July 1939 the Air Ministry issued a contract to Power Jets for a flight engine designated the W.1. Design and manufacture was sub-contracted to BTH, the design team consisting of Whittle, BTH steam turbine engineers and newly recruited Power Jets engineers. Soon after, on 30 August 1939, the Air Ministry awarded a contract to the Gloster Aircraft Company for the design and manufacture of an experimental aircraft to Specification E.28/39 to flight test Whittle's turbojet. Contracts from MAP assisted Power Jets in building up a small engineering team, a sheet metal workshop and combustion testing rigs at Lutterworth.

Thus 1939 was an important year for Power Jets. Following Dr Pye's visit to the company, their work was put on the Secret List and the Air Ministry insisted that the company should endeavour to acquire more private funding. The Air Ministry officials were quite unable to grasp that it was very difficult to encourage private investors when they could not be shown the secret work! From this point on Power Jets could collaborate with the RAE and, in theory, there would be a free exchange of information. Unfortunately such collaboration was never very good, probably due to a needless and destructive rivalry.

Whittle's W.1 turbojet

In making up his engineering and development team, Whittle hired only those qualified to at least a university First Class Honours standard: these included Dr Robert Feilden and Dr Geoffrey Bone, the latter working on combustion testing. Whittle himself moved into Brownsover Hall, a nearby country house where the conditions were very peaceful and more conducive to thinking. Everyone worked extremely hard, seven days a week, and those involved on testing frequently found themselves soaked in fuel! Their boss, Whittle, was perceived as kindly, enthusiastic and a workaholic. On 3 September 1939 Britain was forced into war with Germany and so all over the country hard work and long hours became the norm.

While the W.1 engine was under construction, certain of its major components were considered not to be airworthy and so Power Jets used these, plus other spare components made for the WU engine, to build an early version of the W.1, designated W.1X. The W.1X was ready by about November 1940 and tests with it provided feedback with which to modify the W.1 engine.

The general design of the W.1 engine was very similar to the 'third edition' of the WU engine, the main differences being some lightening, a completely new auxiliaries drive box and seventy-two turbine

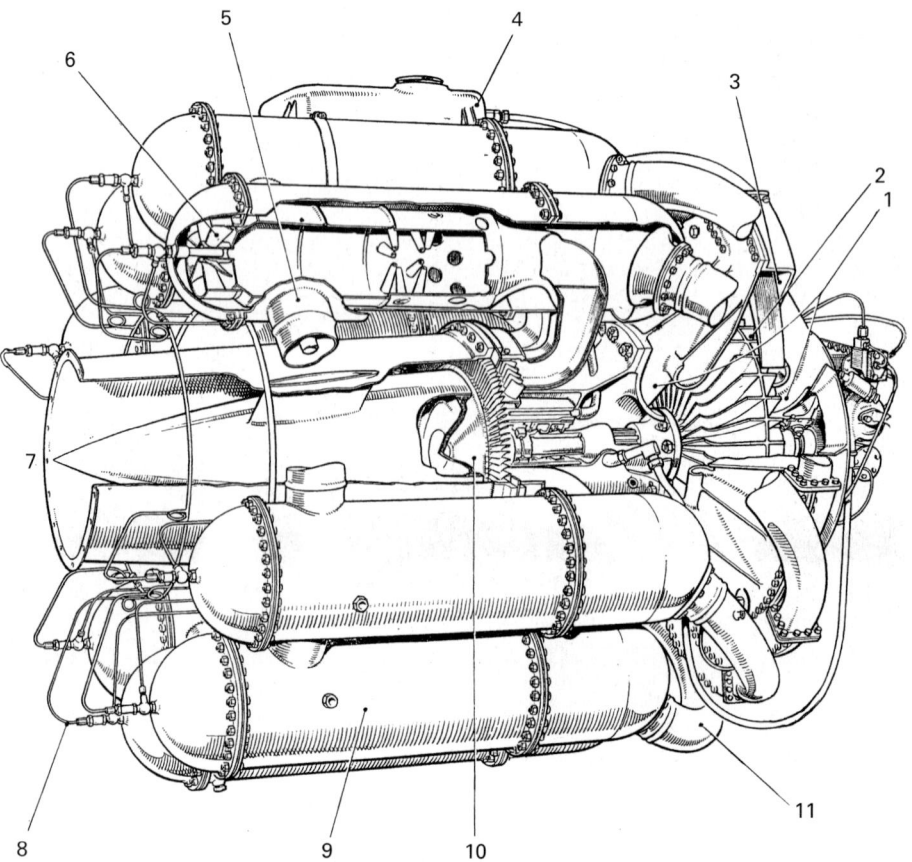

Whittle (Power Jets) W.1 first flight engine: (1) air intakes; (2) double-sided centrifugal compressor; (3) diffuser passage; (4) oil tank; (5) combustion chamber interconnector; (6) swirl vanes; (7) exhaust nozzle; (8) fuel feed pipes; (9) ten reverse-flow combustion chambers; (10) single-stage turbine; (11) air feed duct.

blades, instead of the earlier sixty-six, which now used fir-tree roots. The metal for the turbine blades was changed to Firth-Vickers Rex 78. After preliminary testing, beginning on 12 April 1941, the W.1 was put through a special category 25-hour test to clear it for 10 hours of flight trials before overhauling it. Its static thrust was given as 562kp (1,240lb) at 17,550rpm.

Gloster E.28/39

Meanwhile, at the Gloster Aircraft Company, Chief Designer George Carter had been working up the design of the E.28/39 aircraft, unofficially but aptly named Pioneer. In this work he collaborated closely with Frank Whittle. At the layout stage, consideration was given to adapting a projected twin-boom layout of a fighter intended to have a Sabre piston engine driving a pusher airscrew. It was realized that such a design, with a short jet pipe for the turbojet, would be most efficient, but the idea was discarded since nothing was known about the effects of hot gases on the airflow over the tailplane. Therefore it was decided to use a very long jet pipe so that the jet efflux would exhaust beyond the tail empennage. In any case, the extra length of 1.83m (6ft) in the jet pipe resulted in an estimated loss of only about 1 per cent in the thrust. In practice, the effect of the jet exhaust, just beyond the empennage, resulted in an unexpected reduction in drag and a better performance.

The E.28/39 was a straightforward aircraft of conventional all-metal, monocoque construction with fabric-covered control surfaces, straight flying surfaces and a retractable tricycle undercarriage. The nose air intake led into ducts that bifurcated around the cockpit to meet in a plenum chamber for the engine. This chamber surrounded the engine's air intakes and supplied it with air pressurized by ram effect. Behind the cockpit, which had a semi-bubble, framed canopy, was a 368ltr (81gal) fuel tank, followed by the engine. The top section of the fuselage, from behind the cockpit to the tail, was removable to give easy access to the engine.

For experimental purposes, two sets of wings were constructed for the first prototype, one set having a 'high speed' EC 1240 aerofoil section, the other having an NACA 23012 section. They were both of two-spar, stressed-skin construction, joined at the centreline of the fuselage. Automatic

TOP: The original Whittle (Power Jets) W.1 flight engine outside the Lutterworth works. This historic engine now resides in the Science Museum, London.

BOTTOM: The rotor from the W.1 engine, showing the double-sided centrifugal compressor and the turbine. Ian Whittle

balance tabs were fitted to the fabric-covered ailerons and there were hydraulically actuated flaps operated by a cockpit hand pump. Hydraulic pressure was also used to retract the undercarriage, but the system's accumulator was hand charged on the ground. An unusual feature were the two radiators, one in each air duct flanking the cockpit, for the turbine bearing coolant, of which 16ltr (3.5gal) were carried. This cooling system was later removed when it was found that cooling air bled from the compressor was sufficient.

In mid-1940 work began on the first of two prototypes at Hucclecote in Gloucestershire, where Gloster had its factory and airfield. However, because Hucclecote was building Hurricane fighters and was therefore a likely target for bombing, the construction work on the E.28/39 was moved to Regent Motors' garage in Cheltenham, about 4 miles to the northeast. There the first prototype, W4041/G, with the NACA 23012 section wing, was completed and then transported by road to Hucclecote for testing. While waiting for the W.1 engine to become available, the aircraft was fitted with the W.1X engine, which was not cleared for flight, in order to perform taxiing trials; these began with Flt Lt P.E.G. Sayer at the controls on 7 April 1941.

On 8 April 1941 Sayer continued taxiing tests. During the third test, three short hops were made of about 2m (6ft) off the ground, varying in distance from 90m to 180m (300–600ft). Whittle taxied the aircraft on

George Carter, designer of Britain's first jet aircraft, the Gloster E.28/39, in the early 1930s. Jet Age Museum

The Whittle W.1X turbojet engine installed in the E.28/39 aircraft in April 1941 by Robert Feilden of Power Jets. Jet Age Museum

Jerry Sayer, Gloster's Chief Test Pilot and Britain's first jet pilot. *Flight*

E.28/39 with W.1 engine		
Span	8.84m	(29ft 0in)
Length	7.715m	(25ft 3¾in)
Wing area	13.63sq m	(146.5sq ft)
Wing dihedral	4°	
Height	2.82m	(9ft 3in)
Empty weight	1,309kg	(2,886lb)
Loaded weight	1,700kg	(3,748lb)
Maximum speed (at 4,572m/15,000ft)	589km/h	(366mph)
(at 3,000m/9,840ft)	750km/h	(466mph)
Landing speed	138km/h	(86mph)
Service ceiling	9,754m	(32,000ft)
Range	740km	(460 miles)

The prototype (W4041/G) of Britain's first jet aircraft, the Gloster E.28/39, outside its hangar at Farnborough.
Tony Buttler

The prototype of the Gloster E.28/39 (W4041/G). In the background are camouflaged hangars and, on the right, a barrage balloon winched down. Jet Age Museum

its first run that day and again after Sayer made the hops. After these taxiing tests, the E.28/39 was taken back to Cheltenham, this time to Crabtree's Garage and a new, longer-travel, nosegear was fitted. Undercarriage retraction tests were also made.

The aircraft was then dismantled and taken for flight trials to RAF Cranwell, which had the advantages of long runways and greater secrecy. By 14 May 1941 the W.1 engine had been fitted. (The W.1X engine was sent to GEC at Lynn, Massachusetts,

in October 1941 for demonstration purposes, as related in the USA section.)

At 19.45 hours on 15 May 1941, Sayer took off in the E.28/39 on its maiden flight. This lasted 17 minutes and was an outstanding success. The undercarriage was

The prototype Gloster E28/39 (W4041/G) at Farnborough. Jet Age Museum

A low pass by the E28/39 (W4041/G) at Farnborough with Capt. Eric Brown at the controls. Jet Age Museum

retracted on this first flight. Although the flight was some 21 months later than the world's first turbojet aircraft flight (the German Heinkel He 178), it should be noted that the W.1 engine was technically much more advanced than the He 178's HeS 3b engine, especially as regards combustion and general engineering. Also the E.28/39 was a long-lived, durable experimental aircraft, unlike the He 178, although this observation is not meant to detract from the German achievement. Over the thirteen days following its first flight, the E.28/39 was flown on fifteen test flights, adding up to 10 hours, and during this time there was no need to inspect the engine. Climbs up to 7,600m (25,000ft) were made. Altogether, including bench testing, taxiing and flight tests, the engine ran a total of 39 hours 57 minutes before being dismantled, when it was found to be in excellent condition. Flight testing was then transferred to Edgehill in Warwickshire, since it was located more conveniently for both Gloster and Power Jets.

It was not until 4 February 1942 that the E.28/39 (W4041/G) was again ready for flight tests, although these did not begin until 16 February. The aircraft was now fitted with the 'high speed' wing and the more powerful W.1A turbojet. This engine was fitted with a new barostatic control to reduce the fuel flow with decreasing atmospheric pressure as the aircraft climbed. Various problems were found and corrected with this engine, including a heat-distorted exhaust pipe, a failed turbine blade and a faulty barostat. Turbine blade tip clearances were also reduced. Finally, on 6 June 1942 the main bearings failed due to frozen oil at 9,150m (30,000ft). Back in the air again, towards the end of September 1942, W4041/G again suffered engine oil problems. That was sadly Sayer's last flight in the E.28/39, because on 21 October he was killed when flying a Hawker Typhoon fighter off the Northumberland coast. His place was taken by Michael Daunt and extensive testing of the E.28/39 continued. Originally it was to be fitted with four .303 machine guns as a fighter, but this was never done. Although it was a resounding success in its role of demonstrating Whittle's turbojet, the little aircraft was never going to be a fighter.

On 1 March 1943 the second E.28/39 prototype (W4046/G) made its first flight with Gloster's development test pilot John Grierson at the controls. On 17 April 1943 Grierson made the first cross-country flight in this aircraft, flying it from Edgehill to Hatfield in order to give a demonstration for Prime Minister Winston Churchill, who was most impressed and enthusiastic.

Both E.28/39s finished their flying lives in testing at RAE Farnborough and this included much flying by RAF service pilots. On 30 July 1943 W4046 was destroyed after grease on the aileron controls solidified at 10,000m (33,000ft), causing them to jam. The aircraft went into an inverted spin and its pilot, Sqn Ldr D. Davie, baled out. Fortunately he was able to deploy his parachute and thereby became the first pilot to escape from a jet aircraft.

The highest altitude reached by the E.28/39 (W4041) was 12,853m (42,170ft) on 24 June 1943, when the aircraft was powered by a Power Jets W.2/500 turbojet and flown by John Grierson. For that remarkable flight he inhaled pure oxygen for 30 minutes before take-off and wore a pressure waistcoat. This aircraft's long and useful career came to an end early in 1945 when it was dismantled at Farnborough. It was reassembled and put on display at the Science Museum, London, in April 1946.

CHAPTER FOUR

W.2 Variants and the Spread of Turbojet Technology

At the beginning of 1940, when war had been in progress for a few months, the Air Ministry concluded that there was a good chance of producing a jet aircraft in time for deployment in the conflict. In August 1940 George Carter of Glosters submitted his proposal for a twin-engined jet fighter with wing-mounted engines, a high-set tailplane and a tricycle undercarriage. In November the Air Ministry issued Specification F.9/40, which embodied Carter's proposals, and the design arrangements were finalized in December 1940. This design was to lead to the Meteor fighter.

W.2 turbojets

This fighter was to be powered by two Power Jets W.2 turbojets, a new and more advanced type to be manufactured by BTH under contract to MAP. By then the official decision had been made that Power Jets, under Whittle's direction, should become a research and development organization providing manufacturing companies with all necessary drawings and data to facilitate engine production. This arrangement was also extended to the USA (*see that section*) in autumn 1941 when the W.1X engine, a set of W.2B drawings and a small team of Power Jets engineers were sent to GEC at Lynn, Massachusetts, to kick-start turbojet work there.

Meanwhile, the Rover car company at Barnoldswick went directly to MAP and obtained a contract to build Power Jets engines. Thus, instead of being a subcontractor to Power Jets, they became its competitor but still had the benefit of Power Jets research and development; Whittle maintained that they went behind his back. Also unfortunate was that the W.2 engine was a failure, largely because it was too ambitious. It had eighty diffusers (the W.1A had only ten), which were later blamed for surging problems. The surging manifested itself in a series of loud reports accompanied by violent shaking of the engine, test frame and test house doors when the engine was running at 14,500rpm. Fitting a larger exhaust pipe raised the surge speed to 15,500rpm (design full speed was 16,500rpm) but with lower thrust. At that early stage, a means of driving the compressor separately, as was done later, would have shown that it had a very narrow operating range. A great number of experiments were required to establish the essentials and these were made more troublesome by mechanical failures clouding matters. Compressor surge and high exhaust temperatures made it impossible to run at more than 75 per cent of the design speed.

Therefore work proceeded on a revised, less critical design, the W.2B. Its essential features were again a double-sided centrifugal compressor, ten reverse-flow combustion chambers, a single-stage turbine and a fixed-area exhaust nozzle. Power Jets and BTH between them built the W.2 Mk.IV (or W.2/4) based on the W.2 and gradually developed it as the basis for the W.2B. Unfortunately the W.2/4 engine was wrecked on its first run on 10 October 1941, when its faulty (cracked) compressor burst asunder at about 9,000rpm. Fortunately, four people inside the test cell at the time escaped serious injury.

Development of the W.2B engine took two main paths. At Power Jets the limits of the combustion chambers and airflow were explored, while at the Rover company

Power Jets-Rover W.2B turbojet, predecessor of the Rolls-Royce Welland engine.

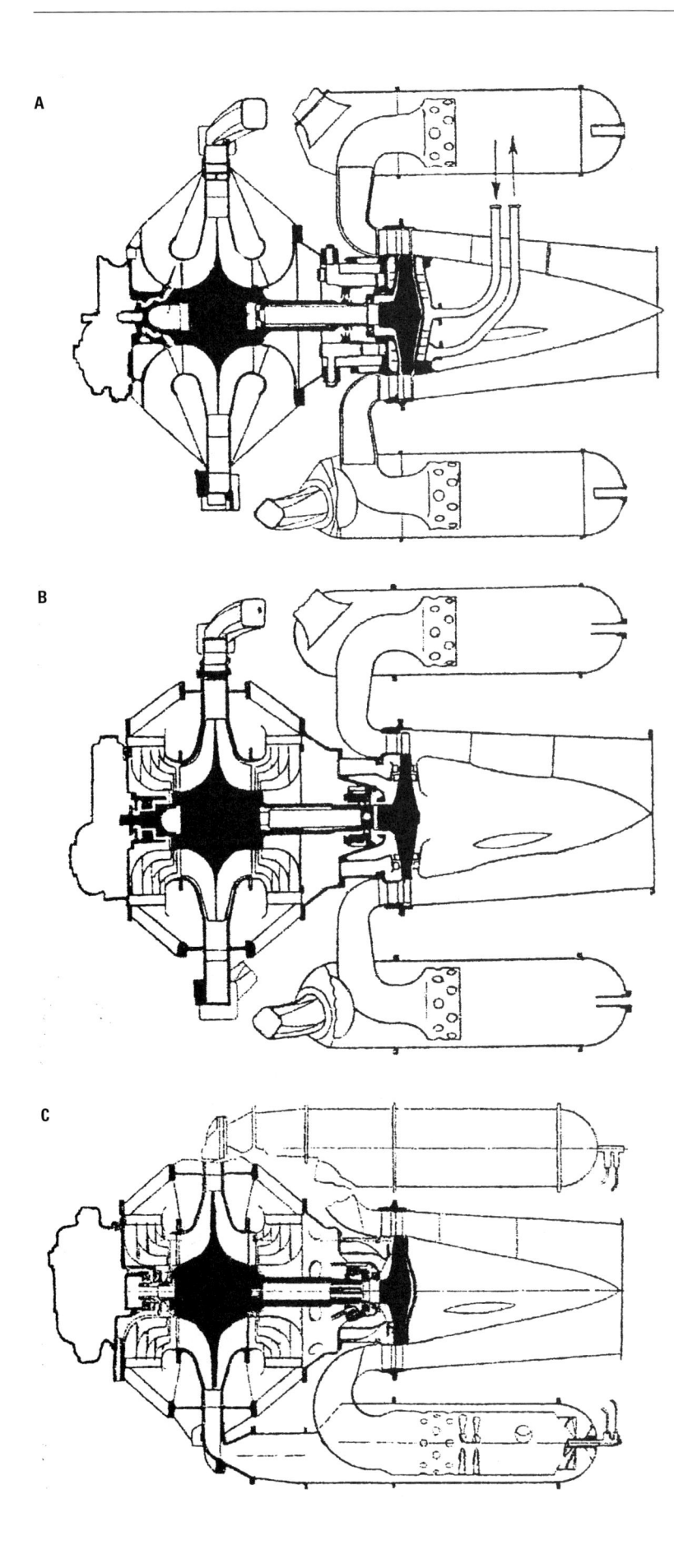

Whittle turbojets: (A) Power Jets W.1; (B) Power Jets W.1A; (C) Power Jets-Rover W.2B.

the engine was modified to have straight-through combustion (as opposed to reverse-flow) with the Lucas primary and secondary air system. The Rover development was not authorized by Whittle.

W.2/500 and W.2/700 turbojets

Power Jets gradually increased the length of the turbine blades to match increased combustion to obtain higher airflow and thrust. This led to the W.2/500 engine, a complete redesign that began running in September 1942. It had a thrust of 796kp (1,755lb) at 16,570rpm, a jet temperature of 606°C and an airflow of 15.75kg/sec (34.73lb/sec). An increase in the length of the turbine blade trailing edges (not the leading edges) produced higher performance figures of 839kp (1,850lb) thrust at 16,750rpm, a jet temperature of 620°C and an airflow of 18.29kg/sec (40.33lb/sec). Specific fuel consumption was 1.12. After improvements to diffuser design were made, performance went up again and, after the leading edges of the turbine blades were increased, the final performance figures were a thrust of 1,127kp (2,485lb) at 16,750rpm, a jet temperature of 647°C and an airflow of 21.39kg/sec (47.15lb/sec). At the same time, the specific fuel consumption went down to 1.05. The development was successful to the extent that the W.2/500 engine was the same size as the W.1, but had three times the thrust for a weight gain of only 50 per cent.

Much of the success of the W.2/500 engine was due to improved turbine blade material, notably Nimonic 80 nickel-chromium alloy developed by the H. Wiggin company. Its full design speed was reached on 2 September 1942, a mere six months from the inception of the design. However, a number of resonance failures of the compressor occurred at around 14,000rpm and it was not until 1943 that the W.2/500 was flight tested in the Gloster E.28/39 and Meteor.

At Rover, meanwhile, work on their 'unauthorized' version of the W.2 engine was not proceeding at all well and they had made various mechanical changes from

Power Jets W.2/500 turbojet wrecked by the failure of the compressor.

The Power Jets W.2/700 turbojet, the last of Whittle's own designs and the last with reverse-flow combustion chambers. Ian Whittle

injection gave appreciable increases in thrust. Rolls-Royce tested afterburning on the Rover W.2B/23 with promising results for short-term power boosting, provided a variable-area exhaust nozzle was used.

By 1944 Power Jets had a factory with facilities to series manufacture turbojets. Having outgrown its Lutterworth site, the company had moved in 1942 to Whetstone, near Leicester, and was employing around a thousand people. They made limited numbers of the W.2/700 engine, which was the last of Whittle's own designs and the last with reverse-flow combustion chambers. From April 1944 this engine was test flown in the tail of a Wellington II (W5518/G), when the thrust was measured at 1,134kp (2,500lb).

In 1943 work began on Whittle's design for a turbofan, which was an idea he patented in 1936. The engine was designated the L.R.1 (for long range), the front fan being of the multi-stage axial type, and it was in an advanced stage of construction when the government inexplicably cancelled it in 1944. A projected version of the W.2/700 with a rear fan and afterburning was also cancelled. Thus, a British lead in an important type of engine, the turbofan, which would later revolutionize civil air transport, was ruined by bureaucrats. However, worse was to come, because the Power Jet's drawings. Their straight-through version of the engine could not be coaxed above a thrust of 454kp (1,000lb) and any advice offered by Whittle was ignored. Quite rightly, Whittle saw the priority as getting an engine ready for production for the Meteor fighter. Rover's W.2B/23 version of Power Jet's engine was used only in taxiing tests in the F.9/40 prototype (predecessor of the Meteor) on 10 July 1942 and the engine was flight tested in the tail of a converted Wellington II bomber (Z8570/G), beginning on 9 August 1942. The thrust at that time reached 567kp (1,250lb). From May 1943 another Wellington II (W5389/G) was also used, when the W.2B/23 thrust reached 725kp (1,600lb). However, Rover's engine was a long way from being ready for production. Eventually their work was turned over to Rolls-Royce.

Experiments to boost thrust were carried out by Power Jets using the W.1/3 and W.2B engines, the earliest being with ammonia injection. Methyl chloride injection was not encouraging but water

36

A Wellington II (W5389/G) flying test-bed with a Rover W.2B/23 turbojet in the tail, May 1943. Philip Jarrett

A rare picture of a Power Jets turbojet and afterburner being demonstrated. Ian Whittle

the MAP nationalized Power Jets in April 1944 and afterwards decreed that it could not make engines in competition with private industry! Renamed Power Jets (R&D) Limited, the company's purpose was to administer the Power Jets patents and this left Whittle without much to do. The company also ran the School of Gas Turbine Technology under D.L. Brown at Lutterworth and, later, Farnborough, the purpose of which was to encourage the growth of gas turbine industries at home and abroad and to provide adequately trained technical staff for such industries. In 1946 the school was taken over by the newly formed National Gas Turbine Establishment (NGTE).

Miles M.52 supersonic aircraft project

The most advanced planned use of the W.2/700 with aft fan and afterburning was the projected Miles M.52 supersonic research aircraft. This highly advanced and revolutionary machine was ordered in autumn 1943, while the war was at its height, to meet the MAP Specification E.24/43, which called for an aircraft capable of flying at speeds up to 1,600km/h (1,000mph) at 11,000m (36,000ft) or Mach 1.5. Actually, the official specification was only six lines long and it stated that the aircraft should be ready in nine months. Strangely, an error had been made in translating a German document and 1,000 mph was written instead of 1,000km/h (621mph). However, everything about the M.52 was reaching far into the unknown and it is extraordinary that such a machine was ordered in the midst of the war.

The programme was launched on 8 October 1943 and the team under the brothers F.G. Miles and George Miles evolved the airframe, designed by Dennis Bancroft, while Frank Whittle worked on the turbojet. Naturally the whole project was guarded in the utmost secrecy. The M.52 had a tubular-shaped fuselage, 1.52m (4ft 11⅞in) in diameter, containing the turbojet and fuel; at the nose there was a partially glazed, cone-shaped cockpit section, leading back to an annular air intake. Flying surfaces were only slightly swept but of very thin, knife-edged sections, the wings and tailplane being attached on the fuselage centreline and the fin on top. How much theory was known or used in those days of piston-engined aircraft is uncertain, but Miles reasoned that the shape that was good for a supersonic bullet might also serve the M.52. (The same reasoning had been used by the designers of Germany's V2 and other rockets.) All possible help was obtained on the aerodynamics, including consultation with the RAE and the NPL. Certainly, shock waves from the sharp lips of the annular air intake and the pointed centre cone would assist in pressure recovery of the supersonic air flow before it entered the turbojet.

The structure of the M.52 was designed to be of high-tensile steel with aluminium alloy covering, but the flying surfaces were almost solid in the manner used by certain aircraft after the war, such as the Lockheed F-104. The pressurized cockpit, in which the pilot assumed a semi-prone position, was designed to be ejected complete in a high-speed emergency and slowed by parachute, the pilot then baling out at lower altitude. Of special interest, and showing much foresight, was that the tailplane was of the all-moving type, another feature that was to become important for supersonic aircraft in post-war years. A narrow-track, tricycle undercarriage, retracting into the fuselage, was to be provided.

Frank Whittle with a 1/4-scale model of the W.2/700 turbojet at Brownsover Hall, 1944. Ian Whittle

In 1939 Frank Whittle moved from the Power Jets factory to Brownsover Hall, a nearby country house where conditions were very peaceful and more conducive to thinking. This photo was taken in 1943. On the desk are models of the F.9/40 jet fighter and a compressor while, by the window, is a model of the W.2/700 turbojet. Ian Whittle

A conference at Brownsover Hall: (left to right) W.E.P. Johnson, R. Dudley Williams, L.J. Cheshire, Mary Phillips, Frank Whittle, J.C.B. Tinling, D.N. Walker and Wg Cdr G. Lees. Ian Whittle

special engine was put in hand and combustion tests on the afterburner section were made before the project was officially cancelled.

To gain knowledge on the characteristics of the thin wing at lower speeds, a full-size wing was built in wood and fitted to a Miles M.3B Falcon light aircraft (L9705), which first flew in this form on 11 August 1944, piloted by H.V. Kennedy. Later a wooden version of the M.52 tailplane was also added. Because of its 'razor-sharp' wings, this aircraft became known as the Gillette-Falcon (after the well-known razor blades). To further initial supersonic research, Miles built their own wind tunnel for the testing of beautiful steel models of the M.52. Meanwhile, a full-size wooden mock-up of the M.52 was built at Woodley.

The Miles M.52 project, which was unquestionably full of promise for supersonic research, was dealt a mortal blow in February 1946 when it was cancelled by MAP. At that stage the aircraft was about 80 per cent complete and was due to fly in another three months. Even the pilot, the redoubtable Eric Brown, had been chosen to make the first flights. Following official instructions, the jigs, tools and mock-up were all scrapped and George Miles was ordered to send the M.52 drawings to the USA. The official reason for this decision was lamely given that supersonic flight would be too dangerous for pilots, although there was no shortage of

The only engine available to power the M.52 at the time was the Power Jets W.2/700 of 907kp (2,000lb) static thrust at sea level. However, it was evident that, even with the M.52 having minimal drag, considerably higher thrust would be required. Therefore it was planned to use one of Whittle's bypass and afterburning schemes (see below). It was decided to mount fan blades as extensions of the turbine blades, giving a bypass ratio of 2:1 at about 3,700rpm. Provision was made for afterburning and it was calculated that a total net thrust of 1,450kp (3,200lb) would be available for take-off, the specific fuel consumption being 1.6. At 1,610km/h (1,000mph) at 10,975m (36,000ft), or Mach 1.5, it was calculated that the thrust would be 1,860kp (4,100lb) with a specific fuel consumption of 2.0. By using a slightly higher combustion temperature, the maximum thrust under these conditions would be 2,720kp (6,000lb). Construction of this

Jubilation at Brownsover Hall, 1944: Cheshire, Williams, Whittle, Tinling, Walker, Lees and Johnson. Ian Whittle

Frank Whittle shakes hands with his Power Jets team in 1945: J. Hodge, unidentified, W. Cooper, A. Simmons, Whittle, R. Feilden, D. Walker, D. Shepherd, R. Voysey and H. Cohen.
Ian Whittle

volunteers. (Much later, doubts were expressed that, following its climb, the M.52 would not have enough fuel for the supersonic mission.) In fact, attention was already turning to swept wings for high-speed flight following the study of German research results discovered in 1945. Interestingly, from 1947 René Leduc in France conducted tests with ramjet-powered aircraft that were similarly configured to the M.52, but it was the Bell X-1, also with thin, straight flying surfaces, that first exceeded Mach 1.0 in the USA later that year. The Bell X-1 had a tailplane that was adjustable on the ground but, following the Miles design, this was then altered to be an all-moving tailplane.

The Gloster E.28/39 (W4041) on display after the war at the Promenade, Cheltenham, Gloucestershire, probably in April 1946. Russell Adams Collection

CHAPTER FIVE

Bypass and Ducted Fan Projects from Whittle and Power Jets

Frank Whittle conceived his ideas on ducted fan and bypass engines as early as 1928/29 and worked on them throughout the life of the Power Jets company. All subsequent engines of this type have been based on this pioneering work, which has proved of great importance in the modern world, especially for long-range transport. From the account below it will be evident that a wide range of schemes were considered.

Whittle's ducted fan patents

Whittle's initial thought in 1928 was for a piston engine driving a low-pressure ducted fan housed in the hollow nacelle fuselage of an aircraft, and he made provision for burning additional fuel downstream of the fan and before the exhaust nozzle. Thus the ideas of the ducted fan and afterburning came to him before his turbojet patent of 1930; he later found that the basic concept had been patented by a Dr Harris of Esher in 1917. By 1929, however, Whittle had calculated that using a piston engine was not worthwhile, since such an arrangement would be no better than using a conventional piston engine and airscrew. Despite this, the piston-driven ducted fan engine was pursued by most countries into the 1940s and was particularly advocated and publicized by Campini in Italy.

Once his turbojet ideas were formed and patented in 1930, Whittle realized that a ducted fan driven by a gas turbine was a very much better prospect. However, on 4 March 1936 his patent for a different type of engine, the bypass turbojet, was applied for and was accepted on 3 September 1937 (Patent No. 471,368). This meant that the ideas in the patent were there for all to see, but no country, not even Germany, took advantage of them at the time.

The wording of Whittle's patent of 1937 makes it clear that he fully appreciated the advantages of the bypass turbojet. Its primary objective was to 'gear down the jet', since it was realized from the start that the pure jet engine had a low propulsive efficiency until speeds of around 800km/h (500mph) could be reached. The patent specification states: 'The primary object is to provide as great a mass-flow of the working fluid as possible . . . The high mass flow is particularly desirable in the case of the propulsion of aircraft by fluid reaction in order that the jet velocity shall be as low as possible.' In the patent scheme, part of the intake air was fed to the intake of a gas turbine, the hot gases from which drove another turbine that, in turn, drove a ducted fan. The air delivered by the ducted fan mixed with the hot gases from the turbine. Thus, the total mass flow of the engine was augmented and the jet velocity reduced.

Augmented turbojets

In March 1940 Whittle applied for a patent to cover another type of thrust Augmentor. This comprised a twin contra-rotating ducted fan mounted ahead of the turbojet, in an arrangement similar to that favoured today. Although mechanically simple and therefore attractive, investigation threw up some difficult aerodynamic problems and so in the same month Whittle applied for a patent for another scheme. In this a shrouded turbine behind the turbojet drove a ducted fan, the turbine blades being extended beyond their shroud to carry the fan blades. The efflux from the turbojet was led through an annular duct to the Augmentor turbine; the ducted fan mounted on this turbine induced a flow of air that joined the hot gases downstream of the turbine. This idea was later revived by Metropolitan-Vickers (*see* Chapter 6) and in 1942 its general arrangement was sent to the USA.

Around 1941/42, Whittle worked out a scheme for a bypass engine using a very large ducted fan driven by a two-spool turbojet. This engine had a three-stage axial high-pressure (HP) compressor driven by a single-stage turbine and a three-stage low-pressure (LP) compressor driven by a two-stage turbine. The LP shaft was extended forward to drive, via gearing, the very large, single-stage ducted fan. This fan was about three times the diameter of the turbojet and its airflow was intended to supercharge the engine but mostly to bypass it. A variant of this scheme proposed a centrifugal HP compressor preceded by a six-stage axial LP compressor and a two-stage ducted fan. Both these engines remained in the project design stage.

In 1941 Whittle applied for a patent on another ducted fan scheme. This unit comprised four shaft-power turbojets enclosed in a nacelle of about 2.59m (8ft 6in) diameter. The engines were arranged radially around a central shaft that was driven through reduction gearing and carried at its forward end a large ducted fan used to supercharge the four turbojets. The large nacelle enclosing this arrangement was intended to form part of the fuselage of a large four-engined bomber, the crew being housed in a conical, pressurized cockpit forward of the ducted fan, thereby forming an annular intake for the engines.

A variant of this scheme comprised four turbojets arranged radially around a central shaft carrying, at its rear end, a large turbine driven by the hot gases from the four turbojets. The hot gases were conveyed to the turbine via an annular duct. At its forward end the shaft carried a large ducted fan, part of its airflow entering the turbojets and the rest bypassing them to meet the hot exhaust gases downstream.

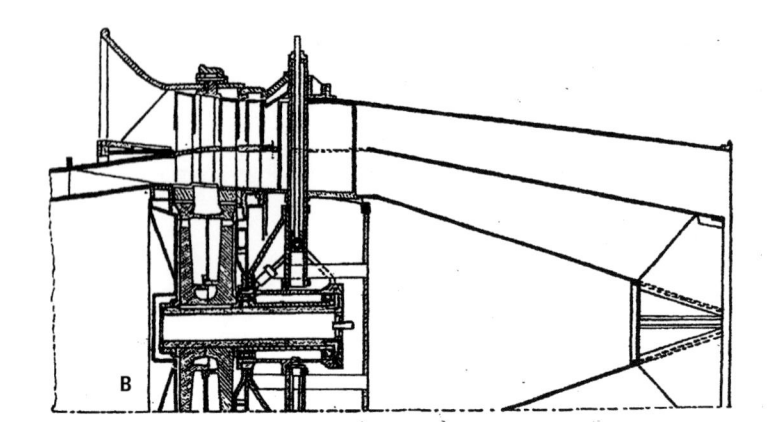

Some of Frank Whittle's schemes for bypass and thrust-augmented turbojets, illustrate the fertility of his mind:

(A) 1940 thrust augmenter, aft of the turbojet. The external ducted-fan blades are attached to inner turbine blades operating in the engine's exhaust.

(B) 1940 thrust augmenter comprising a twin contra-rotating ducted fan mounted ahead of the turbojet and driven through reduction gearing. A second ducted fan at the rear is driven by its blades being attached to a turbine in the engine's exhaust. This second fan re-energized the bypassed air.

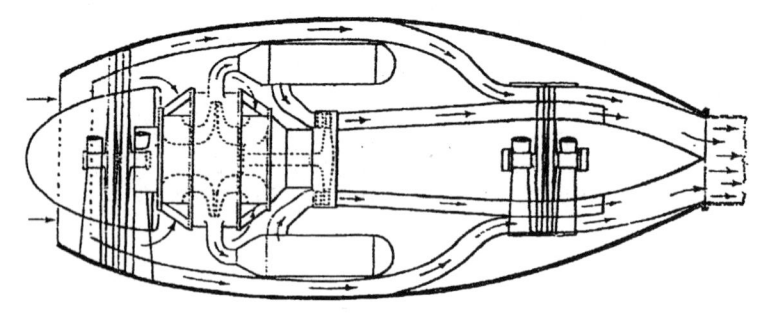

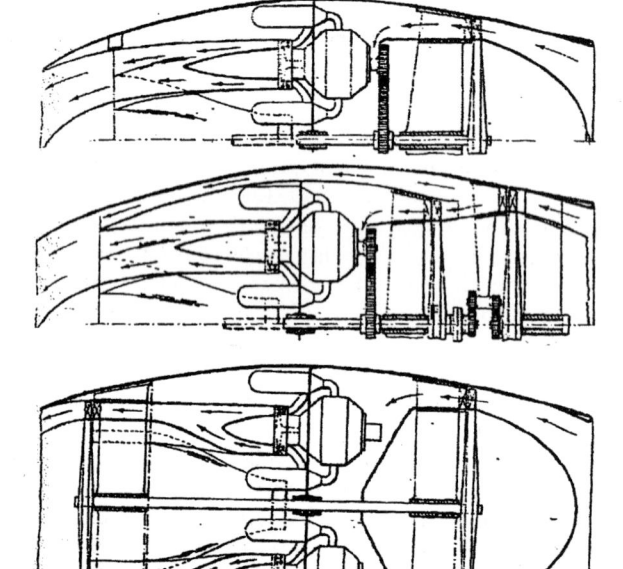

(C) 1941 ducted fan scheme. The unit comprises four shaft-power turbojets (turboshafts) enclosed in a nacelle. These are arranged radially around a central shaft that was driven through reduction gearing and carried at its forward end a large ducted fan used to supercharge the four turbojets.

(D) Four turbojets arranged radially around a central shaft carrying at its rear end a large turbine driven by the hot gases from the four turbojets. At its forward end, the shaft carried a large ducted fan, part of its airflow entering the turbojets and the rest bypassing them to meet the hot exhaust gases downstream.

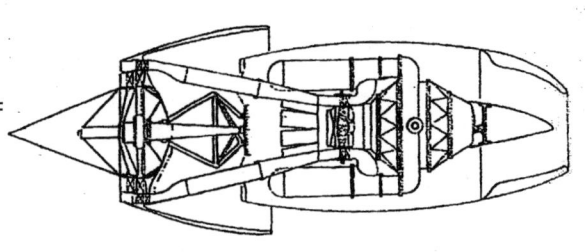

(E) Four turbojets arranged radially around a central shaft and delivering hot gases to a large two-stage turbine mounted on the central shaft. The shaft drove at its forward end a single-stage ducted fan used to supercharge the four compressors via a plenum chamber.

(F) 1943 No.3 Thrust Augmenter arranged at the rear of a turbojet. A turbine annulus surrounds a fan annulus, the turbine blades being carried as extensions of the fan blades.

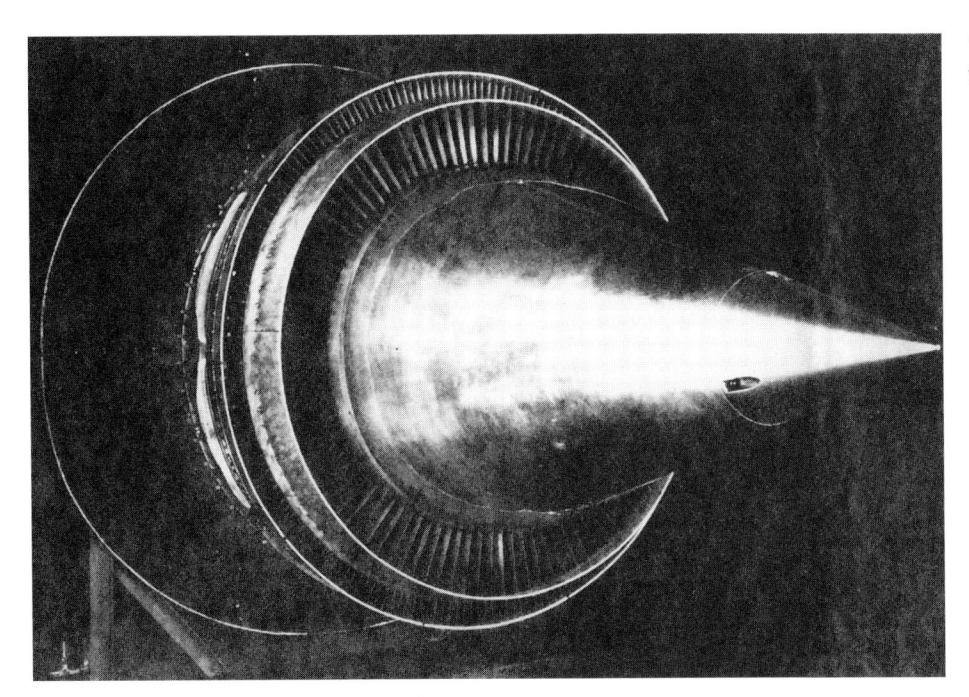

1943 No.3 Thrust Augmenter: aft fan with tip turbine.

Perhaps a final variant of this arrangement consisted of four centrifugal compressors, mounted in a large nacelle and arranged radially around a central shaft, delivering, via a series of combustion chambers, to a large two-stage turbine mounted on the central shaft. The shaft drove the four compressors through gearing and, at its forward end, it carried a single-stage ducted fan used for supercharging the four compressors via a plenum chamber. Ahead of this, the shaft drove, via gearing, a very large ducted fan, the output from which bypassed the remainder of the powerplant and joined the hot gases downstream of the turbine. As before, the crew were housed in a cockpit inside a conical nacelle, surrounded by an annular air intake.

Although Power Jets desired to avoid gearing in favour of mechanical simplicity, the manufacture of double-tiered blading in some of the above schemes ran into difficulties. Initially it was thought that such blades could be made by the lost-wax method of casting but this idea revealed various problems. In any event, all the multiple powerplants mentioned remained as sketches and calculations.

A simpler proposal was put forward in 1943. This consisted of a turbojet driving through gearing at its forward end a ducted fan that supercharged the engine and provided a bypass airflow. In this case, however, the bypassed air did not join the hot gases downstream of the turbine. Instead it flowed through an annular duct surrounding the turbojet and continued aft to the end of the tailpipe. Furthermore, the annular duct was provided with a means of afterburning. Thus the final efflux consisted of two streams – the central turbojet exhaust being surrounded by the bypass efflux – the latter being either cold or hot depending upon whether or not afterburning was used. This was another engine that was not built.

The No.3 Augmentor, however, was built in 1943. In this a turbine annulus surrounded a fan annulus, the turbine blades being carried as an extension of the fan blades instead of the other way round as before. By this means, the hot turbine blades were relieved of the loads previously imposed by carrying and driving the fan. This arrangement also allowed higher, more appropriate blade speeds.

The No.3 Augmentor was fitted to a W.2/500 turbojet and had a bypass ratio of 3. (It was also proposed to use the augmentor with a projected two-spool centrifugal engine, the LP compressor being double-sided and the HP compressor single-sided.) In order to pass the combustion gases to the turbine, it was necessary to pass exhaust ducts through the shroud for the fan. In tests it was found that these ducts interfered too much with the air flowing into the fan, resulting in a poor performance. It was therefore decided to revert to the schemes whereby the fan blades were an extension of the turbine blades. This was the arrangement used in the special version of the W.2/700 turbojet planned for the Miles M.52 (*see above*).

L.R.1 turboprop and turbojet projects

During 1943 Frank Whittle inspired a discussion at the RAF Staff College on the requirements for a future long-range bomber for use in the Far East theatre of war against the Japanese. It was suggested that the probable requirements for such a bomber would be a radius of action of 3,220km (2,000 miles) at 640 to 725km/h (400–450mph) at 12,200m (40,000ft) altitude, carrying a bomb load of 6,800kg (15,000lb). This translated into a still-air range of 8,000km (5,000 miles), for which was envisaged a four-engined bomber with a loaded weight of 54,430kg (120,000lb).

Regarding the choice of engine, it was originally thought that a turboprop would be suitable. This was to use an eight-stage axial compressor followed by a centrifugal compressor. A two-stage turbine was to drive the compressors and, via gearing, a six-bladed contra-rotating airscrew. Designated the L.R.1 (L.R. for long range), this turboprop was expected to give a total thrust (airscrew plus exhaust) of 770kp (1,700lb) at 95 per cent full power at 757km/h (470mph) at 12,200m (40,000ft). For take-off at 8,000rpm, it was estimated that the engine would give 4,670shp plus 1,433kp (3,160lb) of jet thrust. Other data for the L.R.1 included a weight of 1,395kg (3,070lb), a diameter of 1.143m (3ft 9in) and, under cruising conditions, a specific fuel consumption of 0.41.

There were variations of the L.R.1 engine without the airscrew. These included a bypass version, a straight turbojet version and a supersonic version. The bypass version had a two-stage ducted fan at the front, rotating at 2,300rpm and giving a bypass ratio of 2.5. It was planned to weigh 1,256kg (2,770lb) and have a thrust equivalent to 2,860hp. The straight turbojet version was planned to give a static thrust at sea level of 2,495kp (5,500lb) for a weight of 1,180kg (2,600lb) and have a specific fuel consumption some 15 per cent better than that of other turbojets of the time. As for the supersonic version, this was to have a ducted fan and afterburning, the bypass ratio being 2.6. Its maximum net thrust at 1,370km/h (850mph) at 11,000m (36,000ft) altitude or Mach 1.28 was estimated at 4,695kp (10,350lb).

The long-range bomber project, with four L.R.1 turboprops, was first submitted to the MAP late in 1944. In May 1945, the month that war in Europe came to an end, a project was planned for a transatlantic airliner powered by four L.R.1 ducted fan engines. With a loaded weight of 70,760kg (156,000lb), the airliner was designed to cruise at 757km/h (470mph) at 13,700m (45,000ft) altitude. It was to have a still-air range of 8,500km (5,280 miles). The bomber was not taken up and the MAP Powerplants Committee considered that future airliners of the type proposed would not be required and, in any case, developing the pressurized cabin would be too difficult!

However, this did not condemn the L.R.1 engine because in August 1945 Power Jets was awarded a contract to build one engine plus a set of spare parts. Unfortunately a mechanical failure occurred during compressor tests at the NGTE in 1949 and the cancellation of the L.R.1 project soon followed. With the cancellation of all of Whittle's bypass and ducted fan engines, Great Britain's hopes of achieving supersonic flight in 1946, exceeding 1,610 km/h (1,000mph) in 1948 and producing a high-speed, non-stop, transatlantic airliner in 1951 were dashed. All these milestones in aviation did, of course, come to pass, but at a later date. There were a number of reasons why Whittle's ideas were not capitalized upon, chiefly the wartime priority to concentrate on getting the straightforward turbojet developed up to production status. Then, following the 1944 nationalization of Power Jets, the company was no longer allowed to direct its own policy. After the war Whittle's brilliant team was disbanded and the design and development expertise of Power Jets was no more.

Frank Whittle resigned from Power Jets in January 1946, owing to disagreements with official policies. No British company ever hired Whittle. For someone of his stature, there was no place in the industry that his talent and determination had founded. Never one for grabbing the glory for himself, he always acknowledged the important part played by his team and others in developing the turbojet. From 1944, when Whittle's work became more widely known, he was showered with honours, including a knighthood in 1948. The government awarded him £100,000 tax free for his inventions and services to the country; before he left Power Jets, Whittle enquired of his staff, one by one, as to their circumstances and then made out cheques to them from his award.

After the war, Whittle suffered several spells in hospital due to ill health brought on by his struggles, largely against government departments and large companies. The great pressures of working against time while beset with technical and financial difficulties also took their toll. In addition, he was devastated by the cancellation of his W.2/700 turbojet and his bypass and ducted fan projects. Eventually he recovered and went to live in the USA in 1976. He lectured regularly and extensively, visited England and travelled much. Among many awards and accolades, he was awarded the Order of Merit from the Queen in 1986. The great Stanley Hooker, guiding light of Rolls-Royce and Bristol gas turbine development, later recalled of Whittle: 'He laid down the performance and aerodynamic design of the jet engine with the precision of Newton, and even today we still use his formulae for the calculation of the performance of the jet engine. I was very impressed with his work.' Sir Frank Whittle died on 8 August 1996 in Maryland. By the time of Power Jets's nationalization, other British companies had well and truly entered upon turbojet development and production, and these will be looked at next.

CHAPTER SIX

Metropolitan-Vickers Electrical Co. Ltd

In March 1937 Hayne Constant of the RAE proposed a gas turbine-driven airscrew for an aircraft. The following month the Air Ministry authorized drawing work to begin, the project having been recommended by a sub-committee of the Aeronautical Research Committee (ARC). The engineering company of Metropolitan-Vickers Electrical Co. Ltd (MV), in Trafford Park, Manchester, was seen to be well placed to help with this project, due to its long history of successful design and construction of steam turbines. It was

BELOW: **A Metropolitan-Vickers F.2, Britain's first axial turbojet engine. Note that the drawing shows the accessories below the engine.**
RIGHT: **The compressor rotor of the F.2.**

45

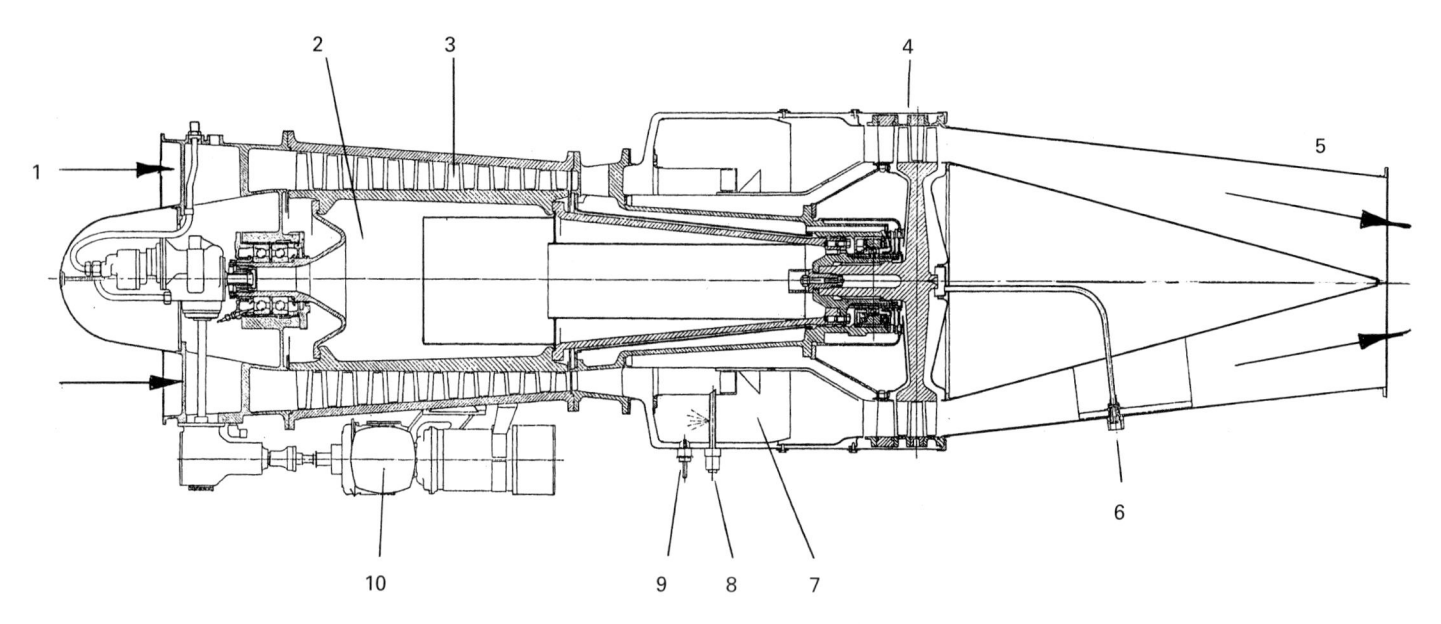

Metropolitan-Vickers F.2/1 turbojet, first batch flight engine: (1) air intake; (2) rotor drum, with double-cone forward end; (3) 9-stage axial compressor; (4) 2-stage turbine; (5) exhaust nozzle; (6) turbine wheel cooling air outlet; (7) annular combustion chamber; (8) fuel burner; (9) igniter; (10) gearbox for fuel pump and accessories.

also an organization at the forefront in the world of mechanical sciences research. As a result this company and the RAE collaborated on the study of several layouts, designated 'B', that featured a low-pressure axial compressor driven by a high-pressure turbine. The complete, dispersed, static test engine, designated B.10 and with a nine-stage axial compressor, was run in December 1940 (see Chapter 1). It gave a remarkable performance and provided invaluable data for subsequent work.

The next static test engine, designated D.11, aimed at reducing pressure losses by using as straight a flow path as possible through the engine (see Chapter 1). It was abandoned when only partially constructed, however, since with the advent of war in September 1939 a much higher priority was given to a simpler turbojet designated the F.2. This axial engine featured an RAE 'Freda' nine-stage axial compressor (hence the 'F' designation), an annular combustion chamber, two-stage turbine and a fixed-area exhaust nozzle. A two-stage turbine was needed in order to drive the compressor at a slower speed and thereby lessen centrifugal forces. The engine's simplicity ensured a much more rapid development than previously projected turboprops.

F.2: Britain's first axial turbojet to fly

By July 1940 MV had assembled a design and development team in a Gas Turbine Department at Barton, under the leadership of Dr D.M. Smith, to work on the F.2 turbojet. Prominent in his team, which collaborated with Hayne Constant at the RAE, was Dr I.S. Shannon. Overall responsibility and technical direction for their work fell to Dr Karl Baumann, Chief Mechanical Engineer of the company. Compressor blade design was the particular province of the RAE and by 1942 the compressor was giving an efficiency of 88 per cent at a pressure ratio of 3.2:1. Static tests of the first F.2 Series 1 engine began in December 1941 and it passed a twenty-five-hour special category test in November 1942 with a take-off thrust of 816kp (1,800lb). Flight tests began on 29 June 1943 by fitting the third prototype into the tail of the prototype Avro 683 Lancaster (BT308/G). An air intake was fitted above the tail section of this Lancaster, its duct curving downwards to feed the turbojet. The F.2 thereby became Britain's first axial turbojet to fly. This particular engine completed 106 hours flying

Test flying a Metropolitan-Vickers F.2 turbojet in the tail of a Lancaster. Note the dorsal air intake for the turbojet. Philip Jarrett

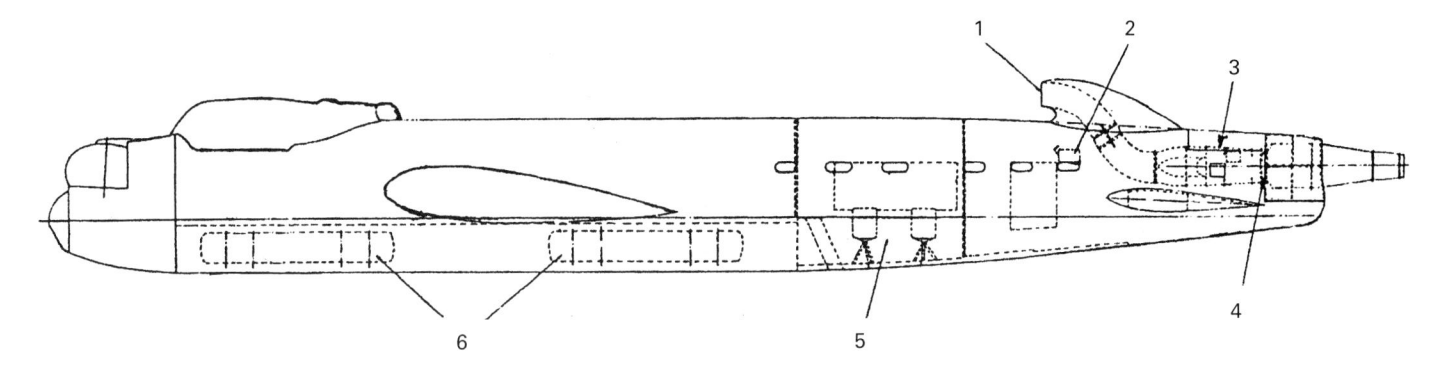

F2/1 turbojet installation in a Lancaster flying test-bed: (1) air intake; (2) oil tank; (3) F2/1 turbojet; (4) engine mounting; (5) test observer's cabin; (6) auxiliary jet fuel tanks.

in the Lancaster plus another 104 hours of static running before completing its programme. No trouble with compressor surging occurred.

Testing of a pair of F.2 Series 1 engines installed in a Gloster F.9/40 prototype (DG 204/G) began on 13 November 1943 but yielded disappointing results due, it was later thought, to the use of short, underwing nacelles that caused excessive air spill at high speeds.

Problems to be overcome during development of the F.2 engine included overheating of the turbine bearing and combustion hotspots that caused distortion of the combustion chamber and fracture and burning of the turbine inlet nozzles. By August 1942 a redesigned engine designated F.2/2 had been produced. Its turbine rotor blade metal had been changed from Rex 75 to Nimonic 75 and the combustion chamber lengthened by 15.2mm (6in). All along, simplicity of construction was the keynote, due to a shortage of skilled labour. The F.2/2 gave a static thrust of 1,089kp (2,400lb), but the combustion chamber still suffered from severe distortions and a build-up of carbon.

During 1943 another redesign, designated F.2/3, was ordered purely for experimentation. In this version the annular combustion chamber was replaced with can-type chambers and a static thrust of 1,225kp (2,700lb) was produced.

Two views of the F.9/40 prototype (DG204/G) at Gloster's Bentham facility in July 1943. This aircraft was tested with a pair of Metropolitan-Vickers F.2 axial turbojets. Tony Buttler

The Saunders Roe SR.A/1 was the world's first turbojet-powered flying boat. Three prototypes were built, each powered by two Metropolitan-Vickers F.2/4 Beryl engines, TG263 being shown here.

Beryl turbojet

In the same year, however, the final design, known as the F.2/4, was produced and the annular combustion chamber was re-introduced. The F.2/4 now had a ten-stage axial compressor and a single-stage turbine. Its length was 4.04m (13ft 3in) and its diameter was only 0.963m (3ft 1.9in). For a weight of 794kg (1,750lb), a rated thrust of 1,588kp (3,500lb) at 7,700rpm was achieved. The specific fuel consumption was 1.05 at maximum thrust and 1.0 at cruising speed. The F.2/4 first ran in January 1945 and passed a 100-hour type test that year. During the programme the first flight tests were made in the tail of a Lancaster II (LL735/G). The engine was then named the Beryl and was built in small numbers during 1945.

The Beryl engine also powered a Meteor F.4 test-bed (RA490) in September 1947 and the three Saunders Roe SR.A/1 flying boat fighter prototypes, the first of which (TG263) made its maiden flight on 16 July 1947 piloted by Geoffrey Tyson. The SR.A/1 was the world's first jet-powered flying boat. The Meteor F.4 test-bed (RA490) showed a fantastic rate of climb with the Beryl turbojets. It reached 3,050m (10,000 ft) in 55 seconds, 6,100m (20,000ft) in 2 minutes 5 seconds, 9,150m (30,000ft) in 3 minutes 43 seconds and 12,100m (40,000ft) in 7 minutes 31 seconds. Major production of the Beryl was not undertaken, despite its great promise, because it was superseded by the F.9 turbojet, later known as the Sapphire.

At the inlet of the Beryl engine were five concentric rings known as an 'anti-Daunt guard' in memory of the incident when the test pilot Michael Daunt narrowly escaped being sucked into an engine. The compressor casing was cast in R.R.50 light alloy, ribbed externally for stiffness, fully machined internally and bolted together. Attachment of the fixed stator blades was by dovetail-section axial grooves with spacers in between. Another light alloy casting extended from the compressor casing to carry the turbine bearing. The annular combustion chamber inlet cover was a one-piece casting, ribbed externally, with internal bosses for twenty fuel burners. A special sheet metal known as 'Immaculate 5' (an austenitic steel developed before the war) was used for the combustion chamber and exhaust details.

The compressor rotor and its hollow extension leading back to the turbine disc were light alloy forgings. The metal, known as R.R.56, was also used for the compressor rotor blades, which were attached by serrated roots. Anti-friction bearings were used at the front of the compressor rotor and in between the rear of the rotor extension and the turbine disc, leaving the latter overhung. A molybdenum-vanadium steel, developed by MV several years before for steam turbines, was used for the turbine disc. All turbine blades were made from Nimonic 80 heat-resisting alloy, the rotor blades being attached by bulb roots.

Fuel was sprayed into the combustion chamber and mixed with the primary air in an upstream direction, the temperature in front of the twenty burners reaching 1,200°C. Secondary air was fed into the combustion chamber through hollow wedges, the temperature of the resulting hot gas reducing to 800°C at the turbine. Air was tapped off from the fourth stage of the compressor for the cooling of bearings, the front face of the turbine disc and the roots of its blades. The rear face of the turbine disc was cooled by air piped from the outlet of the compressor to pass through the struts supporting the exhaust cone and internal tubes that led the air inside a plate covering the turbine disc. For the bearings, oil was mixed with the cooling air to form a mist for lubrication, this method having the advantage of avoiding the drying out of surplus oil when the engine was switched off. Fuel flow was controlled by a throttle valve and barostat, which adjusted fuel flow according to altitude, and there was a governor incorporated in the fuel system for emergency over-speed control only.

Research and development for turbojets carried out by MV was extensive and was largely under a MAP collaboration scheme. Apart from the Gas Turbine Department at Barton, work on compressors was undertaken at the Trafford Works, where the necessary great power and speed variation was provided by geared steam turbine drives. Later, a power station of the Northampton Electric Supply Company was also used and, later still, a power station of the Wigan Corporation. Valuable data from the RAE and their blade cascades was used. Temperature, stress, vibration and creep testing was carried out with special apparatus on turbine discs and blades.

Of great importance was the work carried out by MV in collaboration with High Duty Alloys, from 1940, in developing a process to mass produce precision-forged, light alloy compressor blades. By 1943 this process had been perfected and a major step towards the mass production of the turbojet had been taken.

Because MV had chosen to use an annular combustion chamber, to take advantage of the smaller engine diameter that resulted, it accepted that greater development difficulties lay ahead. The only way that modifications to the combustion chamber could be tested was by testing on the engine itself. This was because very large supplies of high-pressure air were not available at the works.

MV's experience with steam turbines proved to be of great benefit in developing turbojets. The aerodynamic design of the turbine blades and labyrinth gas seals, for example, were derived from steam turbine practice. It was also possible to carry out the difficult machining of turbine blades, as another example, in the steam workshop section.

F.3, the world's first practical turbofan

During 1941 MV began work on a turbofan version of the F.2 engine as a private venture, with the aim of creating an engine more economical for slower, long-range flight. Dr Baumann consulted with the RAE and then commenced design work on this so-called thrust Augmentor. In 1942 the MAP issued a contract for the construction of this thrust Augmentor attached to an F.2/2 turbojet, its designation being F.3. As we have seen, Whittle patented the turbofan in 1936 but his practical work on it was cancelled by the government in 1944.

Baumann's F.3 scheme was to take an F.2/2 turbojet and add, behind the two-stage turbine, another four-stage, contra-rotating turbine, which was connected to a two-stage, contra-rotating fan. The separate intake for the fan surrounded the engine and its cold air left the engine to surround the hot exhaust gases. The discs for the fan turbines were swept forward to allow room for four anti-friction bearings at the hubs while allowing the turbine blades to be close to the compressor turbines. The inlet duct surrounding the fan was supported at the front by streamlined guide vanes and, further aft, by extensions of the fairing spokes supporting the hot gas exhaust nozzle. It is instructive to compare the elegant simplicity of this design with that of the only wartime turbofan built outside Britain, the complex Daimler-Benz 109-007 ZTL engine.

The development of the F.3 turbofan engine proceeded with few problems and the design performance was met during bench testing. Compared with the F.2/2 turbojet, the F.3 gave a static thrust of 1,814kp (4,000lb) (1,089kp/2,400lb for the F.2/2) for a weight of 998kg (2,200lb) (748kg/1,650lb for the F.2/2). Furthermore, the specific fuel consumption went down from 1.05 to 0.65 and the cold, slower fan air surrounding the fast, hot air exhaust resulted in very marked noise suppression. The early development of a turbofan for civilian aircraft seemed to be at hand but, although MAP issued an order to flight test the F.3, this was prevented until after the war when, sadly, the development contract was cancelled. Nevertheless, the F.3 was the world's first practical turbofan engine. This only came about, however, because the government cancelled Whittle's aft fan W.2, which was so close to full development, and his L.R.1, which was almost ready for testing. One of the reasons the government cancelled Whittle's turbofan developments was that they did not wish Power Jets to be a competitor of the established aero-engine manufacturers and decided to sideline the ambitious Whittle commercially.

F.5 turbofan

Following the end of the war, MV explored another scheme using the F.2/4A turbojet to provide, in a similar manner, an open-fan thrust Augmentor that could compete

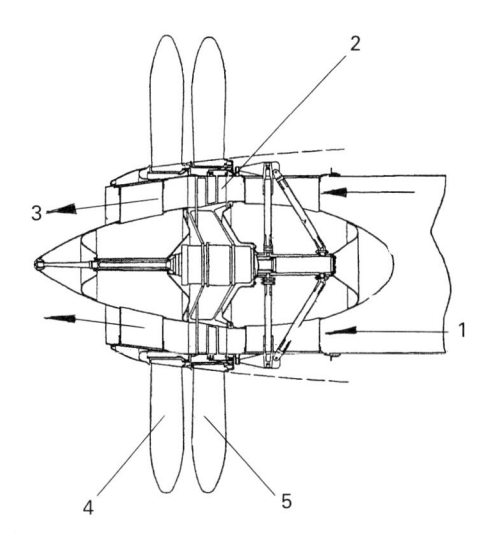

Metropolitan-Vickers contra-rotating, thrust augmenter unit to fit behind a standard F.2 turbojet (to create the F.5): (1) turbojet exhaust gases; (2) turbine; (3) exhaust efflux; (4) 14-blade fan; (5) 12-blade fan.

with the turboprop. The fan section could be added as a pusher fan behind the turbojet, which provided the hot gas to drive a four-stage turbine driving a contra-rotating fan. The fan was 1.676m (5ft 6in) in diameter, of fixed pitch and had fourteen blades on the front row and twelve blades on the rear row. Connection of the fan section or Augmentor unit was by a 0.661m (2ft 2in) diameter hot gas duct, which incorporated a streamlined bullet for the turbine bearings and an exhaust nozzle for the hot gases. In the case of the unit known as the F.5, the fan and duct was intended to be mounted behind the trailing edge of the wing or at the rear end of the fuselage. The F.5 was about 45kg (100lb) lighter than the F.3 turbofan and gave 322kp (710lb) more thrust. It gave a total sea-level fan and jet thrust of 2,136kp (4,710lb) and was designed for optimum performance at 644km/h (400mph) at an altitude of 6,100m (20,000ft). Testing of the F.5 began in 1945 but two years later MV decided to leave the field of aero-engines, remaining exclusively in the field of heavy industrial gas turbines; this was a most curious decision by a great engineering company but made under government pressure. Prior to this, in 1946, development had started on the F.9 turbojet, later known as the world-beating Sapphire, which was passed on to Armstrong Siddeley (*see* Chapter 15).

CHAPTER SEVEN

The de Havilland Engine Company

Whereas Metropolitan-Vickers's work on the turbojet had its roots in the axial studies of the RAE, that of de Havilland started by studying primarily Whittle's work and was therefore concerned with centrifugal engines. In this it was to be more successful than MV or Power Jets in getting its engines into operational aircraft and was to be the first British company to develop a turbojet engine for series production.

In the 1920s and 1930s the de Havilland company had been especially successful with their Gipsy piston aero-engines and various Moth aircraft. During the war its famous Mosquito twin-engined aircraft were to be particularly outstanding in many roles. In January 1941 the ARC, at the instigation of its chairman Sir Henry Tizard, invited de Havilland to design and build a jet fighter, giving them a free hand since no specification was issued.

A small team of about fifty people was brought together at de Havilland, headed by Major Frank Bernard Halford (1894–1955), J.L.P. Brodie and E.S. Moult. Frank Halford learned to fly at Brooklands in 1913, when he was eighteen, and was with the Royal Flying Corps in France during the First World War. He was brought back to England to utilize his experience with engines in the design of new ones. His 230hp B.H.P. engine was evolved into the Puma, thousands of which were made. B.H.P. and Puma engines powered the DH.4, DH.9 and other de Havilland aircraft. From this time began the close relationship between two great men, Halford and Geoffrey de Havilland. Halford designed many engines, including for de Havilland and Napier, but by 1937 he occupied offices in de Havilland's Stag Lane Works, where engines, airscrews and aircraft were designed, developed and built. Halford and his team remained independent until 1943, when they became part of the de Havilland company.

H.1 centrifugal turbojet

After studying the work of Whittle, the RAE and the Gloster company, Halford decided that their first turbojet should have a thrust of 1,360kp (3,000lb) and have a single-sided centrifugal compressor and straight-through flow using separate, can-type combustion chambers. Because of the lack of high-power test facilities, it was decided to use sixteen combustion chambers. The turbine was to be a single-stage type and the whole engine was to be kept as simple as possible. The compressor and turbine were connected by a tubular shaft and only two ball bearing races were used. Cooling of the turbine was by air piped from the compressor's diffuser casing, cooled in a ring or choke surrounding the air intake and then piped to the turbine bearing, and finally passed across the forward face of the turbine disc. The rear face of the turbine disc was cooled by uncooled compressor air, which passed through the streamlined support struts of the exhaust cone and discharged from a central pipe.

Keeping rapid development in mind, the axial compressor was avoided and a single-sided centrifugal compressor was chosen in order to take full advantage of ram effect and to avoid the bulkiness of the ducting and plenum chamber needed to feed a double-sided compressor. With Halford's engine, designated the H.1 and code-named 'Supercharger' (later named the Goblin), it was felt that the compressor chosen could almost touch the skin of the aircraft. To suit an aircraft wing root intakes, bifurcated air intakes were chosen for the engine and the space between these provided a convenient position for the auxiliaries. By April 1941 the essential features and layout of the H.1 had been decided.

The process of selecting the features of the H.1 engine was made in collaboration with the de Havilland aircraft designers. To limit duct losses, the shortest possible air intakes and jet exhaust pipe were envisaged. This indicated a short, pod-like fuselage containing the cockpit, fuel and engine, a wing with root intakes, and twin booms to support the tail empennages. The aircraft design was to become known as the DH.100 Spider Crab, later renamed the Vampire.

The first drawings for the H.1 engine, which by now had a MAP contract, were issued to the factory floor on 8 August 1941 and the prototype engine was run for the first time on a test-bed at Hatfield on 13 April 1942, only 248 days later. Initial problems included difficult starting (because the starting power required had been underestimated), buckling of the tailpipe and the insufficient capacity of the fuel pump. As these problems were overcome, the engine speed was gradually increased, but on 5 May 1942 the compressor sucked the intake flat and the engine stopped. The first run up to full speed was made on 13 June 1942 and results were most encouraging, the thrust being as designed for: 1,360kp (3,000lb) at 10,500rpm. However, another 2½ years of development lay ahead in order to achieve the reliability required and type approval.

On 8 December 1942 Specification E.6/41 was issued to de Havilland for a single-seat jet fighter to be powered by the H.1 turbojet. A minimum speed of 789km/h (490mph) and a service ceiling of over 14,600m (48,000ft) was called for in order to give a worthwhile advantage over the latest piston-engined aircraft. Armament was to be four Hispano 20mm cannon with a total of 600 rounds. An airframe to flight test de Havilland's turbojet was now needed but the company was too busy on essential Mosquito developments and the Hornet long-range fighter. Therefore, a visit was made to the Gloster company to see if its

The Gloster F.9/40 (DG206/G) fitted with Halford H.1 turbojets. Note 'Daunt stoppers' in the air intakes.
Philip Jarrett

projected F.9/40 fighter could be fitted with two of the H.1 engines. Gloster's Chief Designer Carter suggested rotating the H.1 through ninety degrees so that the bifurcated air intakes fortuitously passed above and below his wing spar. Some rearrangement of the engine was easy enough and mainly involved the lubrication system and the oil sump.

The first prototype of the F.9/40 (DG 202), fitted with two H.1 turbojets each rated at a static thrust of 907kp (2,000lb), made its first flight on 5 March 1943 from Cranwell. The pilot was Michael Daunt. By this time the H.1 had completed some 200 hours of bench testing. A second F.9/40 (DG207) was also fitted with H.1 engines, and this and DG202 were used as flying test-beds for the H.1.

Goblin turbojet into series production

The H.1 engine was soon into series production and had been named the Goblin. There is no doubt that it would have been used to power production Meteor fighters had not all production been earmarked for the Vampire. The prototype of the Vampire (LZ548/G) made its first flight at Hatfield on 20 September 1943, the pilot being Geoffrey de Havilland Jr. Its engine was giving 1,043kp (2,300lb) thrust at 9,500rpm and this first flight boded well for both airframe and engine.

By the end of 1943 bench running of the H.1 totalled about 700 hours and much flight data had been obtained at altitudes up to 10,700m (35,000ft) and at speeds close to 805km/h (500mph). An extension of the runway made at Hatfield was found to be unnecessary, even with only 1,043kp (2,300lb) thrust in the Vampire.

At about this time a Goblin engine was sent to Lockheed in California to fit in their prototype of the XP-80 fighter, later to be known as the Shooting Star. Unfortunately, severe damage occurred to this engine during initial ground running when the aircraft's intake ducts collapsed. A second engine was rushed out from Hatfield and this powered the prototype XP-80 (AF44-83020) on its first flight on 8 January 1944, giving a thrust of 1,116kp (2,460lb).

During 1944 development of the Goblin resulted in experimental engines being cleared for flight at 1,134kp (2,500lb) thrust at 10,000rpm, although one particular engine completed three 100-hour runs at 1,225kp (2,700lb) of thrust. By the end of 1944 more than two thousand hours of testing had been completed and plans were being made to put the engine into production. In January 1945 the Goblin became the first turbojet to pass an Air Ministry type approval test, its Type Certificate No. 1 stating the thrust to be 1,225kp (2,700lb). (The Rolls-Royce Welland was approved as reliable enough for RAF operational use by May 1944 but does not appear to have received a Type Certificate.) This point had not been reached without difficulties, including failures of the machined centrifugal compressor. A considerable amount of work involved developing a shrouded tail pipe with induced air flow cooling; this had become necessary after the asbestos lagging previously used became soaked in fuel on one occasion and caught fire following a baulked start.

The Goblin's centrifugal compressor had seventeen vanes, which were gently curved at their central intake portion. The size and shape of this compressor remained basically unaltered apart from improve-

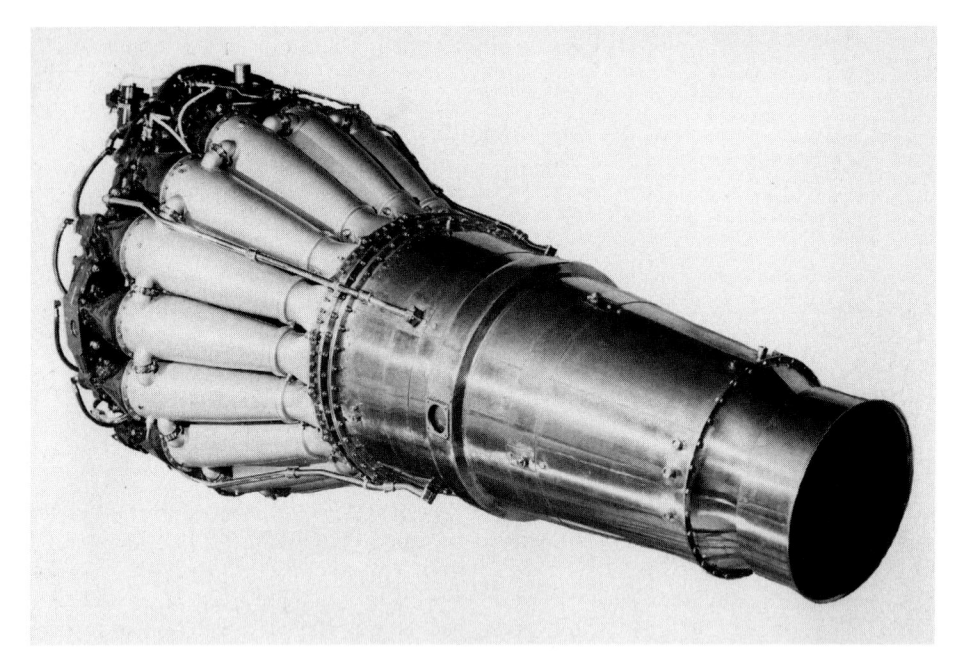

De Havilland Goblin II centrifugal turbojet.
Philip Jarrett

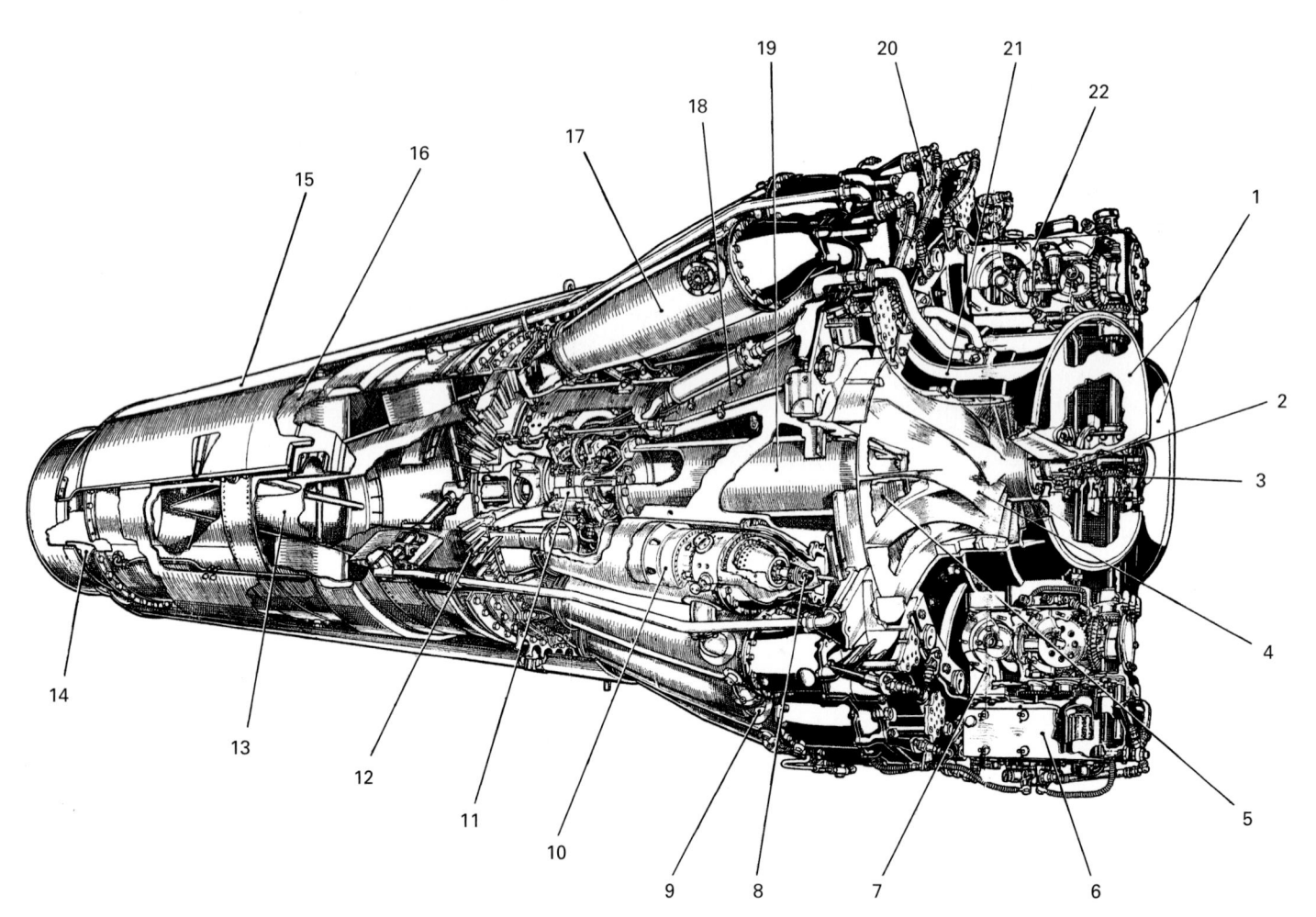

De Havilland Goblin Mk 35 turbojet: (1) bifurcated air intake; (2) front bearing; (3) accessory bevel gears; (4) single-sided centrifugal compressor; (5) back plate with labyrinth seal; (6) oil sump; (7) lower accessory gearbox; (8) fuel injector; (9) interconnector between combustion chambers; (10) combustion chamber; (11) rear bearing; (12) single-stage axial turbine; (13) exhaust duct cone; (14) exhaust nozzle; (15) fire guard; (16) jet pipe; (17) sixteen combustion chambers; (18) centre cone; (19) tubular main shaft; (20) diffuser casing with cascades; (21) intake casing; (22) top accessory gearbox.

ments in material and small blade form changes following vibration tests. The blade changes involved chamfering back the leading edges at the entry portion, which gave an improvement in the Mach number at the exit portion of the entry vanes. Increased thrust and lower air temperature resulted from this modification.

The first compressors were machined from 227kg (500lb) forgings of R-R.59 alloy. Machining reduced the weight of each forging down to 49kg (109lb), which was then heat treated. Using a master compressor to guide the special milling machine, two compressors at a time could be machined out. Testing of each compressor, up to 20 per cent over maximum speed, was carried out by mounting it in a vacuum box and driving it with a car engine.

The continuous and difficult development of the combustion chambers, as all turbojet engineers found, was the most troublesome part. De Havilland had the help of Imperial College of Science (London) and the Lucas company in this work.

For the first Goblin turbines, Firth's Rex.78, the best austenitic alloy steel then available, was used for the disc and its blades. It was then found that the disc temperature was not more than 250°C, far lower than the exhaust gas, and so a ferritic steel could be used. This gave advantages, including lighter construction, easier production and less expense. Accordingly, Jessop's H.3 silicon-chrome steel was used and, for the majority of engines, Hadfield's Hecla 153.

Further development of the Goblin

Early in 1945 the main development effort was focused on bringing the production Goblin up to its original design thrust of 1,360kp (3,000lb), but with improved reliability. This was achieved after something like another 1,500 hours of development testing, mainly on improved combustion chambers, and the design thrust was achieved at only 10,200rpm. After passing a type approval test the engine was known, by July 1945, as the Goblin II. Flight tests of the Goblin II began in the Gloster F.9/40 (DG207/G) on 24 July 1945. The total weight of the Goblin II was 703kg (1,550lb) and its specific fuel consumption was 1.233.

One of the main differences between the Goblin II and the earlier model was the design of the labyrinth seal on the rear face of the centrifugal compressor, which matched grooves on the sealing plate. The rotor shaft was machined from a hollow forging and bolted to both compressor and turbine discs. Roller bearings supported the rotor in front of the compressor, cooled by intake air, and just ahead of the turbine disc, cooled by air bled from the compressor between the sixteen combustion chambers. No thrust bearing was required since the single-sided compressor pulled forward while the turbine pulled aft, the two counteracting. The compressor featured seventeen vanes while the single-stage turbine featured eighty-three blades. For the turbine blades, Nimonic 80 was used and they were joined to the disc by fir-tree roots. The interconnected, can-type combustion chambers used about 20 per cent of the air for primary combustion and the other 80 per cent for secondary combustion, the initial combustion temperature of 2,000°C falling to not more than 790°C at the turbine inlet nozzles.

Further development led to the Goblin IV turbojet of 1,667kp (3,675lb) static thrust. In July and August 1948 a Goblin was given a most severe test on the testbed. It was run for seven weeks to give the equivalent of 462 combat sorties, each of 65 minutes duration. For each 'sortie', maximum power was used for 1½ minutes and 5 minutes to simulate take-off and combat respectively. The engine was still giving full power at the end of this rigorous test and so the whole test was repeated from January to March 1949. By this means, the engine achieved a TBO of 1,000 hours including 100 hours at full power. Halford's original design path, which led from Whittle's work, was truly vindicated.

CHAPTER EIGHT

Rolls-Royce

Development of the turbojet in Britain took a great step forward when Rolls-Royce came into the field and brought its great resources and expertise to bear by spring 1941. This great company was formed in 1906 to produce high-quality cars by Henry Royce and Charles S. Rolls, men of quite different backgrounds. Rolls had a wealthy background and came to engineering through studying at Cambridge, where he gained an honours degree. Royce, on the other hand, had humble origins and struggled to become an engineer. In 1910 Rolls was killed in a flying accident. The following year Royce's health began to fail but, working from his home, he continued to have a strong influence until his death in 1933. In the meantime, the company was run by Ernest Hives and Claude Johnson.

Ernest Hives and Stanley Hooker

With the outbreak of war in 1914, Rolls-Royce began working on aero-engines. Its experimental workshop was headed by Ernest Hives, who was to become of enormous influence in steering the company's business and engineering affairs until well into the jet age. Many great piston engines were developed and produced, helping the country to win wars and Rolls-Royce to prosper, including such famous names as the Eagle, Hawk, Falcon, Kestrel (1934), Merlin (1935), Griffon (1939) and, the last piston engine in the series, the Eagle (1942). Many engines were produced under licence, the most famous being the Merlin, which was essential to the winning of the Second World War.

By 1940 the official view at the Air Ministry was that Whittle's turbojet development taking place at Power Jets would be more effective in the hands of an established aero-engine company, the sooner to bring it into a reliable production form. The Director of Scientific Research favoured Armstrong Siddeley for this work but in due course, as we have seen, the Air Ministry issued contracts to the Rover and BTH companies with the intention that the first of many jet fighters and engines would begin production by summer 1942.

In January 1938 Hives at Rolls-Royce met and engaged Stanley Hooker, who was another figure to become instrumental in the company's success with turbojets and turboprops. Hooker was born in 1907, the son of a Kent farm manager, and showed an early gift for mathematics. He went to Imperial College and Oxford and made his reputation in scientific research at the Admiralty before going on to Rolls-Royce. There, his analytical work on the Merlin engine's supercharger, helping to bring the engine up from 1,000hp to 2,000hp, helped win the Battle of Britain. Interestingly, Hooker described the BMW 801 radial engine in Germany's Focke-Wulf Fw 190 fighter as 'Brute force and bloody ignorance!'

During the summer of 1940 Hooker began visiting Power Jets at Lutterworth and became an early member of the small group of supporters of Whittle's turbojet. It took several attempts to persuade Hives to visit Whittle's workshops, however, and then only after Hooker had stated that Whittle's turbojet was already producing a power similar to that of a Merlin piston engine in a Spitfire when flying at 565km/h (350mph). After hearing from Whittle about the problems in getting engine parts made, Hives soon had Rolls-Royce turning out turbine blades, gearcases and other parts for Power Jets. During 1941 they were also helping to solve compressor surging problems being experienced with the W.2 turbojet. For this, a special test rig was built

Ernest Hives, head of Rolls-Royce. Rolls-Royce Heritage Trust

using a R-R Vulture engine of 2,000hp to spin the compressor. All this activity was conducted as work funded by Rolls-Royce despite its heavy burden of piston engine development and production for the war effort. Late in 1941, however, the Air Ministry directed the company to build a Whittle-type turbojet designated WR.1. This engine was designed with low blade stresses to demonstrate reliability, since weight and size, at that stage, were not of importance. The WR.1 had a maximum diameter of 1.37m (4ft 6in) over the compressor, weighed 499kg (1,100lb) and gave a static thrust of 907kp (2,000lb). Two more WR.1s were built for intensive test work, together with enough parts for a further six.

During 1942 there was a deterioration in the relations between Power Jets and Rover on the one hand and Power Jets and BTH on the other. These companies' technical progress on Whittle's engines had been slow and poor. While Griffith's CR-1 engine could not be made to work at Rolls-Royce, the W.2/500 turbojet at Power Jets was progressing quite nicely, and so Whittle suggested to Hives that Rolls-Royce take over the production of it. At the time Rolls-Royce was testing the W.2B turbojet in the tail of a Wellington II bomber (Z8570/G) and the results helped convince Hives that the future lay with

The WR.1 turbojet, based on Whittle's work, was built by Rolls-Royce to demonstrate reliability; at that stage weight and size were not important. Rolls-Royce Heritage Trust

Sir Stanley Hooker, despite describing himself as 'not much of an engineer', was instrumental in bringing Whittle's turbojet to Rolls-Royce and to the company's success. Later he was the guiding light at Bristol engines. Rolls-Royce Heritage Trust

the turbojet. However, Rolls-Royce was already working flat out on Merlin piston engine production in support of the country's war effort, so additional facilities had to be found.

In December 1942 Hives and Hooker met Spencer Wilkes of Rover for dinner at the Swan and Royal Hotel at Clitheroe. This informal meeting has become famous for the way in which the Rolls-Royce tank engine factory in Nottingham was 'swapped' for Rover's turbojet factory in Barnoldswick, which Wilkes wanted to be rid of anyway. Informal as this swap was, it was carried out with the collaboration of the MAP. To Whittle's delight, the huge resources of Rolls-Royce, including almost 2,000 personnel, could now be focused on bringing his turbojet to production status, although the W.2B/23 engine was chosen for development rather than the potentially more advanced W.2/500.

B.23 Welland turbojet

Rolls-Royce retained Rover's team at Barnoldswick, including John Herriott and Adrian Lombard, and also moved Stanley Hooker (as Chief Engineer) and other personnel there in January 1943. Barnoldswick, a disused cotton mill, was supposed to put the engines into production but in fact was only ever used for development,

leaving Derby as the centre of production. Although he had a splendid relationship with Whittle and greatly admired his abilities, Hives was not happy that development continued at Power Jets's facilities at Lutterworth, and later at Whetstone, and thought that Rolls-Royce should be doing that also. At Barnoldswick, Hooker and his team had a completely free hand to carry out modifications to the engines and the company's great resources meant that such modifications could usually be carried out overnight, ready for fresh tests the following morning. Thus, work on Whittle's W.2B/23 engine, soon redesignated by R-R as the B.23 Welland I ('B' for Barnoldswick and Welland as the first of the 'river' names denoting 'flowing'), was speeded up immeasurably and was further facilitated by being accorded top priority by the MAP.

While Rolls-Royce was largely perfecting the mechanical aspects of the B.23 engine, Joseph Lucas Ltd tackled its combustion problems. The company had been brought into this work as early as 1940 by the then newly formed MAP to undertake basic research and had formed an outstanding research team at Birmingham under Dr E.A. Watson. Originally they used a large compressor from the pilot Dartford tunnel, driven by a 750hp motor. The compressor house and adjoining laboratory were built without windows to preserve secrecy. In

1941 the Lucas combustion organization was moved by MAP to improved facilities near Burnley, which it was hoped would also be safer from enemy attack. The new facilities included greater air flow capacity, sheet metal and mechanical workshops, chemical and physical laboratories, and the ability to build prototype combustion chambers. Initial investigations for Watson's team centred on Whittle's W.2 and W.2B engines. The principle of primary and secondary air supply to the combustion chamber was first embodied in the W.2B engine at the Rover company, and this basic design of combustion chamber and its principle was subsequently applied to all later engines, starting at Rolls-Royce with the B.23 Welland.

The Rolls-Royce B.23 Welland turbojet was first flown in the tail of a Wellington II (W5389/G) in May 1943 and gave a thrust of 726kp (1,600lb). Conversion of the Wellington as a flying test-bed had been undertaken by Vickers by removal of the tail turret to make a position for the turbojet and by fitting a complete set of test instruments in a compartment in the fuselage. This work was carried out early in 1942. Three Wellington IIs were thus converted: Z8570/G, W5389/G and W5518/G. The first two aircraft had a ceiling of 7,600m (25,000ft) but the third aircraft had a ceiling of 10,650m (35,000ft), owing to its Merlin engines being equipped with two-stage superchargers. There is a story, possibly apocryphal, of an occasion when a Wellington, with airscrews feathered and turbojet on, overtook a B-17 Flying Fortress to the amazement of its American crew!

On 12 June 1943 two B.23 engines were test flown in the Gloster F.9/40 fourth prototype (DG205/G), when the rated thrust of each was 658kp (1,450lb). On 24 July 1943 two B.23s were first flown in the first prototype F.9/40 (DG202/G), when the thrust of each was 771kp (1,700lb) using area-reducing inserts in the exhaust nozzles. At that time, there were never fewer than six F.9/40s undertaking test flying. By October 1943 the B.23 turbojet, still using the reverse-flow layout, was flight-cleared at 726kp (1,600lb) thrust and named the Welland I. This engine had a maximum diameter of 1.09m (3ft 7in), weighed 386kg (850lb), and had an air mass flow of 14.70kg/sec (32.4lb/sec), a combustion temperature of 754°C and a specific fuel consumption of 1.178. The design was now considered ready for production and,

The Gloster F.9/40 (DG205/G) fitted with two B.23 turbojets: (left to right) John Crosby, Michael Daunt, Frank McKenna (Gloster's General Manager), Frank Whittle and George Carter; this aircraft crashed at Minchinhampton on 27 April 1944, killing its pilot. Jet Age Museum

The first prototype of the Gloster F.9/40 (DG202/G), precursor of the Meteor fighter.
RAF Cosford Museum

by May 1944, approved as reliable enough for RAF operational use. A batch of 100 engines was delivered from Barnoldswick to power the F.9/40 fighter, now named the Meteor I (*see* Chapter 9).

From the small production batch of Welland I engines, an example was given a type test of 100 hours but went on for another 400 hours under these conditions. Because of this excellent performance, the Air Ministry increased the time between overhaul (TBO) rating from 80 to 180 hours.

Derwent turbojet

There was no doubt that the Welland I engine had great development potential and so a team under Geoffrey Wilde set about redesigning it in April 1943 under the designation of B.37, later to be known as the Derwent I. Many changes were made for the Derwent I, beginning with an increased airflow and changing to straight-through combustion chambers that originated with B.26 chambers from Lucas. The initial aim was to achieve a static thrust of 907kp (2,000lb), but within the dimensions of the Welland I engine in order to fit the Meteor's nacelles. Almost 20 per cent greater air mass flow through the engine was obtained by improving the diffuser shape leading from the compressor to the combustion chambers, and the turbine was redesigned to handle this. Other improvements included a new Lucas fuel system with barometric control to suit different altitudes, a recirculating oil system with oil cooler and expansion joints for the hot sections. The oil cooler was necessary when plain bearings were used on the early Derwent engines, but was dispensed with later on when ball and roller bearings were used instead. Meteors using engines with the oil cooler were easily identified by the vent on the upper, starboard side of each engine cowling. The combined delivery of the two oil scavenge pumps was passed through a cylindrical Serck cooler, mounted vertically on the starboard side of the compressor casing, and discharged into a tank. Air passing through the cooler was ducted from its upper end to the vent on the engine nacelle, but when the aircraft was stationary on the ground a depression inside the nacelle drew air in through the vent. Later there were some catastrophic failures of plain bearings (*see* Chapter 9).

On 29 June 1943, only a couple of months from inception, the first Derwent I made its first run and achieved a static thrust of 816kp (1,800lb). The engine passed its 100-hour type test at the design thrust of 907kp (2,000lb) in November 1943. On 20 December a Derwent I made its first test flight in a Wellington II (Z8570/G). Finally, two Derwent Is made the first flight in an F.9/40 (the eighth prototype, DG209/G) on 18 April 1944. By then the service rating was 816kp (1,800lb) thrust for a weight of 417kg (920lb). The combustion temperature had been increased to 843°C and the specific fuel consumption reduced to 1.083.

Before the end of 1943 John Herriott went from Barnoldswick to a new factory

A Gloster Meteor F.I (EE223/G), one of an initial batch of twenty, powered by R-R Welland I turbojets. 616 Squadron RAF was the first to become operational with jet aircraft, on 27 July 1944, pre-dating Germany's first fully operational jet unit by just over two months. Philip Jarrett

Rolls-Royce Derwent I centrifugal turbojet.
Rolls-Royce Heritage Trust

at Newcastle-under-Lyme to head production of the Derwent I. In the second half of 1944 some 500 of these engines were manufactured for the Meteor III. They improved on the speed of the Welland-powered Meteors by 89km/h (55mph) to give a top speed of 756km/h (470mph).

An intensive and continuous development programme followed upon this success and culminated in a successful 500-hour type test. Much was learned in this work and the following improved Derwent engines appeared:

- Derwent II: 998kp (2,200lb) thrust.
- Derwent III: Special engine providing a suction tap for boundary layer experiments with the Armstrong Whitworth AW.52.
- Derwent IV: 1,111kp (2,450lb) thrust.
- Derwent V: 1,588kp (3,500lb) thrust.

The Derwent V was actually an entirely new design that doubled the thrust of the original Derwents while retaining the maximum diameter of 1.09m (3ft 7in). Using Derwent V engines, a Meteor F.4 (EE549) set a new world air speed record of 991km/h (615.81mph), piloted by Gp Capt E.M. Donaldson, on 7 September 1946. By 1947 the Derwent V was brought up to a static thrust of 1,814kp (4,000lb). Among the many improvements to the engines was the superior compressor housing developed at Power Jets (and introduced on the Derwent II) and an enlarged compressor used for the Derwent IV. The incredibly rapid development of the turbojet, exemplified by the Derwent, could never have been achieved by any piston engine development. The importance of the Derwent is such that a few extra notes on it are worthwhile before moving on to the Meteor's story.

When the Derwent was designed, the double-sided centrifugal compressor was chosen owing to its wide operating characteristics (compared with the axial type), robustness, lightness, capacity for high speeds and resistance to icing. A lesser advantage was that forces on each side of the compressor balance out and so there is no need to make special provision for axial thrust. The double-sided version of the centrifugal compressor gives the maximum intake area within a given diameter. Finally, such a compressor was cheaper to produce than an axial type. (Note that, as we have seen, de Havilland found that their single-sided centrifugal compressor, which pulled forward, was largely and conveniently balanced by the turbine pulling rearwards). The Derwent's compressor was a single forging of R-R.56 from High Duty Alloys and had twenty-nine vanes on each side. After machining this forging, the entry part of the vanes were heated and bent to shape. It required about 4,000hp to drive the compressor at 16,600rpm, the vane tips reaching Mach 1.34.

From the ten Lucas combustion chambers, the hot gases were at about 850°C when entering the turbine. The forty-eight nozzle guide vanes before the turbine were individual precision castings of heat-resisting steel. As cast, they were correct to form within 0.13mm (0.0005in) and required only polishing. They were cast by a modernized, mechanized, Austenol system of the ancient 'lost wax' process. The turbine blades, on the other hand, were fully machined from a special Mond Nickel alloy, their aerodynamic form being the same as Whittle's first design but enlarged. The blades were fitted into the turbine disc (of Jessop G.18B steel) by their fir-tree roots and allowed a slight movement in the radial direction. As the blades accommodated to the heat and loads in operation, so the whole assembly tightened up.

It was found important to balance statically and dynamically, separately, the compressor and the turbine assemblies on their respective shafts. The shafts were connected by a ball spline coupling near the centre bearing of the engine. The reason for this was to avoid the imbalance that would occur if the rotating elements

A Gloster Meteor F.III jet fighter, painted all-white to assist wartime identification over Europe. This was the first mark to go into large-scale production. Philip Jarrett

A line-up of eight Gloster Meteor F.III fighters of 74 Squadron RAF. Philip Jarrett

The polished Gloster Meteor F.4 (EE549) being prepared for an attempt on the air speed record at RAF Tangmere. Gp Capt E.M. Donaldson flew this aircraft to a new speed record of 990.97km/h (615.778mph) on 7 September 1945. It was powered by R-R Derwent V engines. Philip Jarrett

were dismantled and reassembled. The turbine and its shaft, for example, could be removed and replaced without problems. Also, because of this design of rotor shaft, it was no longer necessary to keep the shaft as short as possible and so the longer, straight-through combustion chambers could more readily be used.

The supporting structure of the Derwent was largely made up from intricate castings, located by lips and rebates and bolted together. These castings incorporated such sections as the compresser diffuser vanes and support struts in the air intakes. Altogether, the castings were a tribute to the pattern-maker's art. Sheet metal was largely confined to combustion chamber construction.

Starting of the Derwent was by electric motor and was completed within 30 seconds. Ignition was by means of an electric igniter plug in combustion chambers 3 and 10, the flame then spreading to all ten chambers via the interconnecting, pressure-balancing pipes. Once idling, the engine could be immediately throttled up to full thrust. Before production ceased in 1954, more than 9,700 Derwent turbojets had been built in Britain and, under licence, in Belgium, where the Derwent was built by Fabrique National for its Fokker-built Meteors.

CHAPTER NINE

Meteor: The World's First Operational Jet Aircraft

The first eight prototypes of the F.9/40 twin-jet fighter, designed by George Carter at Glosters, were flown on the power of early examples of Metropolitan-Vickers, de Havilland and Rolls-Royce turbojets. Following on from the F.9/40 prototypes the fighter was ordered into production as the Meteor F.I (although the name of Thunderbolt was originally chosen until it was discovered that the Republic P-47 piston-engined fighter already had that name) and was later dubbed the 'Meatbox' by RAF crews. An initial batch of twenty of these fighters was ordered, to be powered by R-R Welland I turbojets; the first to fly was EE210/G on 12 January 1944, the pilot being Michael Daunt.

The Meteor was an all-metal, stressed-skin aircraft having unswept flying surfaces but of generally rugged construction. Its wing had two main spars and was divided into a centre section and two outer panels. The wing aerofoil section was EC 1240 for the centre section, tapering to EC 1040 at the tips. Each engine was fitted in its nacelle behind the forward wing spar and passing through a ring section of the rear spar. The top half and nose section of each engine nacelle was detachable for engine maintenance. The tricycle undercarriage, spilt flaps and air brakes were hydraulically operated while operation of the brakes and the cocking of the guns were carried out

Meteor nacelle intake, showing the streamlined wing spar passing in front of the Derwent V turbojet. Author

Meteor F.I			
Span		13.11m	(43ft 0in)
Length		12.57m	(41ft 3in)
Wing area		34.78sq m	(374sq ft)
Loaded weight		6,260kg	(13,800lb)
Maximum speed	at 3,050m (10,000ft)	660km/h	(410mph)
	at 9,150m (30,000ft)	716km/h	(445mph)
Time to	9,150m (30,000ft)		15 minutes
Service ceiling		12,800m	(42,000ft)
Range		805km	(500 miles)

pneumatically. Fuselage fuel tankage was of the self-sealing type and four Hispano 20mm cannon were fitted in the nose, the pilot having a gyro gunsight. Internally mass-balanced and aerodynamically balanced control surfaces were used and, after the F.9/40 prototypes, a bullet-type fairing was used at the junction of the fin and high-set tailplane.

Meteor operations and development

By May 1944 deliveries of these first Meteors to the RAF had begun and H.J. Wilson commanded a unit known as the CRD Flight under the auspices of the RAE at Farnborough. This flight had six Meteors by June 1944 and the next month these fighters were, beginning with EE213/G and EE214/G, transferred to 616 Squadron based at Culmhead in Devon. On 21 July 1944 this squadron began its move to Manston in Kent in order to help combat the German V1 flying bombs. These operations, known as 'anti-Diver patrols', began on 27 July 1944 but teething problems with the Hispano cannon led Flying Officer T.D. 'Dixie' Dean to destroy a V1 by tipping it over with his wingtip on 4 August. Soon the cannon problem was rectified and the jets were shooting V1s down at a steady rate. Thus 616 Squadron RAF was the first to become operational with jet aircraft and predated Germany's first fully operational jet unit by just over two months (Kommando Nowotny of the Luftwaffe beginning operations on 3 October 1944 with about twenty Messerschmitt Me 262A-1a jet fighters).

A number of RAF pilots visited Manston to try out the Meteor Is but many were not impressed. The faults they found included slow acceleration and deceleration (there were no air-brakes then), very short endurance, a long take-off and flat climb-out and heavy controls. All these characteristics were noted by pilots used to flying Spitfires, Tempests, Mustangs and so forth. Other faults found were snaking, which affected gun aiming, and insufficiently rapid response from the Welland engines should an overshoot off a landing be necessary. The snaking became severe at Mach 0.68 and violent at Mach 0.72. These first Meteors were fast but by the time the maximum speed of 772km/h (480mph) was reached it was time to think about landing. An American pilot invited to fly a Meteor declined, stating that, with the fuel available, he was normally thinking about landing, not taking off!

Despite all these frustrating shortcomings, the Meteor Is were a useful introduction to jet fighters and much better

A Gloster Meteor F.I (RA435) fitted with R-R Derwent turbojets to conduct afterburner flight tests.
Philip Jarrett

performance was not far away. In the meantime, the USAAF 8th Air Force had met with some of the enemy's first Me 262 jet fighters and Me 163 rocket interceptors over Germany and wanted some anti-jet practice. This was provided in October 1944 by some of 616 Squadron's Meteor Is detached to Debden. Unfortunately, the 8th Air Force tactics proved ineffectual against the Meteor 'attacks' since the jets flew into the bomber formations at will, 'destroyed' two or three bombers and easily evaded the escorting fighters. The future clearly lay with the jet.

Although the V1 launching sites in the French Pas de Calais had been destroyed or overrun by the Allies by the end of August 1944 and the Meteors had nothing left to shoot down, 616 Squadron was kept at Manston until the end of the year when their first Meteor IIIs were received. (The Meteor II was cancelled since its proposed de Havilland Goblin turbojets were all earmarked for the company's Vampire fighter). The first fifteen of the 280 Meteor IIIs built (EE230 to EE244) used Welland I engines but then the more powerful R-R RB.37 Derwent Is were fitted. Also, because of the more powerful engines, these aircraft had a slightly larger main fuel tank plus provision for a ventral tank of 455ltr (100gal). The Meteor III was the first mark to go into large-scale production.

616 Squadron moved to Colerne on 17 January 1945, and on 4 February a detachment of four Meteors (painted white to assist identification), led by Wg Cdr McDowall, flew to Melsbroek in Belgium. There, due to an absence of Luftwaffe aircraft to tangle with, this detachment joined the 2nd Tactical Air Force to carry out ground attack sorties in support of the Allied forces advancing into Germany. These sorties began on 16 April. The detachment was joined by 504 Squadron and a mixture of Meteor Is and IIIs were used. Already, on 1 April, the rest of 616 Squadron had arrived from England and the two Meteor squadrons progressed to airfields in the Netherlands and, later, Germany. As the Allies advanced the only Luftwaffe aircraft encountered were shot up on the ground and, much to the RAF's disappointment, no jet-versus-jet encounters took place.

The main problems found with these early Meteors were the jamming of cannon in the Mk I aircraft (this being caused by the airflow preventing the ejection of empty shell cases from the fuselage slots), overheating of the plain bearings in the Welland and early Derwent turbojets and compressor surging. The drill was to keep an eye on the exhaust gas temperature (EGT) gauge and to shut down an engine if its temperature rose too high. Landing on one engine was fairly common as also were forced landings (mainly in England) due to fuel exhaustion. The Meteor was easy to fly except for flying on one engine, which was very tricky and tiring since the wide spacing of the engines caused a strong turning moment and required full opposite rudder when overshooting. The minimum safe single engine speed was as high as 278km/h (173mph) and quite a number of pilots lost their lives landing on one engine. One pilot was killed after an engine failure when he was flying inverted at low level. The problem of surging was especially bad above 6,100m (20,000ft) and could happen at any time, but depended greatly on air temperature. Surging could be stopped by partially closing the throttles.

In order to raise the Meteor III's limiting Mach number, the engine nacelles were lengthened fore and aft, thereby increasing their fineness ratio. The last thirty Meteor IIIs were modified in this way and their limiting Mach number was thereby raised to 0.79. This version had the manufacturer's designation of G.41E. With Derwent I engines of 907kp (2,000lb) static thrust, the Meteor III could reach 737km/h (458mph) at sea level and 793km/h (493mph) at 9,150m (30,000ft).

By the end of the war 616 Squadron with its Meteor IIIs was based at Lübeck in Germany and the opportunity arose to compare its aircraft with the Luftwaffe's Messerschmitt Me 262 jet fighter. Two Me 262s were flown to Lübeck by Wg Cdr Schrader and Flt Lt C. Gosling. Gosling's view was that the Me 262 was a much better fighter than the Meteor III with regard to its better, lighter controls, greater firepower and excellent pilot's view. However, the Me 262's Junkers 109-004 turbojets were totally unreliable and not in the same league as the Meteor's Derwent engines. Also, the Me 262 was more difficult to fly and no aircraft for the novice pilot; the all too regular failure of one of its engines invariably led to a spiral dive and crash, unless the pilot took immediate corrective action. Of course, the ideal fighter at the time would have been an Me 262 with Derwent engines! Interestingly, the fatter engine nacelles of the Meteor III, in their lengthened form, were found to compare aerodynamically very favourably with the nacelles of the Me 262 with its axial engines.

Soon after the end of the war the RAF had two fighter Wings of Meteors, comprising six squadrons. The improved Meteor IV (now designated Meteor F.4 in the post-war style) was introduced in 1948 with the more powerful Derwent V engines, decreased wingspan to improve roll rate, the longer engine nacelles mentioned above and a longer forward fuselage section. With these improvements, the F.4 (Gloster designation G.41F) could reach over 644km/h (400mph) at sea level, flying level, on one engine! It was the first mark to have a fully operational pressurized

A Gloster Meteor F.3 (EE360/G) with R-R Derwent turbojets. Philip Jarrett

Meteor T.7 two-seat trainers: VW440 (nearest), VW416 and VW422.

cockpit, the air being bled from the engine compressor casings. It also had air brakes. In addition to a ventral fuel tank, there was the further option of fitting underwing drop tanks (normally only for ferry flights, because they degraded performance). Typically, when a Meteor F.4 was started quite long flames would initially issue dramatically from the exhaust nozzles of its Derwent V engines. Some 657 Meteor F.4s were built and a considerable number were exported, including 100 to Argentina, 48 to Belgium, 12 to Egypt and 65 to the Netherlands.

During 1945 various attempts were made on the world speed record using Meteors. These culminated on 7 September 1945 with a speed of 990.97km/h (615.778 mph), set by Gp Capt E.M. Donaldson flying an F.4 (EE549) from Tangmere. As a matter of interest, a day or two before this, Frank Whittle flew EE549 to entertain the waiting journalists, the record attempt having been delayed to wait for suitable weather conditions. Afterwards, when Whittle arrived home, his son Ian asked him how fast he had flown. When Whittle said 'About 450mph', Ian said 'Oh, is that all?' and turned away to carry on with whatever he was doing – leaving his father somewhat deflated!

A lot of high-altitude test work was carried out with Meteor F.4s (RA397, RA417 and RA438) by the Aircraft and Armaments Experimental Establishment (A&AEE) at Boscombe Down during 1947/48. Amazingly, these aircraft had no pressure cabins and the pilots were only sustained with oxygen and pressure waistcoats. The highest altitude reached was 15,300m (50,200ft) with RA417 in February 1948. Progressively less time was spent at altitudes above 12,200m (40,000ft) until only seconds could be spent above 14,000m (46,000ft). An F.4 became the first jet aircraft to appear on the British civil register: G-AIDC, used by the Gloster company largely as a demonstrator on overseas sales tours, was painted red with cream lettering and a fuselage flash.

A training version, the Meteor T.7, with a lengthened forward fuselage to accommodate a second seat, was introduced into RAF service and the Royal Navy in the 1950s. (Although Meteor F.3s had been used in deck landing trials in 1948, and had proved to be excellent for the purpose, the Meteor did not go into production for carrier use.) The T.7's cockpit was unpressurized and was flown on oxygen at all heights. Because of this, it was not permitted to fly above 10,700m (35,000ft) and pilots were restricted to spend only ten minutes above 9,150 m (30,000ft).

In 1949 a replacement fighter for the F.4 was introduced in the shape of the Meteor F.8, which became the RAF's

FACING PAGE: A Meteor F.8 (VZ460), standard day fighter of the RAF in the 1950s, in a dive.
Russell Adams Collection

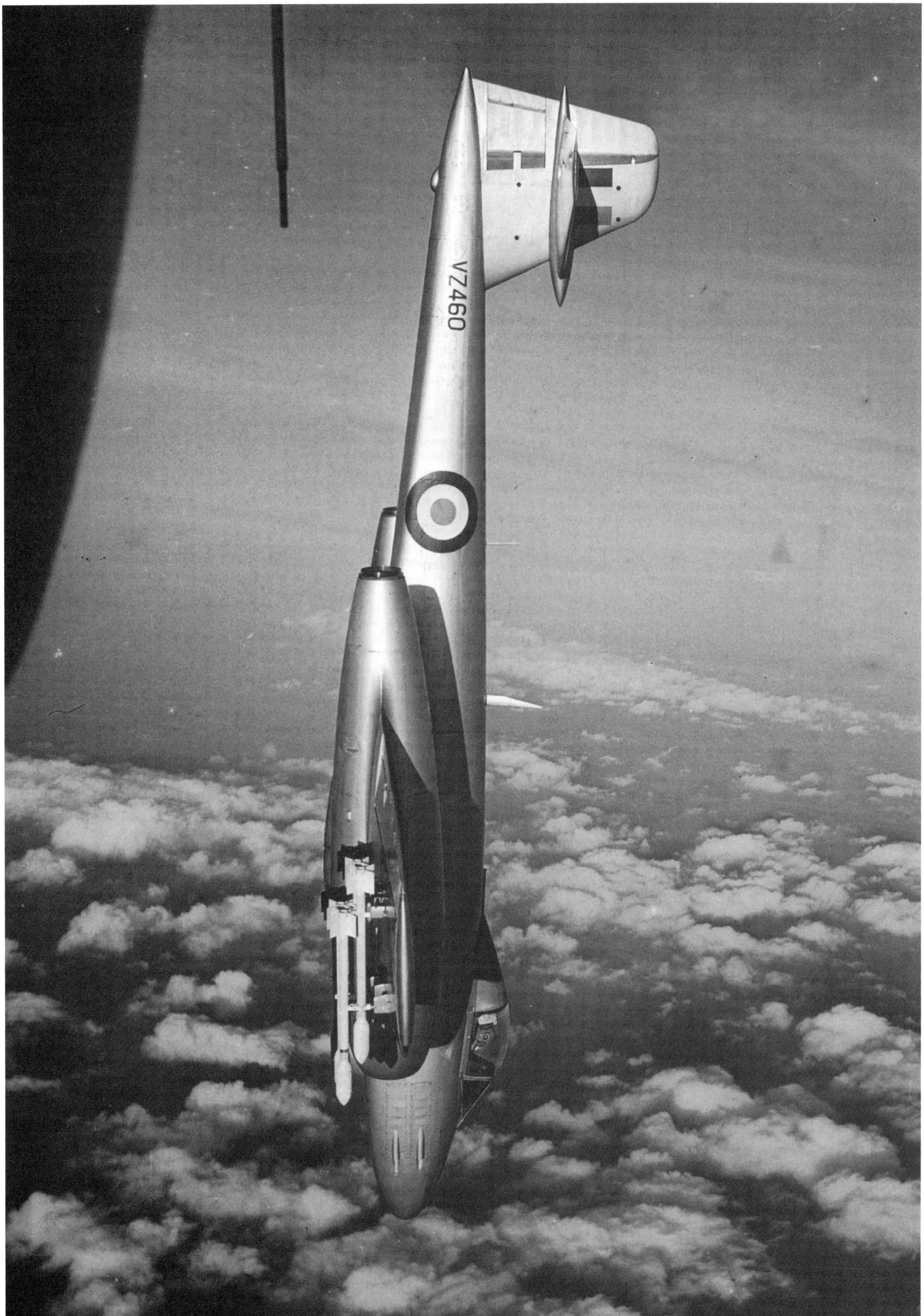
VZ460

Meteor F.8		
Span	11.33m	(37ft 2in)
Length	13.59m	(44ft 7in)
Wing area	32.55sq m	(350sq ft)
Height	3.96m	(13ft 0in)
Empty weight	4,820kg	(10,626lb)
Loaded weight	8,664kg	(19,100lb)
Maximum speed at sea level	950km/h	(590mph)
Service ceiling	13,411m	(44,000ft)
Range	1,578km	(980 miles)

standard day fighter of the 1950s. A very necessary increased fuel tankage was accommodated in a lengthened fuselage. Although as many F.4 components were used as possible, the F.8 required a longer tail unit. Power was provided by two R-R Derwent 8s of 1,633kp (3,600lb) static thrust. The F.8 was a great improvement over previous models, having a limiting Mach number of 0.82, and remained easy to handle. Care was needed not to exceed the limiting Mach number even in level flight, but snaking was considerably reduced. At Mach 0.82 the controls became ineffective and it was necessary to use the speed-brakes to reduce speed in order to regain control. However, Mach 0.82 was only achieved in a shallow dive and with plenty of power on. The F.8 was fitted with the Martin-Baker ejection seat. The only known successful bale-out from a Meteor without an ejection seat was made on 8 January 1950 from a Meteor F.3 (EE472) by Sqn Ldr Dennis Barry of 616 Squadron. His aircraft suffered total engine failure and was icing up in poor weather. After ejecting the canopy, Barry managed to avoid the tail of the Meteor by stalling it and, when the port wing dropped, pushing himself out into the centre of the ensuing spin.

As aircraft speeds built up, thanks to the turbojet and advancing aerodynamics, so it became imperative to provide an automatic and practical means of emergency ejection. In this sphere the Martin-Baker Aircraft Co. Ltd was to reign supreme. Under the guidance of its Chief Designer and Director James Martin, a development programme was begun in 1944. The first British in-flight ejection, using a dummy, was made on 11 May 1945 from a two-seat Boulton Paul Defiant (DR344), the pilot being Brian Greenstead. The work, including ground tests on rails, culminated in the first live in-flight ejection on 24 July 1946 when Bernard Lynch was ejected from a Meteor F.3 (EE416) piloted by J.E.D. Scott. (Pioneering work on ejector seats was carried out in Germany by the Heinkel company from about 1944 and they were first fitted operationally to its He 219A Uhu night fighters, in which the seats saved lives. The first jet aircraft to receive an ejector seat was the barely operational Heinkel He 162 Salamander,

Gloster Meteor NF.12 night fighters forming up over their airfield; nearest is WD603. Philip Jarrett

owing to the close proximity of its dorsal turbojet to the cockpit.)

Beginning in June 1950, the F.8 entered RAF service and eventually equipped thirty-one fighter squadrons. Photo reconnaissance versions, the FR.9 and PR.10, eventually equipped six RAF squadrons. A considerable number of the 1,183 F.8s built were exported. Most countries that had the F.4 replaced them with F.8s; Australia, Brazil, Belgium, Syria and Israel also bought F.8s. Others were built under licence in Belgium and the Netherlands.

In regular exercises against the Sea Fury, F-84 Thunderjet and F-86A Sabre fighters, the Meteor F.8 was only superior at low level, when it had better turning, climbing and acceleration performance than even the F-86A, which had greatly inferior air brakes. However, once above 7,600m (25,000ft) the Meteor was outclassed and its normal maximum operational ceiling was 10,700m (35,000ft). It was also hard pressed to intercept jet bombers such as the Canberra and B-47 Stratojet; 245 Squadron RAF had the embarrassment of being unable to intercept the prototype Comet jet airliners being flown on test flights. Nevertheless, the Meteor F.8 was the favourite mount of many aerobatic teams due to its excellent handling qualities. The short range of the Meteor could be increased by flying on one engine, re-lighting for landing being easy and reliable. The single-engine range of an F.8, for example, was increased by 23 per cent at sea level but less so as height was increased until, at 7,600m (25,000ft), the increase was zero.

As the Gloster company at Hucclecote became occupied with work on the Javelin delta night fighter, so the W.G. Armstrong Whitworth company at Coventry was brought into Meteor production, beginning with the last forty-five F.4s (VZ386–VZ429 plus VZ436) and continuing with some 430 of the F.8s. This company was then given the task of developing a night fighter version of the Meteor as a stopgap measure pending the introduction of the Javelin. Work began at Armstrong Whitworth's experimental works at Bitteswell in January 1948, using a converted Meteor (VW413), which was the fourth production T.7 machine. Flight trials showed up few prob-

Meteor NF.14			
Span		13.11m	(43ft 0in)
Length		15.65m	(51ft 4in)
Wing area		34.78sq m	(374sq ft)
Height		4.24m	(13ft 11in)
Empty weight		5,398kg	(11,900lb)
Loaded weight		9,616kg	(21,200lb)
Maximum speed	at 3,050m (10,000ft)	930km/h	(578mph)
Service ceiling		12,200m	(40,000ft)
Range		1,480km	(920 miles)

lems and the first Meteor night fighter, the NF.11, was produced to the official Specification F.24/48 issued on 12 February 1949. It was based on the two-seat T.7 trainer in order to carry the pilot and navigator/radar operator. The fuselage was extended by 1.53m (5ft 0¼in) in order to accommodate AI Mk.10 radar (of wartime vintage) and extra fuel. Because of the nose radar, four Hispano Mk.5 20mm cannon were mounted in the wings. The prototype (WA546) made its first flight on 31 May 1950 and production began at Coventry on 13 November 1950. The NF.11 entered service in August 1951, with 29 Squadron RAF at Tangmere, finally giving a replacement for the Mosquito night fighter.

The NF.11 night fighter was followed by other versions: the NF.12, with a longer, cleaner nose accommodating Westinghouse APS-57 radar; the NF.13 tropicalized version of the NF.12; and the last, and finest, version, the NF.14 with a clear-view, one-piece, sliding cockpit canopy and general improvements. Beginning with the NF.12, the engines were the final production version of the Derwent, the Mk.9, which gave a static thrust of 1,724kp (3,800lb) that compensated for the increases in weight.

The NF Meteors were stable and rock steady with no tendency to snake but were rarely used for ground attack sorties. They could reach a maximum ceiling of 16,150m (53,000ft) but were then totally unmanoeuvrable. Both NF and F.8 types could intercept bombers of the Lincoln, B-29 and B-36 type, but had great difficulty intercepting Canberras and B-47s in exercises.

Meteors would prove inferior when pitted against Soviet jet aircraft in Korea. The last built, an NF.14 (WS848), left the production line in May 1955. In RAF service the retirement of the F.8 was delayed until April 1957, owing to problems with the next generation of Swift and Hunter swept-wing fighters.

More Meteors were built than any other British jet aircraft, the number being 3,947 including 547 night fighters. From this total, 225 were later converted into target tugs and target drones. Many were also used for engine and weapon test-beds and much experimental work. In the latter category, of particular importance to this history is the Meteor Mk.I (EE227), which became the world's first turboprop-powered aircraft. It was fitted with two R-R Trent turboprops, which were Derwent I turbojets each modified to drive a 2.41m (7ft 11in) diameter Rotol five-bladed airscrew through reduction gearing. This Meteor first flew on 20 September 1945, piloted by Eric Greenwood, each Trent developing 750shp plus 567kp (1,250lb) of thrust. The aircraft had increased fin and rudder area, including two extra fins added to the tailplane, to help counteract asymmetric drag in the event of one engine failing (*see* Chapter 14). Other Meteor Mk.Is (EE215 and RA435) were used in 1944 to conduct the first British afterburner test flights.

The cost of a standard, fully armed Meteor F.4 in 1949 was £30,468 and this rose, for example, to £42,810 in 1952 for an export version of the Meteor F.8.

CHAPTER TEN

Vampire and Venom

Before the war ended, the MAP invited de Havilland to produce a more powerful turbojet than its Goblin II, which at that time was producing 1,360kp (3,000lb) static thrust. The company decided that the best way forward was to produce a scaled-up version of the Goblin II and the first drawings were made in April 1944. The new engine, designated DGT/40, was to produce a static thrust of 1,814kp (4,000lb) and have an overall diameter of 1.448m (4ft 9in). This centrifugal engine had fourteen combustion chambers, each supplied with air from a single diffuser outlet. Two DGT/40 experimental engines were built; the thrust was later increased to 1,996kp (4,400lb), indicating that the design was along the right lines. However, the MAP considered the engine's diameter too large for any aircraft then in prospect and so a redesign was undertaken in 1945.

The diameter of this redesigned engine, designated DGT/50, was, at 1.35m (4ft 5in), only 76mm (3in) greater than the Goblin. The reduction in diameter was achieved by using cascade vanes to deflect air into an axial direction as soon as it left the compressor's diffuser vanes. Only ten combustion chambers were used, but of larger volume, in order to increase efficiency. To provide sufficient air to these larger combustion chambers, each was fed by two outlets from two diffuser ducts.

The first test run of the DGT/50 (later to be named Ghost) was made in October 1945 and the engine reached its design static thrust of 2,268kp (5,000lb) early in 1946. Its first test flight was made in a DH.100 Vampire F.1 (TG278) on 8 May 1947, when it was giving a thrust of 1,996kp (4,400lb). Beginning on 31 July 1947, an Avro 691 Lancastrian (VM703) was used as a flying test-bed, the aircraft being fitted with two Merlin piston engines and two DGT/50 or Ghost turbojets (engines 1 and 4), the full thrust being developed in due course.

DH.100 Vampire jet fighter

Simultaneously with the development of turbojets at de Havilland, the company was also working on a single-engine jet fighter that was closely tailored to the H.1 or Goblin turbojet. This fighter was designed to meet the Air Ministry Specification E.6/41, which called for an experimental prototype with a pressurized cockpit and provision for an armament of four of the new Hispano 20mm cannon. A top speed of 790km/h (490mph) and a service ceiling of over 14,630m (48,000ft) was called for. Design work began as early as 1942 under the designation DH.100 and the code name of Spider Crab. The fighter later became known as the Vampire and was extremely successful, introducing many of the world's air forces to the jet age. Never has Britain exported more jet aircraft of one type.

The design team for the DH.100 was led by R.E. Bishop who, back in 1938, had been prominent in Geoffrey de Havilland's team that designed the outstanding Mosquito. Bishop was able to create a small airframe by virtue of the H.1 turbojet, with its single-sided centrifugal compressor, only needing to be fed by ram air from the front (whereas Welland and Derwent engines, derived from Whittle's work, required air fed to their double-sided compressors to come from a sealed plenum or pressure chamber, which was larger than the engine). The H.1 engine was designed with a bifurcated intake that matched the very short wing root intakes of the DH.100, while the small pod-like fuselage also gave a short exhaust duct for the engine; this

A de Havilland Goblin turbojet with J.L.P. Brodie, Maj Frank B. Halford (Chairman and Technical Director of the de Havilland Engine Co. Ltd) and E.S. Moult.

The first de Havilland DH.100 (Vampire) prototype (LZ548/G), with Halford H.1 turbojet. Philip Jarrett

In April 1946 the de Havilland Vampire jet fighter became operational. This formation of six F.1s was part of 247 Squadron RAF, based at RAF Odiham. They were powered by the de Havilland Goblin 1 turbojet. Nearest is TG/311. Philip Jarrett

DH.100 Vampire F.1 fighter (with Goblin 2)		
Span	12.19m	(40ft 0in)
Length	9.37m	(30ft 9in)
Wing area	24.74sq m	(266sq ft)
Height	1.91m	(6ft 3in)
Empty weight	2,890kg	(6,372lb)
Maximum loaded weight	4,754kg	(10,480lb)
Maximum speed at low level	869km/h	(540mph)
at 9,150m (30,000ft)	821km/h	(510mph)
Initial climb rate	1,311m/min	(4,300ft/min)
Range	1,175km	(730 miles)

all added up to considerable duct efficiency. Due to this efficiency, light construction and clean lines, only one engine was needed to obtain the requisite performance. Since the effects of jet exhaust on the tailplane were then unknown, it was decided to use twin booms with twin fins to support the tailplane above the exhaust line. Even this arrangement was checked by mounting twin booms and a tailplane behind an engine on the test bench.

A mock-up of the DH.100 was essentially complete by September 1942 at the experimental section at Hatfield and this was soon followed by the commencement of construction of three prototypes. Drawing upon experience with the Mosquito, the DH.100 had a smooth fuselage pod constructed largely from balsawood sandwiched between plywood. This was made in two halves that were fitted out before being brought together. The pilot's cockpit (unpressurized in the first three prototypes) was situated well forward, affording a superb view, and was enclosed by a bubble canopy. The tapered wings, booms and tail empennage were all of flush-riveted, stressed-skin, aluminium construction and the simple controls were manually operated. Access for maintenance was exceptionally easy since the aircraft sat on a short tricycle undercarriage, the main wheels retracting outwards and the nosewheel retracting rearwards between the cannon (although the first prototype was unarmed). Actuation of the undercarriage, flaps and dive brakes was hydraulic, while actuation of the brakes and canopy seal inflation was pneumatic. To access the engine, the top and bottom cowlings and the exhaust pipe were removable. Internal fuel capacity was 917ltr (202gal).

The first DH.100 prototype, LZ548/G, made its maiden flight from Hatfield on 20 September 1943 piloted by Geoffrey de Havilland Jr, and this was followed by the other two prototypes, LZ551/G and MP838/G. The flight trials that followed threw up very few problems, although by January 1941 the tops of the fins had been flattened off. By early 1944, with a Goblin engine of 1,225kp (2,700lb) thrust, the DH.100 was flying at speeds up to 853km/h (530mph). On 8 January 1944 Lockheed's XP-80 fighter prototype (AF44-83020) first flew, using the engine taken from the second DH.100 prototype (thereby delaying it for eleven weeks). These aircraft were the first to exceed 805km/h (500mph), apart from Germany's wartime jet and rocket aircraft.

On 13 May 1944 de Havilland received an initial order for 120 (later increased to 300) of the fighters, now officially named Vampire F.I (F.1 later). Due to the company being fully committed on Mosquito work (and later on the even more remarkable Hornet), production was turned over to the English Electric Company at Preston. The first production Vampire F.1 (TG274/G) made its maiden flight on 20 April 1945 from Salmesbury where final assembly and test flying took place. Test flying was also undertaken at Hatfield and official establishments and the first sixteen production Vampires joined the prototypes in this flight development programme. The first prototype (LZ548/G) was destroyed following an engine failure during a take-off at Hatfield but fortunately de Havilland's test pilot Geoffrey Pike survived.

The second prototype, LZ551/G, became the first jet aircraft to land and take off from an aircraft carrier, following trials at RAE Farnborough and RNAS Ford. To make the tests, the aircraft was modified with 4 per cent more flap area, an arrester hook and lengthened undercarriage oleos. The first carrier landing was made on HMS *Ocean* on 3 December 1945, the pilot being the very skilful Capt Eric M. Brown. This successful landing was made despite the inclement weather, considerable pitching of the ship and the problems associated with the slow spool-up of the engine. The landing was soon followed by a very short take-off and further successful trials. Nevertheless, the Vampire was considered unsuitable for naval use because of its short endurance and the slow spool-up of the engine militating against a baulked landing. Other shortcomings with early Vampires were that the Goblin engine could not be relit in flight (normal starting on the ground being by a plugged-in 24-volt battery starter) and, until ejector seats were fitted, baling out was problematical.

The endurance and power shortcomings were addressed firstly by increasing the internal fuel tankage to 1,480ltr (326gal) and then by making provision for a pair of 454ltr (100gal) underwing drop tanks.

DH.113 Vampire NF.10 night fighter (with Goblin 3)		
Span	11.58m	(38ft 0in)
Length	10.54m	(34ft 7in)
Wing area	24.27sq m	(261sq ft)
Height	2.01m	(6ft 7in)
Empty weight	3,931kg	(8,666lb)
Loaded weight	5,942kg	(13,100lb)
Maximum speed at low level	866km/h	(538mph)
Initial climb rate	1,372m/min	(4,500ft/min)
Ceiling	12,800m	(42,000ft)
Range	1,964km	(1,220 miles)

The second prototype (TG/306) of the de Havilland DH.108 Swallow, one of three Goblin-powered research aircraft based on the Vampire fuselage. It was intended for tests up to Mach 1.0, but TG/306 crashed on 27 September 1946, killing its pilot, Geoffrey de Havilland Jr. An early Vampire (TG285) is also on view. This photograph was taken at the 1946 Society of British Aircraft Constructors Show at Handley Page's Radlett airfield. Philip Jarrett

DH.112 Sea Venom FAW.22 (with Ghost 105)		
Span	13.08m	(43ft 11in)
Length	11.15m	(36ft 7in)
Wing area	26.02sq m	(279.75sq ft)
Height	2.60m	(8ft 6¼in)
Empty weight	4,733kg	(10,435lb)
Maximum loaded weight	7,167kg	(15,800lb)
Maximum speed at low level	960km/h	(596mph)
at 9,150ft (30,000ft)	894km/h	(555mph)
Initial climb rate	1,783m/min	(5,850ft/min)
Ceiling	12,802m	(42,000ft)
Range	1,135km	(705 miles)

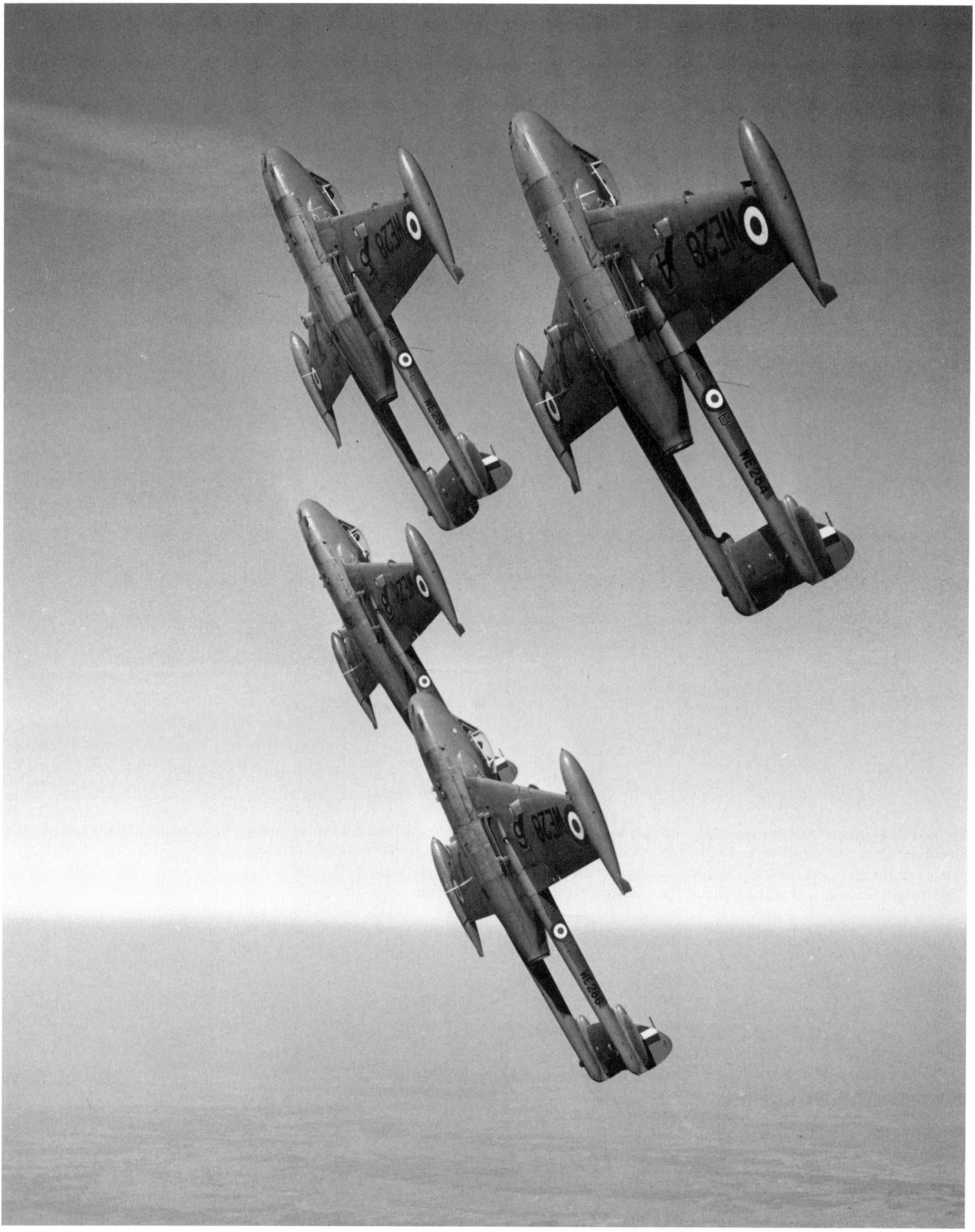

Also, various engines of greater thrust (e.g. the R-R Nene) were tested in 1946, TG276 and TG280 being sent to Rolls-Royce at Hucknall for this purpose. TG275 was tested with a Goblin 2 engine of 1,406kp (3,100lb) thrust and was the prototype for the Vampire F.3. TG278 was built with a larger engine compartment and extended wingtips to test the DH Ghost engine. Up to 2,268kp (5,000lb) thrust was promised with these engines.

A great deal of Vampire testing was carried out including service evaluation to ready the fighter for RAF service and export to other air forces. However, it was too late to see wartime service, the first deliveries of F.1s being in April 1946 to 247 Squadron RAF at Odiham. This squadron was able to fly their new jets in the VJ Day fly-past over London on 8 June 1946.

Through Mach 1.0

In the meantime, de Havilland was working on three Goblin-powered prototypes of a research aircraft, the DH.108 Swallow, designed by John Wimpenny. This small, tailless aircraft consisted of an adapted Vampire fuselage, with an extended tailpipe section, a single fin and rudder and a wing swept back at 43 degrees. It was produced in accordance with Specification E.18/45 to provide swept wing data for the DH.106 Comet jet airliner (the world's first) and the DH.110 jet fighter. The first DH.108, TG283, made its maiden flight on 15 May 1946 from RAF Woodbridge. Used chiefly for low-speed tests, TG283 crashed following a stall on 1 May 1950, killing the pilot Sqn Ldr Genders. The second DH.108 prototype, TG306, first flew on 23 August 1946 from Hatfield and was intended for tests up to Mach 1.0. This aircraft soon exceeded 965km/h (600mph) and was set fair to break the world air speed record. Tragically, however, during a practice flight for this on 27 September 1946 it apparently hit unexpected turbulence and broke up, killing the pilot Geoffrey de Havilland Jr. The final, modified, prototype of the DH.108, VW120, made its first flight from Hatfield on 24 July 1947, piloted by John Cunningham. On 12 April 1948 John Derry flew this aircraft to a new 10km international closed-circuit speed record of 974km/h (605.23mph). John Derry also took this aircraft through Mach 1.0, the first in Britain to successfully do so, by diving it from about 12,200m (40,000ft) on 9 September 1948. On this and a subsequent supersonic flight the aircraft went out of control until it entered the denser air of lower altitude. This last DH.108 Swallow was finally destroyed on 15 February 1950 during an RAE test, probably owing to oxygen failure rather than to aerodynamic problems.

Venom jet fighter

Most of the fighter, fighter-bomber and night fighter versions of the Vampire were powered by Goblin engines, but the upgraded version, known as the Venom, was powered by the Ghost 103 turbojet of 2,200kp (4,850lb) thrust. The Venom had various other new features including a thinner-sectioned wing with a leading edge sweepback of 17° 6' and the option of carrying wingtip fuel tanks. The prototype Venom FB.1 (VV612) made its maiden flight on 2 September 1949. The Venom proved to be a useful combat and ground attack aircraft and could climb significantly faster than the earlier Vampire.

Various marks of Venom were produced and also night fighter versions of Vampire and Venom. A design team under W.A. Tamblin at Christchurch produced the Ghost-powered Sea Venom, a naval all-weather fighter based on the night fighter Venom NF.2 and fitted with folding wings and a V-frame arrester hook. It was introduced into FAA service in March 1954, pending more advanced types.

Apart from prototypes, production of these de Havilland fighters amounted to 3,042 Vampire fighters, 94 Vampire night fighters, 965 Vampire trainers, 858 Venom fighter-bombers, 281 Venom night fighters (including 250 to Switzerland) and 349 Sea Venoms (including Aquilon versions built for the French Navy), making a grand total of 5,589. These large numbers include many exported all over the world to Sweden, Switzerland, Canada, Australia, Norway, India, France, South Africa, Italy, Egypt, Venezuela, New Zealand, Finland, Iraq, Lebanon, Southern Rhodesia, Portugal, Chile, Finland, Burma, Indonesia, Japan, Jordan, Ireland, Ceylon, Syria and Austria. In addition, many were built under licence abroad. The large numbers of aircraft and customers reflects the economy and success of the de Havilland products. Regarding economy, a Vampire could be bought for about 70 per cent of the cost of a Meteor, a Goblin-powered Vampire F.5, for example, being sold for £22,000 in 1949. Particularly popular were the Vampire trainers that introduced so many pilots to the techniques of flying jet aircraft. Because the de Havilland turbojets were so reliable, they had no means of relighting following a flame-out, resulting in the loss of some Vampires and Venoms when this occurred.

FACING PAGE: **Powered by the more powerful de Havilland Ghost engine, the Venom replaced the Vampire in many RAF squadrons. These Venom FB.1 fighter-bombers, entering a loop, are from the 2nd Tactical Air Force. They are WE284 (nearest), WE285, WE286 and WE288.** Philip Jarrett

A formation of de Havilland Sea Venom FAW.21s of the Royal Navy's No. 890 Squadron, with the witches emblem, in 1956. Power was provided by the 2,245kp (4,950lb) static thrust Ghost 104 turbojet; from the top are WW150, WW217 and WW201. Philip Jarrett.

CHAPTER ELEVEN

Britain's First Jet Aircraft at War

As we have seen, combat between German and Allied jet aircraft did not take place during the Second World War, but RAF Meteors were in action against V1 flying bombs and attacking ground targets on the Continent in the absence of the Luftwaffe. However, the outbreak of the Korean War on 25 June 1950, which began with piston-engined aircraft, soon saw jet aircraft in action. American F-80 and F-84 jets were among the first in action but were joined by Meteors of Australia's RAAF in the early months of 1951. UN forces soon found that these straight-wing jets were no match for the swept-wing Soviet MiG-15s, which first swept into North Korea on 1 November 1950. Notwithstanding this, the first jet-versus-jet victory occurred on 8 November 1950 when an F-80C Shooting Star shot down a MiG-15, but combat with the MiG-15 on more equal terms had to wait for the introduction of the USAF's swept-wing F-86 Sabre in December 1950.

The low wing loading of the Meteor enabled it to outturn the MiG-15, but the MiG could outclimb and fly much faster than the Meteor and had a higher limiting Mach number. Therefore, the RAAF's Meteors were unable to dogfight with the MiG-15s and they were found to be more useful in the ground attack role. On 1 December 1951, however, some fourteen Meteor F.8s of No.77 Squadron heading north into 'MiG Alley' encountered about forty MiG-15s high above them. In the ensuing fight both sides scored hits and two MiGs and three Meteors went down. In this encounter the first aerial victory for a Meteor was chalked up when the Meteor F.8 (A77-15) flown by Fg Off Bruce Gogerly shot down a MiG-15. Pilots ejected from two of the downed Meteors and went into captivity.

In early May 1952 another 77 Squadron Meteor F.8 shot down a MiG-15, but was itself shot down during a ground attack mission in April 1953. Such victories for the Meteor were usually only possible due to its tight turning circle and superior air brakes and were only won with difficulty. In fact, only the fourth (and last) Meteor victory over a MiG-15 occurred on 27 March 1953 near Pyongyang, the North Korean capital: on this occasion the RAAF pilot Sgt George Hale was flying his Meteor F.8 A77-851 'Halestorm'.

Many other MiG-15s were damaged by Meteors, some recorded as probables. In general the Australian pilots were superior to the North Korean and Chinese pilots flying the MiGs, but it should be noted that the MiG-15 was a 'tough cookie' and could sustain a lot of damage. A number of Meteors were lost to the deadly ground fire during ground attack sorties. In 37 months of combat in Korea, the RAAF lost 42 pilots, 32 of them in Meteors, during 18,872 individual sorties covering escort and ground attack missions.

Vampire, Venom and Meteor around the world

Although the Royal Navy flew carrier-based piston-engined aircraft in Korea, the RAF did not take the Meteor or Vampire to Korea and RAF pilots usually flew USAF jets as 'observers' in combat. It was not until 1955 that the Vampire was blooded in combat. On 29 August 1955 two Israeli Air Force Meteor F.8s were patrolling at 6,100m (20,000ft) over the Negev inside Israel when their pilots spotted four Egyptian Air Force Vampire FB.52s below. In a diving attack, the Israelis shot down one of the Vampires, the other three escaping in a long dive over the Mediterranean. Two days later, a dogfight between two Israeli Meteors and four Egyptian Vampires saw two of the Vampires crash in flames. On 12 April 1956 another Vampire FB.52 was shot down by the Meteors.

When Israel acquired more modern French fighters (as Egypt acquired MiG-15s) its Meteors were relegated to the training role. Pilots found the Mystère IVA much easier and more pleasant to fly than the Meteor. However, both sides used their Meteors during the Sinai Campaign of 1956 and the Egyptians also used their Vampires. The victories of the Meteors over the Vampires were really a result of the better training of the Israeli pilots compared with that of the Egyptians, since in Britain the Vampire F.5 had already been assessed superior in combat to the Meteor F.4. Compared with the Meteor, the Vampire was found to be more manoeuvrable, lighter on the controls and a steadier gun platform: in the right hands a Vampire would have no difficulty in getting onto a Meteor's tail during a dogfight.

The Venom was even more manoeuvrable and could even outturn and outclimb the F-86 Sabre, which in exercises could not reach Venoms above 12,200m (40,000ft). Carrier-borne Sea Venoms and Cyprus-based Venoms took part in the Suez Crisis. Large numbers of Meteor F.4s were sold to Argentina, where both government and rebel factions flew them during the Revolution of 1955.

Meteors and Vampires, sometimes called first-generation jets, appeared in conflicts all over the world. The French, for example, used Sud-Est-built Vampire fighters (known as Mistrals) in the Algerian war from 1954. Both sides in the India/Pakistan conflict in the 1950s also used Vampires. Once the Second World War was over, Britain set about regaining Malaya from the Communists, a costly business that dragged on until the late 1950s and saw the use of Meteor PR.10s, Vampire F.5s and Venoms, mostly flying from Tengah in Malaya. Various marks of the aircraft also appeared in the 1950s with the RAF protecting oilfields in Iraq (flying from Habaniyah) and protecting oil shipping plying between

Instead of using an electrical accumulator trolley to start the engine, a cartridge starter is used by these de Havilland Venoms. They are with 142 Squadron RAF at Eastleigh, Nairobi, in 1959. Philip Jarrett

the Persian Gulf and the entrance to the Red Sea (flying from Khormaksar in Aden).

From 1953 the Middle East Air Force of the RAF sent Vampires and PR Meteors to assist in quelling the rebellion in Kenya. At the end of 1956 the Suez Crisis broke and Vampires, Venoms and Meteor night fighters were deployed from Cyprus and Malta. The Royal Navy also deployed Sea Venoms from three aircraft carriers in this conflict. The actual Suez war began on 29 October 1956 with French, British and Israeli forces opposing Egypt. Both sides employed a mixture of piston aircraft and various jets. The Egyptian Air Force included Meteor fighters and night fighters and Vampires, often escorted and covered by their MiG-15s. The Israeli Air Force used Meteors, usually in the ground attack role. It was already evident that the first-generation jets were usually outclassed by later aircraft, such as the MiG-15 and F-86 Sabre.

The RAF soldiered on with Vampires, Venoms and Meteors into the mid-1950s, bolstered up by Canadian Sabres, until the Hunter entered service. The last operational sortie by an RAF Meteor was flown by an NF.14 night fighter of 60 Squadron based at Tengah, Singapore, in September 1961. The last six Meteors in operational service anywhere were six of the twelve FR.9s in service with the Fuerza Aerea Ecuatoriana (Ecuadorean Air Force) until 1979. A last few Sea Venoms, FAW.53s, were in service with the Royal Australian Navy until January 1970, but Venom FB.4s stayed on in Swedish Air Force service until the latter half of 1983. Such Venoms continued beyond their service life, showing no signs of wooden structure failure but requiring strengthening work on the metal structure. In Zimbabwe, ex-Southern Rhodesian Vampires were still giving service in 1982.

CHAPTER TWELVE

Further de Havilland Developments

As related, the company's successful Goblin turbojet lent itself readily to redesign, leading to development of the more powerful Ghost turbojet, which was giving its designed thrust of 2,268kp (5,000lb) early in 1946. The Ghost was to power Venoms and Sea Venoms and its reliability was soon proved. It was designed to the same basic formula as the Goblin and also incorporated many of the lessons learned from its predecessor.

As the H.2, the Ghost fitted into the forward-looking development programme planned by Frank Halford in 1944, which covered the following projects:

- H.3 A 500hp turboprop.
- H.4 A very large axial turbojet (later named the Gyron).
- H.5 Further development of the H.2 Ghost.
- H.6 Smaller version of the H.4 (Gyron Junior).
- H.7 Gas producer for helicopter rotor tip drives.

Of these, the H.3 turboprop was only bench run. The H.2 Ghost was built in large numbers and also under licence in Sweden, Switzerland and Italy. However, the greatest moment for the Ghost engine was in the powering of the world's first true jet airliner designed as such, the de Havilland DH.106 Comet. The Comet was preceded by the Avro Lancastrian, which had two of its engines replaced with turbojets, the first of six flying in October 1945, but these were flying test-beds. Also preceding the Comet was a Vickers-Armstrong Viking (G-AJPH), which had its two Hercules piston engines replaced by two R-R Nene I turbojets; in this form it first flew on 6 April 1948. This one-off conversion was ordered by the Ministry of Supply (MoS) to provide early jetliner data. It was also used (as VX856) as an executive transport for government officials.

DH.106 Comet, world's first true jet airliner

The history of the pioneering Comet has its origins in late 1941, even as war was raging, when a committee considered what civil aircraft were likely to be needed by Britain after the war. One of its recommendations was for a jet-powered mail carrier. In May 1943 a second committee, under the chairmanship of Lord Brabazon, changed the mail carrier proposal to one for a fourteen-passenger jet airliner with the same capacity for mail and a range of about 1,300km (800 miles), this proposal being designated Type IV. The aim was to break the airliner dominance of the USA by a bold technological leap forward. Strangely, jet bombers were being considered in the USA but not jet airliners.

The Brabazon jet airliner proposal soon became associated with the de Havilland company since it had strong airframe and turbojet design teams under R.E. Bishop and F.B. Halford, respectively. Early in 1944 the de Havilland teams found the time to study many airliner layouts, including twin-boom aircraft and a tailless one, but eventually decided on a more conventional low-wing monoplane layout with a tail. It was recognized that, at that time, a useful transatlantic aircraft would not be possible and so the design was settled as a short-range, fourteen-passenger airliner. Because the project was shrouded in secrecy, the only customer, initially, was to be the British Overseas Airways Corporation (BOAC). Designing began in September 1946 and the MoS ordered two prototypes. The first production order, for eight aircraft, came from BOAC in January 1947 and in December that year the name Comet was announced.

Meanwhile a great deal of work had to be done to develop the Ghost into an engine suitable for the civilian market. Halford realized that this market needed an engine that was particularly reliable, safe and possessed of long life. Accordingly, more than 60 per cent of the military Ghost had to be redesigned and this led to a programme of intensive work. This included 13,000 hours of bench running and over 5,200 hours of flight testing in two Lancastrian aircraft and, later, the prototype Comet. On 23 March 1948 a Ghost engine powered a Vampire (TG278), piloted by John Cunningham, to a record height of 18,124m (59,446ft), which was higher than it was planned to fly the Comet.

Even at the planned operating height of 12,200m (40,000ft) the Comet was to be pressurized to an altitude of 2,450m (8,000ft), which was double the pressure of any previous airliner. There was the added stress of an outside air temperature of –70°C. Bleed air from the engines was used to pressurize the fuselage and also to prevent icing of the wings and tail unit. Since de Havilland chose to use Redux metal-to-metal bonding on a previously unheard of scale, it was necessary to conduct many tests, including pressure testing of large fuselage sections. On one occasion this resulted in the explosive destruction of a section to an extent that made the tracing of the original cracking impossible. This led to underwater pressure testing subsequently.

As the design of the Comet proceeded, many changes were made. The sweep-back of the wing was approximately halved to 20 degrees on the leading edge and the fuselage was lengthened to accommodate more passengers, up to thirty-two. This led to an increase in wingspan and weight and a reduction of the design speed from 861km/h (535mph) to 813km/h (505mph).

The large-area wing had most of its volume taken up by fuel tanks except where the main wheel wells and control surfaces were and where the engines were buried. A new feature of pressure refuelling was adopted, large diameter hoses attaching beneath the wings. By this means, the 27,467ltr (6,050gal) of fuel could be pumped aboard in 20 minutes. Large, plain flaps were fitted, inboard and outboard, with split flaps beneath the jet pipes. There were slim, perforated air brakes just before the outboard flaps. Two Ghost 50 engines of 2,293kp (5,050lb) thrust were buried in each wing root, each being fed from a plain, oval air intake and exhausting via a long jet pipe that extended just beyond the wing's trailing edge. Provision was made in the wing structure for the installation of rockets between each pair of engines in case assisted take-off was needed in hot or high-altitude locations overseas. The fuselage had a circular cross-section of 3.05m (10ft) diameter. Aft of the four-crew cockpit were the galley, forward toilets and then the passenger seating, which extended as far as the wing trailing edge, followed by the rear toilet and mail compartment. There was a slender fin and tailplane, the latter having dihedral.

On 25 July 1949 the first of the two Comet prototypes was rolled out and taxiing tests were made on the following two days. In the late afternoon of 27 July this prototype, with the 'B' registration of G-5-1 (later registered G-ALVG), made its first flight. The pilot was John Cunningham. Exactly one year later, on 27 July 1950, it was joined by the second prototype, G-ALZK, and extensive flight tests were made with them both. Both prototypes had single, large mainwheels but in December 1950 G-ALVG was tested with the four-wheel main bogies that were to be used for production aircraft. The same aircraft also tested, from 7 May 1951, a de Havilland Sprite rocket motor between each pair of jet pipes. The Sprite was a neat, stored-liquid rocket unit of 2,268kp (5,000lb) thrust. From the flight tests it was established that a full load, with up to thirty-six passengers, could be carried over a range of 2,818km (1,750 miles), cruising at 789km/h (490mph). Many records were broken on these test flights and a certificate of airworthiness was issued on 21 April 1950.

Production of the Comet 1 was shared between de Havilland's plants at Hatfield and Chester. The first production aircraft, G-ALYP, made its maiden flight on 9 July 1951. A certificate for passenger carrying was issued on 22 January 1952. Regular passenger service began on 9 May that year on the route from London to Johannesburg,

The prototype de Havilland DH.106 Comet (G-5-1), the world's first true jet airliner, landing after its first flight on 27 July 1949. Philip Jarrett

via Rome, Beirut, Khartoum, Entebbe and Livingstone. Piston-engined aircraft times were halved. Passengers immediately noticed, and soon loved, the smooth, almost silent flight that was well above any poor weather. A favourite pastime was to balance a pencil on its end on a table where it usually stayed for a long time. Apart from BOAC's nine Comet 1s, other customers to order the Comet included Canadian Pacific Airlines (CPA), Air France, UAT and the Royal Canadian Air Force (RCAF).

For these new customers de Havilland upgraded the airliner to the Comet 1A, which had extra fuel capacity, forty-four seats and water/methanol injection, which was supposed to maintain the take-off thrust of the Ghost engines at hot or high-altitude airports. For the Comet 1A the weight increased to 52,164kg (115,000lb).

Sadly, tragedy stalked the Comet. The first two accidents happened as a result of rotating the aircraft too early during take-off, thereby stalling the wing and reducing airflow into the engines by the high angle of attack. The first such accident happened with BOAC's G-ALYZ on 26 October 1952 at Rome, but luckily without any fatalities. However, the second accident occurred with CPA's CF-CUN on 3 March 1953 at Karachi, when all on board died. The cure for such happenings was to introduce the drooped leading edge to the wing and to ensure correct rotation procedure was followed.

More disasters, with far more serious technical flaws, followed. On 2 May 1953 (almost the first anniversary of BOAC Comet services) G-ALYV crashed when climbing out of Calcutta; this was mistakenly attributed to monsoon turbulence. On 10 January 1954 G-ALYP disappeared after leaving Rome and precisely the same thing happened to G-ALYY on 8 April 1954. The remainder of BOAC's Comet 1 fleet was grounded and there followed an immense and ground-breaking investigation into the causes of the crashes. From February 1954 the Royal Navy located and began salvaging most of the wreckage of G-ALYP near Elba while the fuselage of G-ALYU was put through hundreds of simulated flight cycles in a giant water tank. All Comet 1s and 1As were returned for other tests, or scrapped, except those of the RCAF, which remained in service after modification.

The cause of the disaster was found to be the explosion of the fuselage following a fatigue crack at the corner of one of the rectangular cut-outs for the ADF aerials in the top fuselage skin. The cure was to redesign the fuselage with all cut-outs, windows and doors with oval shapes or with rounded corners and to use thicker skin for many sections. Also, if bonding could not be used, rivet holes were to be drilled and not punched out. The results of the tests were made generally available to the world of aviation and the redesigning resulted in the Comet 2, deliveries of which began to the RAF from June 1956.

The time was also right to find a more powerful replacement for the reliable Ghost engines and two Comet 2s were delivered to Rolls-Royce to flight test the Avon turbojet (see Chapter 13). By the time greatly improved Avon-powered Comet 3s and 4s entered service in the late 1950s, the airline world was dominated by Boeing 707s and Douglas DC-8s, but the Comet went on to some success and to prove what a pioneering technical achievement had been made by de Havilland in producing both airframes and engines.

Gyron turbojet

The two years needed to investigate the early Comet disasters meant that the Ghost engine was overtaken by other designs and developments: the H.5 development of the Ghost was not built and the de Havilland Engine Company decided to concentrate on the very large H.4 axial turbojet. Known as the Gyron, this engine was the first British design aimed at high supersonic speeds of between Mach 2.0 and 3.0. Although it was not destined for large-scale production figures, it introduced many new features.

Its seven-stage axial compressor had the modest compression ratio of 6:1 and most of the compression was to be obtained from ram from about Mach 2.0 onwards. The Gyron had a two-stage turbine and an annular combustion chamber with sixteen Duplex burners. It was aimed at a static thrust of 9,080kp (20,000lb) or 12,258kp (27,000lb) with afterburning. The engine was envisaged as being highly suitable as part of the rocket-plus-turbojet mixed powerplant system advocated by de Havilland and proposed for use in projected designs of a long-range supersonic bomber, an all-weather supersonic fighter and a cruise missile, to name a few.

On 15 January 1953 the Gyron made its first test bench run. Later that year it passed a type test at 6,810kp (15,000lb) thrust and had reached 9,080kp (20,000lb) thrust by September 1954 (as the D.Gy.2).

De Havilland D.Gy.2 Gyron. This turbojet reached a static thrust (dry) of 9,080kp (20,000lb) by September 1954. Philip Jarrett

Comets in formation in 1951: from the front, G-ALYP (first production aircraft), G-ALVG (first prototype) and G-ALZK (second prototype). Philip Jarrett

On 7 July 1955 the first test flight of a Gyron (of 15,000lb thrust) was made by replacing the port lower R-R Avon (6,500lb thrust) engine in the first of the two prototypes of the Short Sperrin heavy bomber (VX158). The Sperrin had first flown on 10 August 1951, piloted by Short's Chief Test Pilot Tom Brooke-Smith, as the SA.4. It had a shoulder-mounted straight wing to give a capacious bomb bay and its four engines were installed vertically in two nacelles. It was later overtaken by the successful Vickers Valiant, the first of Britain's V-bombers, and relegated to a flying test-bed role. The test flights with the Gyron were mostly made from Hatfield and from 26 June 1956 two Gyrons were installed in place of two Avons in the lower positions. By early 1957 the Sperrins had been scrapped.

In 1957 the British Government's White Paper on Defence made it clear that it saw the future of offence and defence being with the guided missile, leaving little future for manned aircraft. The MoS therefore terminated its financial support of the Gyron engine. Thus, after more than four years of development work and over 20,000 hours of bench and flight testing on the Gyron, the engine had no immediate prospect of powering an aircraft unless these were privately sponsored. Therefore, because aircraft companies such as Hawker had projects for the Gyron in the pipeline (e.g. the Hawker P.1121 replacement for the Hunter), de Havilland decided in March 1957 to continue with Gyron development at its own expense, such was the faith in the engine.

The successful de Havilland policy of obtaining reliability through simplicity, prevalent in the Goblin and Ghost engines, was continued in the Gyron. Some of its more interesting features will now be mentioned. Its design was biased towards lightweight construction and this was largely achieved by adopting only a moderate pressure ratio, which required only a seven-stage compressor, and an annular combustion chamber; both these features led to a short engine that in turn only needed a two-bearing layout. This helped in keeping the weight down and was a new departure in an engine the size of the Gyron. For it to succeed, an exceptionally rigid compressor rotor and main shaft was required and this was achieved by the use of thin discs carrying the compressor rotor blades and drums to join with adjacent rotor discs. Centrifugal loads in the rotor blades were utilized to tighten the discs and thereby increase their axial stiffness. The main shaft, attached to the rear of the compressor rotor, consisted of a thin-walled trumpet shape that tapered to a stub shaft at its rear end where it carried the two turbine discs. Because the complete rotor assembly of the engine had its maximum diameter in the centre, tapering down towards the front and rear, an exceptionally rigid assembly resulted.

To obtain stall-free acceleration, the compressor was provided with variable-incidence inlet guide vanes, which were each fitted at their inner end with a short swivel extension arm located in a spherical bush in a circular actuator ring carried at the rear of the intake centre housing. The actuator ring was rotated by means of a pair of hydraulic jacks fitted within the ring itself. The system operating these jacks was under the command of an automatic control unit responsive to selected temperatures, pressures and speeds.

For the intake and compressor casings, a change from aluminium alloy to steel was only made as dictated by temperature. However, all the compressor blades were machined from stainless steel forgings because of the good vibration damping and other qualities of this metal. Mounting of the rotor blades on pins to allow limited tip rock added a further, successful means of damping vibration. The bearing at the front of the compressor rotor comprised two ball races and one roller race.

To prevent the escape of compressor air between the combustion system inner chamber and the main turbine shaft, an ingenious sealing system was used that employed spring-loaded stationary carbon segments and a rotating labyrinth seal. It was designed in such a way that compressed air prevented any more than light rubbing of the carbon and so the system had a long life.

In the quest for reducing weight, the Halford Laboratory experimented with thin compressor blades and blades and rotors made from titanium. A Gyron engine built with these features was tested to a thrust of 10,433kp (23,000lb) with afterburning.

The Gyron had the first large British annular combustion chamber with an atomizing fuel injection system and this was largely developed by the Joseph Lucas company. The system used eighteen Lucas Duplex 3 burners giving upstream fuel injection. A final choice of seventeen burners was made in order to eliminate resonant coupling in the turbine blades.

The two turbine wheels were forgings of heat-resisting ferritic steel and the turbine blades were machined from Nimonic alloys, the rotor blades being attached by fir-tree roots. The forward bearing, from which the turbine wheels were overhung, were of the roller type. Cooling air, tapped from the inner casing of the combustion chamber, was used for items such as the turbine inlet guide vanes, exhaust cone and supporting spokes.

Starting of the engine was either by electric motor or de Havilland hydrogen peroxide turbo-starter. Also developed were several types of variable-area exhaust nozzle, including a mechanical multi-petal type and a combined petal and aerodynamic type. By 1955 the PS.26-6 Gyron with afterburner was giving a thrust of 10,840kp (23,900lb) at a speed of 6,400rpm. By 1957 the Gyron recorded a thrust of 12,700kp (28,000lb) using a simulated intake of Hawker's projected P.1121 fighter.

Gyron Junior turbojet and the Blackburn NA.39 Buccaneer

Given the Government's waning interest in future manned military aircraft, however, de Havilland had little choice but to put the promising Gyron on the back burner and concentrate on the Gyron Junior turbojet, which was a 0.45 scaled down version of the Gyron. This smaller engine was foreseen as the powerplant of Blackburn's NA.39 low-level bomber. Later famous as the Buccaneer, this was the world's first specifically designed high-speed strike aircraft, the design team being led by B.P. Laight. It was intended to meet an Admiralty requirement for a naval aircraft capable of delivering a nuclear weapon by

A de Havilland D.G.J.10 Gyron Junior turbojet, scaled down from the Gyron.
Philip Jarrett

Buccaneer S.1		
Span	13.41m	(44ft 0in)
Length	19.33m	(63ft 5in)
Wing area	47.89sq m	(515sq ft)
Height	4.95m	(16ft 3in)
Loaded weight approx.	20,410kg	(45,000lb)
Maximum speed at sea level	Mach 0.95	
Tactical radius of action	800–965km	(500–600 miles)

penetrating beneath enemy radar at a speed in excess of Mach 0.9.

The Gyron Junior turbojet was designed to be produced more economically than the Gyron having, for example, cast turbine blades. The PS.43 or Gyron Junior 101 version had a take-off rating of 3,320kp (7,100lb) and also a large air bleed flow for boundary-layer control for the Buccaneer's wing. This latter feature was an essential element in the design of the wing, which was small for low-level manoeuvrability and for storage on aircraft carriers. A small wing with boundary-layer control had already been used on Supermarine's Scimitar fighter. Other features of the Buccaneer were the twin engine nacelles alongside the fuselage with plain, circular intakes, high-mounted tailplane, a large split tailcone-type air-brake and two seats enclosed by a bubble canopy. Another important aerodynamic feature was the area-ruled fuselage, which stemmed from the wartime work of the German Junkers company attributed to Prof. Hertel, among others. In this concept, which was to be employed in many post-war, high-speed aircraft, the cross-sectional area of the aircraft, including the fuselage, wing and engine nacelles, was made to conform to an equivalent but streamlined longitudinal shape. Thus, because the wing and engine nacelles were, apart from their position, fixed, the cross-section of the fuselage had to be altered and this led to a narrowing of the waist in the vicinity of the wing. Inevitably it became known as the 'Coke bottle' fuselage.

The maiden flight of the Buccaneer prototype (XK486) took place on 30 April 1958 from Thurleigh (RAE Bedford), the pilot being Lt Cdr D. Whitehead and the observer B.J. Watson. Deck landing trials aboard HMS *Victorious* began in January 1960 and the Buccaneer S.1 went into Royal Navy service that July. It went on to have a long and useful career, beyond the timescale of this history, including export to South Africa, and plugged the gaps left by the problematical F-111 and cancelled TSR-2. However, with the introduction of the Buccaneer S.2 in 1964, the Gyron Junior engines were replaced with R-R Spey engines, which gave more thrust while using less fuel.

Bristol Type 188 research aircraft

The other highly interesting employment of the Gyron Junior turbojet, in a much modified form and with an afterburner, was to power the Bristol Type 188, which was that company's first turbojet-powered aircraft. The project originated in February 1953 when Bristol was invited to tender for an experimental aircraft capable of sustaining a speed of Mach 2.0 long enough for study to be made of the consequent kinetic heating effects on the structure. The requirements were embodied in the official Specification ER.134 and required a maximum speed of Mach 2.50. By 1955 a layout had been decided upon and comprised a minimal, sleek fuselage with thin wings and a broad fin supporting a high-mounted, all-moving tailplane. Two Gyron Junior DGJ.10 engines were enclosed in long nacelles, each mounted halfway along the span of each wing. Pressure recovery or air compression and slowing down of the intake air to the engines was achieved by passing it through shock waves generated by the sharp lip and movable, conical centre body formed at each nacelle intake. The development of this so-called supersonic diffuser was another German wartime development and was carried out under Dr Klaus Oswatitsch and his assistant H. Böhm at the Kaiser-Wilhelm-Institut at Göttingen from 1941.

The intake of the port engine nacelle of the Bristol Type 188 research aircraft is a prime example of the German Oswatitsch conical shock wave diffuser for supersonic flight. On view at the RAF Cosford Museum. Author

The stainless-steel Bristol Type 188 (XF923), powered by two modified de Havilland Gyron Junior turbojets with afterburners (D.G.J.10s). The aircraft was designed to investigate kinetic heating effects of sustained speeds up to Mach 2.0. Philip Jarrett

The structure of the Type 188 was primarily of stainless steel and included many specially made bolts, rivets and screws to ensure compatible expansion at the high temperatures expected. Many manufacturing problems had to be overcome and here the assistance of Armstrong Whitworth was enlisted. The oval-sectioned fuselage, no larger than the cockpit dimensions, largely contained the fuel but also instruments, telemetry and cockpit refrigeration equipment.

In May 1960 the first of three prototypes (c/n: 13518, XF923) was delivered to the RAE at Farnborough. Following adjustments to the air intakes and taxiing trials, the prototype made its first flight on 14 April 1962. Although among the world's fastest turbojet aircraft at that time, the Type 188s were unable to deliver much useful data on kinetic heating since the excessive fuel consumption of the 6,350kp (14,000lb) static thrust Gyron Junior engines prevented flights of sufficient duration. At altitude, maximum speeds of over 1,930km/h (1,200mph) were reached.

In addition to turbojet development, de Havilland developed a number of successful rocket motors that were employed chiefly for assisted take-off. They ran on high-test peroxide for the oxidant and kerosene for the fuel. These rocket motors included the Sprite for the Comet 1, Super Sprite for the Valiant V-bomber, and the Spectre for the Victor V-bomber and SR 53 interceptor. The company also built, under licence, General Electric's T58 turboprop engine as the Gnome, which was first run in 1959. Finally, in 1961 the de Havilland Engine Company, along with Blackburn, was absorbed into the Bristol Siddeley organization.

De Havilland built General Electric's T58 turboprop under licence as the Gnome turboshaft engine. Here, a senior technician attached to the Royal Norwegian Air Force 330 Squadron inspects one of two Gnome turboshafts that power that unit's Sea King search and rescue helicopters.
Rolls-Royce Heritage Trust

CHAPTER THIRTEEN

Rolls-Royce: Nene and Avon

Early in 1944, by which time Rolls-Royce was well along with the development of the Welland turbojet, which was developed into the Derwent, the company was asked by the MAP to produce a more powerful engine of about 1,905kp (4,200lb) thrust. A maximum diameter of 1.40m (4ft 7in) and weight of 998kg (2,200lb) was specified. This job, undertaken by Stanley Hooker and Adrian Lombard, a brilliant young engineer from Rover, benefited from the early tests on the first Derwent engines. The new engine, designated RB.40, was of a similar layout with a double-sided centrifugal compressor. Following a visit to the USA in mid-1944, where Hooker found that engines of 1,815kp (4,000lb) thrust were already at the test stage, he decided that Rolls-Royce should aim for a 2,268kp (5,000lb) thrust engine. He therefore instructed Lombard to redesign the RB.40 and the result was the RB.41, later named the Nene. This engine was to become the world's most powerful turbojet in production at the time. Also, along with the Derwent, it was the basis for early turbojet production all over the world, especially in the Soviet Union.

Nene turbojet

It would be wrong to suppose that the Nene was merely a scaled-up version of the Derwent since, if that had been done, the engine diameter would have increased from 1.09m (3ft 7in) to nearly 1.52m (5ft). (In fact, the later Derwent V engine was virtually a scaled-down version of the Nene.) By a complete redesign, it was found that the specification could be satisfied within a diameter of 1.257m (4ft 1½ in), while the compressor diameter was 0.732m (2ft 4.8in). In only five-and-a-half months the design of the Nene was complete and the first engine was built, by which time the Derwent IV was giving a thrust of 1,090kp (2,400lb).

The first run of the Nene took place at the Barnoldswick factory on 27 October 1944. After some difficulty with ignition, the engine ran up to 1,815kp (4,000lb) thrust. Then, after the installation of inlet guide vanes, which had been provided for, the engine achieved its design thrust of 2,268kp (5,000lb) on its second run. The first test flight of a Nene was made in March 1946 using a Vampire F.1 (TG276). Once installed in an aircraft the Nene's thrust was liable to drop to 1,588/1,815kp (3,500/4,000lb) due to intake losses. This did not happen to a de Havilland Goblin engine due to its single-sided compressor, but the Nene could have a smaller nacelle with less drag to offset its reduction in thrust. Subsequent flight testing of the Nene was carried out with two of them replacing the outboard Merlin engines of a Lancastrian, beginning on 14 August 1946 with VH742. Later this was joined by a second, similar Lancastrian and flight testing took place from the Rolls-Royce Flight Development and Testing Establishment at Hucknall, the Chief Test Pilot being Capt R.T. Shepherd. The performance of a Lancastrian on the power of two Nenes alone was very good but then their combined power was more than that from the normal four Merlin piston engines. Although the Lancastrians were equipped with Nene engines purely to gather flight data, large numbers of people were able to experience jet flight in them for the first

Rolls-Royce Nene centrifugal turbojet. Rolls-Royce Heritage Trust

An Avro Lancastrian (VH742) flying test-bed, fitted with two R-R Nene turbojets in the outboard positions, replacing the piston engines. This aircraft is shown at the Rolls-Royce aerodrome at Hucknall, being prepared for a flight. Philip Jarrett

time and so they could be called the first jet transports. With the Merlins shut down, the very quiet, vibration-free flight, accompanied only by the whistling of the airflow, was noted by the passengers.

The Nene had two shafts, one for the compressor and one for the turbine, connected by a readily detachable, spherically seated coupling. This permitted the withdrawal of either shaft for inspection without disturbing the main structure of the engine. The compressor was of the single-stage, double-sided, radial flow type that required a plenum chamber for the intake air in order to distribute it to both sides of the compressor. Also on the same shaft was a smaller, single-sided, centrifugal compressor or impeller to deliver cooling air to the turbine bearing and disc. The compressor delivered air at a pressure ratio of 4:1 to the nine large Lucas combustion chambers, which had downstream fuel injection from Duplex burners.

The single-stage turbine disc was forged from Jessop's G.18B heat-resisting steel and its blades were forged from Nimonic 80 steel. These rotor blades were fixed by fir-tree roots. Precision casting by the lost-wax process was used to produce the turbine inlet nozzle guide vanes. Anti-friction bearings consisted of a roller type for the ends of the two shafts and a ball type at the centre. These were lubricated by a high-pressure oil system. Interestingly, at speeds up to 8,000rpm the axial thrust on the shafts was directed forwards, but above that speed it was directed rearwards. Axial thrust was transferred by the ball of the shaft coupling while torque was transmitted by its interlocking teeth. The maximum speed was 12,300rpm. The exhaust cone was of a fixed length but the jet pipe could be varied, within limits, to suit the installation. All auxiliaries were mounted to the front of the engine, along with an oil sump. The drive for these was from a simple wheelcase of spur gears.

Attacker and Sea Hawk fighters

The engine went into production in September 1946 at the Derby factory as the RD.45 or Nene 1 of 2,200kp (4,850lb) thrust. A Nene 1 could be accelerated from idling to full power in 4.5 seconds and its dry weight was 708kg (1,560lb). Its first applications in Britain were to power prototypes, notably the Supermarine Attacker and Hawker Sea Hawk naval fighters, and experimental aircraft. The Attacker originated in the Type 392 designed to

Specification E.10/44 calling for an experimental fighter to be powered by the new Nene turbojet. This aircraft had a straight wing (based on that of the piston-engined Spiteful), a circular-section fuselage with side air intakes alongside the cockpit and a tailwheel undercarriage. The prototype (TS409) made its first flight on 27 July 1946, piloted by Jeffrey Quill; its Nene gave only 1,815kp (4,000lb) of thrust due to duct losses. Since the RAF showed no inclination to replace its Meteors or Vampires, the Type 392 was navalized as the Type 398 with a Nene 3 of 2,495kp (5,500lb) thrust. The first flight of the Type 398 (TS413), now named the Attacker, was made on 17 June 1947, piloted by Mike Lithgow. This was followed by the first deck landing on 5 October 1947 aboard HMS *Illustrious*.

Just before this, on 2 September 1947, the first flight was made of Hawker's first jet, the P.1040 (VP401), which was powered by a Nene giving 2,040kp (4,500lb) of thrust. The pilot was T.S. Wade. This aircraft was built to meet Specification N.7/46 for a naval interceptor with folding wings and arrester hook. It was a conventionally shaped, straight-wing aircraft but with wing root air intakes and bifurcated exhaust pipes at the wing trailing edges for the single Nene engine. The thinking behind this unusual arrangement was that it allowed an extra fuel tank to be fitted in the rear fuselage. A bubble canopy near the nose and a tricycle undercarriage was provided. This aircraft was modified as the P.1072 by the addition of an AS Snarler rocket motor in the tail; the first flight in this form was made on 20 November 1950. This method of boosting, however, was stopped in favour of afterburning development.

A swept-wing version of the P.1040 was then built as the P.1052, powered by the Nene 2 of 2,268kp (5,000lb) thrust. This prototype (VX272) first flew on 19 November 1948, piloted by Sqn Ldr Wade. A second prototype of this aircraft (VX279) was converted into the P.1081, which differed in having a straight jet pipe exhausting beneath the tailplane. This machine first flew on 19 June 1950 but crashed the following year, killing the pilot Wade.

Development of Hawker's P.1040 prototypes led to the Sea Hawk, which began service trials in 1950, and the type was accorded super-priority status for the Royal Navy following the start of the Korean War. This was despite the fact that its performance was inferior to swept-wing fighters and it seems inexplicable why the swept-wing derivative of the P.1052, which had flown supersonically, was not chosen for production. The Sea Hawk's range of 463km (288 miles) was especially poor but it saw action in the Suez campaign of 1956. The usual armament was four 20 mm cannon and, in ground attack versions, underwing bombs. The Sea Hawk also served the forces of Germany, India and the Netherlands.

While the Sea Hawk was being developed, the Attacker had been ordered into

A Vickers 618 Viking (G-AJPH) with two R-R Nenes, in place of the usual Hercules piston engines, for flight tests to obtain data for future jet airliners. On 25 July 1948 this aircraft made a pioneering passenger-carrying flight from Heathrow to Villacoublay, Paris, in 34 minutes 7 seconds. This aircraft was rebuilt in 1954 as a standard freight-carrying Viking. Philip Jarrett

An example of the Hawker FGA.6 Sea Hawk fighter bomber (XE456) of the Royal Navy. Built by Armstrong Whitworth, the type was powered by a single R-R Nene 103 turbojet, exhausting through bifurcated jet pipes flanking the rear fuselage. It was without vices and a joy to fly. Philip Jarrett

production for the Royal Navy, the first production example flying in April 1950. With an armament similar to that of the Sea Hawk, the Attacker was none too successful and the only three front-line navy squadrons equipped with it began receiving Sea Hawk replacements in 1954. Abroad only the Royal Pakistan Air Force bought the Attacker. Both the Attacker and the Sea Hawk types were equipped with Martin-Baker ejector seats and could use rocket assistance for heavily laden take-offs.

During 1947 mass production of the Nene and its derivatives began in earnest in the Soviet Union, unlike the situation in Great Britain. From 1948, however, licence production of the Nene was undertaken in France, Canada, Australia and, as the J42, in the USA. Most development of the Nene was probably undertaken in France by Hispano-Suiza. In Britain, the Nene was used to power sundry experimental aircraft such as the Armstrong Whitworth AW.52 flying wing, the Gloster G.42 fighter, the Canberra B.1 second prototype (VN813), the Boulton Paul BP.111A and BP.120 delta-wing research aircraft, the Handley Page HP.88 crescent-wing research aircraft and the G.41F Meteor F.4 (RA490) used for jet deflection trials. Various foreign prototypes such as the AIME Pulqui (Arrow) in Argentina and Grumman XF9F-2 Panther in the USA also used the Nene. Outside of Britain and the Soviet Union, production aircraft to be powered by the Nene included the Canadair T-33A-N Silver Star trainer from 1952, the Fokker S.14 'Mach Trainer II' from 1953, and the Fiat G.82 trainer from mid-1954.

A great deal of test flying was carried out with the Nene. Flying test-beds, apart from the Lancastrians already mentioned, included a Vickers 618 Viking (G-AJPH) with two Nenes, an Avro 688 Tudor (VX195) with four Nenes, a Supermarine Attacker (TS409), an Avro 706 Ashton

Hawker Sea Hawk FGA.6 with Nene 103 turbojet of 2,359 kp (5,200 lb) thrust		
Span	11.88m	(39ft 0in)
Length	12.09m	(39ft 8in)
Wing area	25.85sq m	(278sq ft)
Height	2.66m	(8ft 9in)
Loaded weight	7,348kg	(16,200lb)
Maximum speed at 3,048m (10,000ft)	844km/h	(524mph)
Service ceiling	13,564m	(44,500ft)
Range	463km	(288 miles)

Mk. 2 (WB491) with four Nenes plus an underslung Avon, and another Ashton (WB493) with four Nenes and an underslung Olympus (and also with two Olympus engines).

Perhaps the most interesting use of the Nene was in the Rolls-Royce 'Flying Bedstead' (XJ314) VTOL test rig, which first flew on 3 August 1954, piloted by R.T. Shepherd. The TMR (Thrust Measuring Rig), as the machine was officially known, used two Nene engines mounted horizontally in the rig with their intakes facing outwards and their jet exhausts ducted through 90 degrees to point downwards. Each downpipe was divided into two halves so that, should one engine fail, the TMR could still descend evenly. Control and balancing of the rig was by means of air jets at the ends of four long arms, the air for each being shared by bleed air from the two engines, again guarding against engine failure. The undercarriage for the TMR consisted of four long tubes at the corners of the rig, culminating in small wheels, the appearance inevitably leading to the appellation of 'Flying Bedstead'. The pilot sat perched on top of the whole ensemble, in the open.

Tay turbojet

As already mentioned, a scaled-down version of the Nene I was developed to produce the Derwent V. At just over 85 per cent of the Nene's size, the Derwent V went on to give 1,588kp (3,500lb) of thrust and powered a Meteor to world speed records in 1945 and 1946. The final development of the Nene, and for Rolls-Royce of the centrifugal turbojet, was the RB.44 Tay. This development was carried out at Derby and encompassed many refinements of details and an increase in air mass flow to initially produce a thrust of 2,835kp (6,250lb). The Tay first ran on 27 September 1948. Beginning on 15 March 1950, flight tests began with two Tays in place of the four turboprops on a Vickers Viscount airliner (VX217), type V.663. Each Tay was installed in a long nacelle underneath the wing. The aircraft was used to develop powered flight controls and also fly-by-wire systems. This was, in fact, the only British aircraft to use the excellent Tay turbojet. Partly this was due to the perception that the centrifugal turbojet would soon be outmoded and that only the axial turbojet was worth pursuing. This mistaken view was encouraged by A.A. Griffith and Hayne Constant and accepted by Ernest Hives. The Tay was therefore merely licensed to Hispano-Suiza in France and Pratt & Whitney in the USA for production in those countries. In France it became known as the Verdon (e.g. Verdon 350 of 3,500kp/7,720lb thrust), which powered, for example, the Dassault Mystère IVA fighter from 1953. In the USA the Tay was designated the J42 in military use or Turbo-Wasp JT-6B in commercial use.

Avon turbojets, RA.1 to RA.3

Early in 1945 Chief Research Engineer A.A. Griffith at Derby was instructed to begin the calculations and layout of a straightforward, relatively simple axial turbojet of around 2,950kp (6,500lb) thrust. The thrust was chosen as the highest likely to be eventually reached by a single-stage centrifugal engine. By June 1945 this preliminary work had been completed and the results were sent to Barnoldswick for working up into hardware by Hooker's team. This was the origin of the AJ.65, later known as the Avon, turbojet. By that time, work was also proceeding at Barnoldswick on turboprops (see below).

As the engineering design of the Avon proceeded, under Adrian Lombard, it soon became clear that Griffith had underestimated its weight by a large margin. Unlike Rolls-Royce's previous centrifugal engines, which had been developed rapidly, the Avon axial engine's development was troublesome and protracted, although its

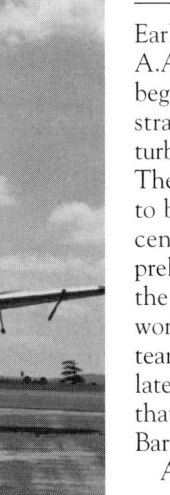

This Gloster Meteor IV (RA490), originally flown with Metropolitan-Vickers F2/4 Beryl turbojets, was fitted with R-R Nene 101 engines in July 1952 to perform jet deflection experiments. The Nenes were fitted well forward so that, by the use of valves, their exhausts could be deflected downwards at the centre of gravity to reduce the landing speed, or rearwards for normal flight. Inexplicably, from April 1957 this historic aircraft finished its days being used for firefighting practice. Philip Jarrett

Another flying test-bed, this Avro 688 Tudor (VX195) has had its piston engines replaced with four R-R Nene turbojets, a pair to each nacelle. Philip Jarrett

Rolls-Royce TMR (Thrust Measuring Rig) or 'Flying Bedstead' (XJ314) VTOL test rig. First flown on 3 August 1954, it was powered by two horizontal R-R Nene turbojets with their exhausts ducted through 90 degrees to point downwards. Balancing jets or puffer pipes were at the ends of the long arms. Rolls-Royce Heritage Trust

This view below the Rolls-Royce TMR 'Flying Bedstead' in the Science Museum, London, shows one of the Nene turbojets with its right-angled exhaust ducting and the two exhaust nozzles underneath the machine. Author

The final development of the Nene was the RB.44 Tay, which was Rolls-Royce's last centrifugal turbojet. Philip Jarrett

eventual great success led to it powering very many aircraft of a wide range of types. The first prototype of the Avon, designated RA.1, was ready for testing in spring 1946 but refused to accelerate and shed the first stage of its compressor blades. Two young engineers at the Derby factory, Geoffrey Wilde and Geoffrey Fawn, redesigned the compressor and turbine and the engine was gradually coaxed up to a thrust of 2,268kp (5,000lb). Gradual improvement of the compressor was made and variable inlet guide vanes and blow-off valves were added to facilitate starting and acceleration. An automatic starting and acceleration control system was also developed. Finally, a run of the RA.1 on 25 March 1947 produced the design thrust of 2,948kp (6,500lb).

Sadly, these early problems with the Avon led to an argument between Hooker and Hives, the latter calling the Barnoldswick factory 'that bloody garage!' Hives was especially keen to get the axial turbojet into production and decided that Avon development should be moved to Derby. He told Hooker to go home for three months but Hooker, who had been promised the position of Chief Engineer of Rolls-Royce, resigned in December 1948 and went to the Bristol company, where his great expertise was badly needed.

In the meantime, improvements to the Avon produced the RA.2, which, for a

To flight test the R-R Tay engine, two were fitted in place of the four turboprops on this Vickers Viscount airliner (VX217). Philip Jarrett

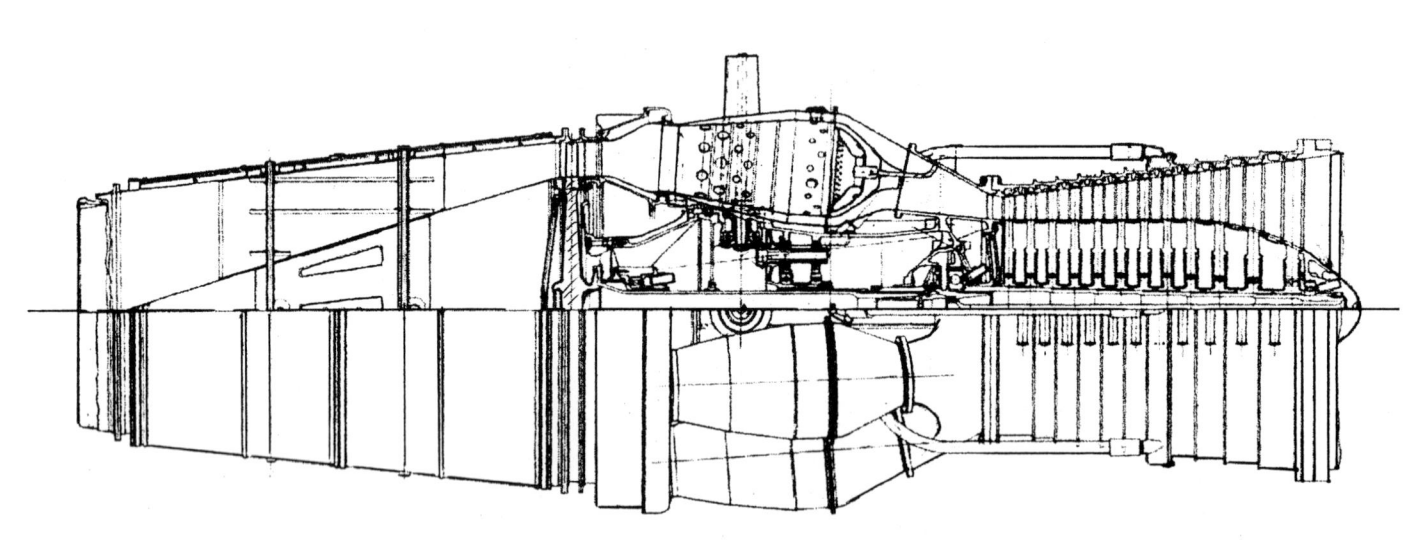

Half-section of the precursor of the Rolls-Royce Avon turbojet.

Rolls-Royce RA.2 Avon axial turbojet. The Avon powered large numbers of aircraft during the 1950s.
Rolls-Royce Heritage Trust

Rolls-Royce RA.3 Avon turbojet.
Rolls-Royce Heritage Trust

weight of 1,157kg (2,550lb), reliably produced a thrust of 2,495kp to 2,630kp (5,500–5,800lb). The RA.2 was basically similar to the RA.1 except for a reduction in the compressor's mass flow and the addition of two-position inlet guide vanes. It made its first run in January 1948. Test flying of this engine began by using the Avro 691 Lancastrian (VM732), two RA.2s replacing the outboard Merlin engines. Its first flight was made on 15 August 1948, piloted by Capt R.T. Shepherd. This aircraft gave the first public demonstration of the Avon when it flew at Farnborough the following month.

The Avon was designed with a twelve-stage axial compressor delivering 54.43kg/sec (120lb/sec) of air at a pressure ratio of 6.5:1, but experiments with fewer stages were made in early prototypes. Eight Lucas combustion chambers fed a large, single-stage turbine. In the RA.3, a new, smaller diameter, two-stage turbine was fitted; this and other improvements gave a thrust of 2,948kp (6,500lb) for a reduced weight of 1,020kg (2,250lb). Most Avons had cartridge starting and air bleed for cockpit pressurization. Beginning on 27 October 1949, two RA.3s were test-flown in Lancastrian VL970. Previously, from 29 April 1949, two RA.2s had been test-flown in a Meteor F.4 (RA491). In August 1950 two RA.3s were given a 25-hour special category test to clear them for 3,538kp (7,800lb) thrust and these were fitted into the Meteor F.4 test-bed, outclimbing anything known at the time. At the Farnborough Air Show in September that year, this Meteor took off in less than 300m (984ft) and climbed away at about 60 degrees. It could climb to 12,200m (40,000ft) in less than 3 minutes. Naturally, this did nothing for the Meteor's short endurance.

Canberra bomber

During 1949 Avon-powered Canberra bombers began appearing and these were the first aircraft that the Avon became involved with in a major way. The maiden flight of the Canberra, a B.1 (VN799), took place on 13 May 1949 with R.P. Beamont at the controls. The Canberra soon proved to be extremely agile. Its powerplants were two RA.2s and few could have suspected that this small, straight-wing bomber was destined to become one of the greatest aircraft of all time.

The Canberra was designed by W.E.W. Petter soon after he joined The English Electric Aircraft Division at Preston in 1944. At first the requirement was for a replacement for the Hawker Typhoon and other ground attack fighter-bombers,

Making a low pass at Farnborough is the Avro 706 Ashton (WB491) flying test-bed fitted with four R-R Nene engines in wing nacelles and one underslung R-R Avon. Philip Jarrett

but then changed to a two-seat, high-altitude bomber with a Mosquito bomb load (1,800kg/4,000lb) and a Lancaster radius of action (1,300km/800 miles). These requirements were covered by Specification B.3/45 in October 1945. Petter's first thoughts were for a single turbojet of 5,897kp (13,000lb) thrust housed in the fuselage, but this caused difficulties in accommodating the bomb load. He therefore decided on two engines of 2,948kp (6,500lb) thrust each, hence another reason for the Rolls-Royce Avon. What resulted was an aircraft of elegant simplicity and utility but outstanding adaptability and longevity. The fuselage was of circular cross-section with a pressurized cockpit in the nose, covered by a single-piece, sandwich Perspex, domed canopy. Access to the cockpit was via a small door on the starboard side of the nose. Fuel tanks were in the roof of the fuselage, with the bomb bay below. The large-area wing had a straight centre section with equi-tapered outer panels outboard of the two engine nacelles. There was a short, tricycle undercarriage and a tailplane with pronounced dihedral to clear the wing wake. Controls and flaps, as with virtually all of the features of the Canberra, were conventional, although the air-brakes were unusual in being multiple fingers that extended above and below the wings. Conventional former frame and stringer construction was used for the fuselage with a pressure bulkhead behind the cockpit section and another bulkhead and fuselage frame for connection to the rear and main spars, respectively, of the wing. The wing and tail surfaces were lightly constructed, using multiple ribs and stringers.

The first Canberra production model was the B.2 with RA.3 engines, the first example (WD929) making its maiden flight on 8 October 1950. The B.2 had a transparent nose for a bomb aimer (since the radar bombing system was not ready) and it entered RAF service early in 1951. With tip tanks, the B.2 had a range of 5,650km (3,510 miles) with a bomb load of 2,700kg (6,000lb) and it could carry a nuclear weapon. It proved to be very difficult for the fighters of the day to intercept, thanks to its low wing loading, which conferred excellent high-altitude performance.

A number of records were set by the Canberra B.2, including the Cape to London in 13 hours 16 minutes on 19 December 1953 (728.69km/h or 452.8 mph), London to Baghdad in 4 hours 51 minutes on 31 July 1954 (842.47km/h or 523.5 mph) and Ottawa to London in 6 hours 42 minutes on 27/28 June 1955 (799.50 km/h or 496.8mph). A Canberra B.2 also made the first non-stop, direct crossing of the Atlantic without refuelling by flying from Aldergrove to Gander in 4 hours 37 minutes on 27 February 1951 (724.18km/h or 450mph). A Canberra also made the first Polar Inter-Continental jet flight by flying from Bardufoss (Norway) to Fairbanks (Alaska) in 6 hours 20 minutes in June 1955 (816.07km/h or 507.10mph). Such records proved the reliability of the Avon.

The Korean War brought about super-priority orders for the Canberra and

First roll-out, on 8 May 1949, of the English Electric B.1 bomber (VN799) prototype, fitted with Rolls-Royce RA.2 Avon engines. This small, straight-wing jet bomber was destined to become one of the greatest aircraft of all time, the Canberra. Philip Jarrett

Representing the first Canberra production model is this B.2 (WD952) with Rolls-Royce RA.3 Avon turbojets.
Philip Jarrett

production took place at the companies of A.V. Roe, Handley Page and Short Brothers. At the same time, the RA.3 Avon engine went into production at the Rolls-Royce Derby and Hillington (Glasgow) factories and also at a new one built at East Kilbride in Lanarkshire. Production was also subcontracted to Bristol, Napier and Standard Motors. Subsequently, the Canberra was produced in a large number of variants and by many licensed manufacturers so that it appeared in service all over the world. In the USA, Martin produced its version of the Canberra as the B-57. The Canberra proved to be especially good in the photo reconnaissance role, a PR Mk 3 winning the London to New Zealand Air Race in October 1953, covering the 19,746km (12,270 miles) in 23 hours 50 minutes 42 seconds. Canberras went on to set another 21 records for speed and altitude. Theatres of war in which they have been used include India, Malaya, Vietnam and the Falklands. Of all types, including trainers, about a thousand Canberras were built, of which many were later converted for other roles. An amazing 103 Canberras were converted as flying testbeds for engines, missiles and many systems. The engine test-beds were used by Rolls-Royce, Bristol, Armstrong Siddeley, de Havilland and Napier. Most Canberras were powered by Avon engines.

After 55 years of service with the RAF, the Canberra finally bowed out in August 2006 when the last PR9s were retired from 39 (1PRU) Squadron at RAF Marham.

Avon 100-series, RA.7, RA.7R, RA.9, RA.21 and RA.25

A modified RA.3 engine was type-tested at a thrust of 3,606kp (7,950lb) and work began on a new Avon, the RA.7, to provide a more powerful engine for the pending Hunter and Swift fighters. This was the first of the important 100-series. It incorporated modifications developed in the RA.3 and did not have the pressure-balancing piston behind the compressor. Anti-icing equipment was developed for the RA.7 and, during ground tests, ice cubes as large as 38mm (1½in) square were fed into the intake without causing damage. To assess how much ingress of ice the new Avon could stand, one was mounted in the nose of the second Lancastrian and surrounded with a water spray grid for flight tests.

The first RA.7s were delivered in 1950 and were soon giving a thrust of 3,402kp (7,500lb). By August 1952 the RA.7 was sufficiently developed to pass a 150-hour official type test and went into production.

English Electric Canberra B.2 with RA.3 Avon engines			
Span		19.49m	(63ft 11½in)
Length		19.96m	(65ft 6in)
Wing area		89.28sq m	(960sq ft)
Height		4.75m	(15ft 7in)
Loaded weight		20,865kg	(46,000lb)
Maximum speed	at 12,200m (40,000ft)	917km/h	(570mph)
Service ceiling		14,630m	(48,000ft)
Range		4,280km	(2,660 miles)

Its specific fuel consumption was 0.92 and its diameter was unchanged from the RA.3 at 1.07m (3ft 6.20in). Weight increased slightly to 1,116kg (2,460lb) due to general all-round increases in engine stresses to cope with supersonic speeds.

Beginning in May 1951, the RA.7 was test-flown in a Canberra B.1 (VX185). From April 1951 development running was started with an afterburning version of the RA.7, designated the RA.7R, in order to ensure forthcoming fighters had the thrust needed for a satisfactory rate of climb and altitude performance. By June 1952 the RA.7R was ready for testing in the Canberra test-bed and by March 1953 it had passed an official 150-hour type test. With a variable-area exhaust nozzle, the RA.7R became the first British afterburning turbojet to go into production. Using the afterburner, this Avon gave 4,310kp (9,500lb) of thrust, but its afterburner and extra equipment caused the weight to rise by 227kg (500lb) to 1,343kg (2,960lb). Its length was 3.048m (10ft) compared with 2.59m (8ft 6.1in) for the basic RA.7.

Development of the RA.7 then took two paths. First was the RA.9, a 2,948kp (6,500lb) thrust civil version for the Comet 2X airliner (pre-production aircraft for the Comet 2). Designated the Mk. 502, it flew in the Comet 2X on its first flight in February 1952. A later version of the RA.9, the RA.25 (Mk. 503), was a 3,220kp (7,100lb) thrust engine for the Comet 2 production aircraft, the first of which (G-AMXA) flew on 29 August 1953. Later RA.25s for the Comet were designated Mk 504s and 505s. Both the RA.9 and RA.25 Avons differed from the RA.7 largely in details of installation and in the higher standard of anti-icing systems. This latter was necessary for the flight let-down requirements for civil aircraft where the engines operated at lower speed (rpm).

The second development path taken by the RA.7 led to the military RA.21 of 3,629kp (8,000lb) thrust, which was introduced to compete with the more powerful AS Sapphire turbojets. The RA.21 entered production as the Avon Mks 113 and 115. However, it was seen that even more thrust would be needed in the future and so by 1950 work had begun on a redesign leading to the 200-series.

Avon 200-series, RA.14

The first of the 200-series was the RA.14. It had a new, larger fifteen-stage compressor that owed some of its aerodynamic design to the AS Sapphire and delivered air at a mass flow of 68kg/sec (150lb/sec) and a pressure ratio of 7.45:1. This was followed by a new, cannular combustion system with eight flame tubes and a two-stage turbine. Accessories for the engine were neatly grouped in the space afforded by the tapered, high-pressure section of the compressor casing.

The RA.14 was first run on 17 November 1951 and gave a thrust of 4,223kp (9,310lb), which was close to its design thrust, and it had a specific fuel con-

Rolls-Royce RA.25 Avon turbojet. Rolls Royce Heritage Trust

sumption of only 0.84. Two of these engines were first flight tested in February 1953 in the Canberra B.2 test-bed (WD930). Soon developed to give a thrust of 4,309kp (9,500lb), the RA.14 gave the highest official rating of any British turbojet at the time. Its diameter was only 1.05m (3ft 5½in), the same as the early RA.1, and its weight was 1,297kg (2,860lb).

Some selected design details used in Avons are of interest. In the intake casing, the six streamlined support struts for the centre housing were tangentially offset to reduce stresses set up during cooling and to provide a semi-tangential wake. By this means the compressor rotor front stages passed through the disturbed air flow progressively rather than all together and so blade vibration was reduced. The Avon was one of the first turbojets to use pinned root mountings for the compressor rotor blades; this allowed each blade to rock to avoid root failure caused by flutter. Steel for the rotor blades was avoided as much as possible since a failed steel blade could burst through the compressor casing. So-called self-clearing labyrinth seals were used for the compressor: should a rotor blade rub at its tip, the frictional heat generated caused diagonally mounted, fixed, outer strips of aluminium to expand radially out of blade contact. Thus, very close tolerances could be used with consequent gains in efficiency.

Meanwhile, RA.2 Avons were used to power the prototype (18101) of Avro Canada's highly successful CF-100 twin-jet fighter, which first flew on 19 January 1950, and the prototype of a four-jet bomber intended as insurance in case the V-bombers should fail, the Short SA.4 Sperrin (VX158), which first flew on 10 August 1951. RA.3 Avons powered a number of important prototypes during 1951 including the Vickers 660 Valiant bomber, the Hawker P.1067 Hunter fighter, the Supermarine V-S 541 Swift fighter and the de Havilland DH.110 fighter. Thus, 1951 was an important year for Rolls-Royce, the Avon promising a bright future. Early in 1952 the Comet 2X (G-ALYT),

This Vickers V.660 (WB210) was the prototype of the Valiant, first of the V-Bombers. It was powered by four Rolls-Royce RA.3 Avon engines. Philip Jarrett

fitted with four RA.3 Mk 501 Avons of 2,948kp (6,500lb) thrust each, was brought into the test programme as a flying testbed.

Valiant V-bomber

The Vickers Valiant became the first strategic jet bomber to enter RAF service and was the first of three types of V-bomber. Designed to deliver a nuclear weapon over a long range, two prototypes were ordered in 1949 to meet the MoS Specification B.9/48. The first of these, designated the Vickers Type 660 (WB210), was powered by four RA.3 engines and made its maiden flight on 18 May 1951, the pilot being J. Summers. Early demonstrations of this prototype, especially at the 1951 Farnborough Air Show, showed its fighter-like performance, which far outshone the pedestrian performance of the stopgap Sperrin bomber. Unfortunately, during engine shut down and start tests on 12 January 1952, one of its Avons caught fire and the crew had to eject. Sadly the co-pilot, Sqn Ldr B.H.D. Foster, was killed when he struck the aircraft's fin. Within three months, the programme was able to continue with the second prototype.

This second prototype, designated the Vickers Type 667 (WB215), first flew on 11 April 1952 and was powered by four, more powerful, RA.7 engines. On the face of it the Valiant, as the type became known, was a fairly simple concept (compared with the other V-bombers, the Vulcan and Victor) but this overlooked the many advanced techniques used in its construction. It had a shoulder-mounted wing, with moderate sweep on the leading edge only, and the engines were buried in the wing roots with bulges below them. The only protuberance on the fuselage was a moderate bulge with windows above the pressurized cockpit. Tail surfaces were slightly swept on their leading edges, the tailplane being mounted halfway up the fin, and there was a tricycle undercarriage. The overall finish of the airframe was extremely clean, permitting a maximum speed of Mach 0.86 and a cruising speed of Mach 0.82.

In April 1951 the Valiant was ordered into production, initially as five pre-production aircraft with RA.14 engines and then twenty heavier B.1s with new RA.28 engines, rated at 4,536kp (10,000lb) thrust each. The installation of the RA.28s required longer jet pipes. In the meantime, Britain bridged the bomber gap with American B-29s (from March 1950) and development of the country's first nuclear bombs went ahead. This development was without the support of the USA in spite of Britain's major contribution to the USA's atomic bomb development during the Second World War. Eventually the USA reluctantly contributed some engineering data only and Britain's first nuclear test weapon was detonated aboard the decommissioned frigate HMS *Plym* off the Monte Bello Islands (Western Australia) on 3 October 1952.

The first Valiant B.1 (WP199) flew on 21 December 1953 and the first operational squadron, No. 138, received the Valiant in February 1955. Once production was in full swing, Valiants left the Weybridge factory at the rate of one per week and

The Vickers Valiant B.1 began reaching operational RAF squadrons in February 1955. The Valiant was used to test-drop Britain's first A- and H-bombs. Philip Jarrett

Valiant B.1		
Span		34.85m (114ft 4in)
Length		33m (108ft 3in)
Wing area		219.67sq m (2,362sq ft)
Height		9.80m (32ft 2in)
Loaded weight		63,504kg (140,000lb)
Maximum speed	at 9,144m (30,000ft)	912km/h (567mph)
Service ceiling		16,459m (54,000ft)
Range		7,242km (4,500 miles)

eventually ten RAF squadrons were operational with the type. One of the records set by the Valiant B.2 was a flight from Singapore to Darwin in 4 hours 1 minute on 7 August 1955, using RA.14 engines (834.26km/h or 518.40mph).

Valiants were used to test-drop Britain's first nuclear weapons. On 11 October 1956 an A-bomb was dropped over Maralinga, South Australia, and on 15 May 1957 an H-bomb was dropped south of Christmas Island over the Pacific. During the Suez Campaign in 1956, Valiants dropped conventional iron bombs.

Hunter fighter

The first of the second-generation British fighters powered by the Avon was the Hunter, which was without doubt the country's most successful jet fighter of the twentieth century. The Hunter was derived from the straight-wing Hawker Sea Hawk (*see above*), but with a 35-degree swept wing and the more powerful, axial Avon engine. Following various design studies, Sydney Camm's team settled on the P.1067, which more than met the MoS Specification F.3/48 of March 1948. Three prototypes were ordered, the first two to be powered by the RA.3 Avon and the third by the AS Sapphire. The new fighter was not only to replace the Meteor, but the appearance of North American's swept-wing XP-86 fighter (Sabre) urged the RAF to seek something equally as advanced.

The MoS ordered the production of the Hunter even before the first prototype flew, spurred on by the Korean War. The prototype (WB188) first flew on 20 July 1951 from Boscombe Down, the pilot being Hawker's Chief Test Pilot Neville Duke. This was followed by the second prototype (WB202) on 30 November 1952.

Soon sonic booms were being heard all over southern England and, in those days, there was virtually no murmuring in opposition. Despite the general soundness of the design and its beautiful, aesthetic lines, all was not well with the Hunter and various problems took years to resolve. These problems included surging of the RA.3 Avons, exacerbated when the guns were fired, poor longitudinal control at high speeds, misting up of the canopy, low endurance, damage caused by ejected ammunition links and incorrect dive brakes. All the problems were gradually rectified, but at the expense of delaying introduction into service. Almost 2,000 Hunters were produced. Because of its teething troubles, fighters from other countries were able to steal a march on it and penetrate the markets, ironically using Avon engines, such as Canadian-built CA-27 Sabres with RA.7 engines in 1953 and, in France, the Dassault Mystère IVB, also with the RA.7 engine. Also powered with the RA.7R (with afterburner) was Sweden's SAAB A-32 Lansen, the prototype of which (32001) first flew on 3 November 1952.

Following its introduction into RAF service in July 1954, the Hunter went on to serve the air arms of twenty-one other countries. The sound engineering of the Hunter's structure allowed almost four hundred machines to be refurbished for the export market. The F.6 was the most successful version of the Hunter, favoured by RAF aerobatic teams, and powered by the RA.14 Avon of 4,223kp (9,310lb) thrust. The first F.6 (XF833) made its maiden flight on 22 January 1954. The Hunter was also demonstrated to be potent in the ground attack role during the Suez crisis.

Swift fighter

In contrast to the Hunter, the competing Supermarine Swift was initially an abysmal

The prototype (WB188) of Hawker's Hunter fighter, powered by a Rolls-Royce RA.3 Avon engine. Philip Jarrett

This example of the Hawker Hunter (XG168) represents the most successful version, the F.6. Powered by a Rolls-Royce RA.14 Avon, the aircraft is pictured on 4 September 1960 on its way to Farnborough in readiness for the annual air show. Philip Jarrett

Hawker Hunter F.6			
Span		10.26m	(33ft 8in)
Length		13.98m	(45ft 10½in)
Wing area		31.62sq m	(340sq ft)
Height		4.01m	(13ft 2in)
Empty weight		5,788kg	(12,760lb)
Loaded weight		8,051kg	(17,750lb)
Maximum loaded weight		10,886kg	(24,000lb)
Maximum speed	at low level	1,143km/h	(710mph)
	at 11,000m (36,000ft)	Mach 0.95	
Initial climb rate		2,560m/min	(8,400ft/m)
Service ceiling		15,700m	(51,500ft)
Maximum range	with four drop tanks	2,965km	(1,840 miles)

failure, despite the fact that, too late, it was developed into an excellent fighter. Following on from the previous straight-winged Attacker and the experimental types 508 and 529, the Supermarine company experimented with the swept-wing fighters types 510, 517 and 541. Modification of the 541 design led to the first prototype of the Swift (WJ965), which made its first flight on 18 July 1952 powered by an RA.7 Avon of 3,400kp (7,500lb) thrust.

Already, in November 1950, orders had been awarded for production Swifts at the same time as the prototypes: the first production Swift F.1 (WK194) with an

The first true Supermarine Swift fighter (WJ965) first flew on 18 July 1952, powered by a Rolls-Royce RA.7 Avon turbojet. Philip Jarrett

Supermarine Swift FR.5			
Span		9.85m	(32ft 4in)
Length		12.87m	(42ft 3in)
Height		4.14m	(13ft 6in)
Empty weight		5,806kg	(12,800lb)
Loaded weight		9,707kg	(21,400lb)
Maximum speed	at sea level	1,102km/h	(685mph)
Low level range	with drop tanks	772km	(480 miles)

RA.7 engine made its first flight on 25 August 1952, only five weeks after the first prototype. If the Swift was superior in any way to its rival, the Hunter, then it was only in range and the fact that the firing of its two 30mm cannon did not precipitate engine surging. Nevertheless, the Avon fitted in the Swift was extremely unreliable, the cause being eventually traced to an alteration in compressor rotor blade attachment that had been made by one of the sub-contractors. Also much development work was needed on the Swift aerodynamics, especially on the swept-back wings: on 18 February 1953 a Swift was dived beyond Mach 1.0.

After entering RAF service with various flight restrictions, the Swift was grounded during 1954 following two crashes. The Swift F.2s, the first of which (WK214) flew in December 1952, were little better. An afterburning RA.7R was flown in the first F.3 (WK247) early in 1953, but was not accepted for operational use. Finally the main production variant, the F.4, appeared and featured many improvements including an all-flying tailplane, powered ailerons and a ventral fuel tank to compensate for the thirsty afterburner. The F.4 prototype (WK198) first flew on 27 May 1953 and, on 25 September that year, was flown at low level by Mike Lithgow over the Libyan Desert to a new world speed record of 1,187.18km/h (737.70mph). At that year's Farnborough Air Show, Lithgow and Morgan in two Swift F.4s dived in formation at supersonic speed and then flew just above the runway to the edification of an amazed crowd. Despite the much improved performance of the F.4, the conclusion of the Korean War brought about the cancellation of Swift fighter production in February 1955.

On 24 May 1955 the first production example of a fighter-reconnaissance Swift, the FR.5, made its maiden flight with an RA.7R engine. This aircraft served the RAF well in the low-level role and could carry rockets and bombs. The FR.6 was not proceeded with, but the FR.7 turned out to be excellent. First flown in April 1956, the FR.7 prototype (XF113) was powered by the RA.21R, which had a dry thrust of 3,225kp (7,110lb) and an afterburning thrust of 4,420kp (9,745lb). Although the FR.7 was considered to be the best Swift, only a small number were built.

Sea Vixen fighter

Turning now to de Havilland's efforts in the second-generation fighter field, we see that they continued with the twin tail-boom theme begun with the Vampire and continued up to the Sea Venom. Then came the very advanced DH.110 twin-engined, long-range fighter intended for

The first prototype (WG236) of the de Havilland DH.110 was powered by two Rolls-Royce RA.3 Avon engines and reached Mach 1.11 in a shallow dive on 9 April 1952. Later that year this aircraft fatally broke up at low level at the Farnborough Air Show. The type was, however, developed into the very successful Sea Vixen. Philip Jarrett

BELOW: The prototype Fairey F.D.2 (WG774) on a test flight. Because of the high angle of attack required during the landing of this delta, the nose was arranged to depress by 10 degrees to improve the pilot's view. Powered by a Rolls-Royce RA.14 Avon with afterburner, the F.D.2 held the world air speed record when it was flown at 1,821.95km/h (1,132.136mph). Philip Jarrett

the Royal Navy (as the Sea Vixen) and the RAF and which was a potential world-beater. It was a swept-wing aircraft with wing root intakes for the side-by-side engines and had twin booms leading back from the wings, each ending in a swept-back fin. A rectangular tailplane joined to the tops of the twin fins and there was a tricycle undercarriage. Curiously the pilot's cockpit, with bubble canopy, was off-centre to the left of the nose and the radar observer sat inside the right side of the nose with no direct external view. Armament was to be four 30mm cannon mounted under the nose. Although the DH.110 had been mooted back in 1946, official vacillation meant that no orders were issued until 1949 and the first prototype did not fly until 1951. Orders for production were cancelled, however, and only prototypes were proceeded with.

The first DH.110 prototype (WG236), with two RA.3 turbojets, reached Mach 1.11 in a shallow dive on 9 April 1952. Tragedy struck in September that year when this aircraft disintegrated at low level at the Farnborough Air Show; the crew of two were killed and the Avon engines broke free, one ploughing into the spectators and killing many. Despite this disaster, the flying programme continued, the next to fly being Neville Duke in a Hunter fighter. It was not until 25 July 1952 that the second DH.110 prototype (WG240) flew, now powered by two more powerful RA.7 engines. This prototype was used for touch-and-go deck-landing trials aboard HMS Albion in autumn 1954. A navalized, interim aircraft called the Sea Vixen Mk. 20X began full arrested-landing trials aboard HMS Ark Royal from 5 April 1956. Production FAW.1 Sea Vixens began appearing for the Royal Navy in March 1957 (the RAF having lost interest long before) and eventually went into service, initially on HMS Ark Royal. The Sea Vixen was further developed and gave good service until 1972.

The prototype (XA847) of the English Electric P.1B, predecessor of the Lightning Mach 2.0 interceptor. It was powered by two Rolls-Royce Avon 200 R engines with afterburners. Philip Jarrett

Avon 200-series, RA.24, RA.26, RA.28 and RA.29

During the 1950s development of the Avon 200-series continued. From the RA.14 was developed, via an intermediate engine, the RA.28. This had the same diameter as the RA.14 but produced a thrust of 4,536kp (10,000lb) for a weight increase of only 14kg (30lb) to 1,310kg (2,890lb). The RA.28s specific fuel consumption was 0.86 and it was readily identified by the air blow-off volute added to the front half of the compressor casing. In September 1954 the RA.28 entered production as the Mk.204. A civil version of the RA.28 was designated the RA.26 and it was produced as the Mk.521 for the Comet 3. The civil RA.29 of 4,650kp (10,250lb) thrust was produced for the Comet 4 and the French Caravelle airliners. It had an extra compressor stage added before the first stage (a so-called '0' stage) and had a specific fuel consumption of 0.786. A further development, the RA.29/6, had an extra two stages added before the first compressor stage and a three-stage turbine to produce a thrust of 5,715kp (12,600lb), again with a specific fuel consumption of 0.786. In July 1956 the RA.24 was type tested at a thrust of 5,103kp (11,250lb). This was the first Avon with production air-cooled turbine blades and was used to power the production Supermarine Scimitar F.1 twin-engined naval fighters, the first of which (XD212) first flew on 11 January 1957.

Fairey F.D.2 and English Electric P.1 research aircraft

The Avon was used to power a number of research aircraft, most notably the single-engined Fairey F.D.2 (the F.D.1 was a small, stubby, delta-wing research aircraft powered by a Derwent) and the twin-engined English Electric P.1. These arose from the MoS Specification E.R.103, which followed on from discussions in 1947 on how best to advance into the supersonic field. This, in turn, had followed on from the extraordinary and disastrous decision in February 1946 to cancel the Miles M.52 supersonic research aircraft (see Chapter 4). Two prototypes each of the F.D.2 and P.1

English Electric P.1B			
Span		10.61m	(34ft 10in)
Length	including pitot tube	16.84m	(55ft 3in)
Wing area		42.64sq m	(458.5sq ft)
Height		5.97m	(19ft 7in)
Empty weight		11,256kg	(24,816lb)
Loaded weight		15,485kg	(34,140lb)
Maximum speed	at 12,200m (40,000ft)	Mach 2.10	
Initial climb rate		15,240m/min	(50,000ft/min)
Ceiling		18,300m	(60,000ft)

were ordered, but work on them was delayed due to concentration on super-priority orders precipitated by the Korean War.

Therefore, construction of the first F.D.2 did not begin until the end of 1952. Designed under H.E. Chaplin and R.L. Lickley, it was a needle-nosed, all-metal, delta-winged aircraft with a single, swept fin and sharp-lipped wing root air intakes. A slight bulge, largely of metal, covered the cockpit. Because of the high angle of attack adopted by a landing delta, the forward fuselage with cockpit was designed to depress 10 degrees to improve the pilot's view (the so-called 'droop snoot' later used by Concorde).

Powered by an RA.14R with afterburner, the first F.D.2 (WG774) made its first flight from Boscombe Down on 6 October 1954, piloted by Peter Twiss. On 17 November 1954 an engine failure occurred at 9,150m (30,000ft) and, although Twiss managed to glide back to Boscombe Down, the high-speed landing at 400km/h (250mph) resulted in some wing damage. Further test flying was delayed until August 1955. Mach 1.0 was first exceeded on 28 October 1955, without recourse to the afterburner. Then, in pursuance of a world air speed record, the F.D.2 achieved an average speed of 1,821.95km/h (1,132.136mph) at 11,600m (38,000ft). This exceeded the previously held record of the North American F-100C Super Sabre by more than 483km/h (300mph). At that time, Britain held the 'Triple Crown' of world speed records on land, water and in the air.

The second F.D.2 (WG777) made its first flight on 15 February 1956 and exceeded Mach 1.0 on this flight. Both aircraft went on to turn in much valuable research data under the RAE at Bedford. The first machine was rebuilt by Bristol with wings of ogival delta planform to assist in research for the future Concorde and began flying in 1964.

As for the P.1 research aircraft, this was conceived by W.E.W. Petter's team at English Electric with an eye on its future development as a high-peformance interceptor. Its advanced features included a wing and tailplane sharply swept back at 60 degrees. The ailerons were transversely mounted, joining the leading and trailing edges, and the low-mounted tailplane was of the all-flying type. The simple, oval-sectioned fuselage had a sharp-lipped nose intake and a cockpit mounted well forward, its canopy blending into a long fairing leading aft to the triangular fin. The two engines were arranged one above the other, this avoiding any horizontal asymmetric thrust in the event of a flame-out, and the wing structure passed between the engine air ducts. Fuel was carried in the wings and there was a tricycle undercarriage, the nosewheel retracting into the fuselage and the thin mainwheels into the wings.

The two P.1 prototypes, which reached Mach 1.5, were powered by AS Sapphire engines. The third P.1 prototype had two 30mm cannon mounted in the sides of the nose but was used for static tests only. From late 1954, the P.1 was redesigned as the P.1B under F.W. Page (Petter having departed to join Folland) and was enlarged to take two Avon RA.24R engines, each of 5,148kp (11,350lb) dry thrust or 6,658kp (14,680lb) with four-stage afterburning. Everything was now more complex since the P.1Bs were prototypes for the all-weather interceptor, the Lightning. The sharp-lipped, oval air intake was replaced with a circular one that also had a conical centre body in it. This was after the work of the German engineer Oswatitsch and it provided pressure recovery to the supersonic airflow by providing shock waves to slow it to subsonic velocity before the engines. The conical centre body also housed Ferranti AI.23 monopulse radar. Using shock waves for intake air pressure recovery, whether generated by a conical centrebody or sharp lips, became universal for turbojet aircraft flying at high supersonic speeds. Such a system seems to have already been planned for the Miles M.52 research aircraft.

Air-brakes were fitted to the sides of the rear fuselage, above wing level. Armament was to be two 30mm cannon and air-to-air missiles mounted on pylons beneath the nose and it was intended to fit the world's first head-up-display (HUD). However, the P.1B prototype (XA847) was unarmed and first flew on 15 February 1956. It exceeded Mach 1.0 on 4 April 1957 and Mach 2.0 on 25 November 1958. Although the official view from 1957 favoured the abandonment of manned aircraft for unmanned missiles, the P.1B was eventually turned into the successful Lightning with great development potential, chiefly serving with the RAF.

Other research aircraft powered by the Avon included the Short SC-1, which carried on VTOL research from the R-R Flying Bedstead, and the Ryan XF 109, also for VTOL research. In the period of this history, probably more aircraft were powered by Avon turbojets than any other. Apart from those already mentioned, it will suffice to mention the SAAB J-35 Draken and Dassault Super Mystère B-1 fighters, Sud Ouest S.O. 4050 Vautour 012 bomber and Hurel-Dubois H.D.45 airliner. Because of its widespread application, many records were made using the Avon and it became, as Ernest Hives had predicted, the 'Merlin of the future'. RA.7 Avons were built under licence in Australia by the Commonwealth Aircraft Corporation to power Canberras and Sabres. In Belgium the RA.21 and 23 Avons were built by Fabrique National to power Dutch and Belgian-built Hunters.

CHAPTER FOURTEEN

Rolls-Royce: Turboprops, Bypass and Lift Engines

Rolls-Royce was also especially successful in the turboprop field. Its first experiments with turboprops were with the readily-produced Trent turboprop which was simply a Derwent turbojet modified to drive an airscrew through reduction gearing (*see above*). Two Trents were fitted to a Meteor Mk.I (EE227), which first flew on 20 September 1945. The Trent was run for 633 hours on the test bench and 298 hours in flight (including 47 hours on the Meteor) and provided much valuable data regarding control and gearing problems for later turboprop work. A problem that nearly turned into a disaster occurred during the first landing of the Trent-powered Meteor. Because the five-bladed airscrew on each Trent was designed to be virtually pitchless to facilitate starting, when the pilot Eric Greenwood throttled back during the final approach for landing the great drag of the airscrews in almost zero pitch caused an extremely rapid descent and only quick application of full power, a natural reaction, prevented a crash.

Rolls-Royce's first turboprop, the Trent, which was produced by fitting a reduction gearbox for the airscrew to a modified Derwent turbojet. Rolls-Royce Heritage Trust

A Meteor Mk I (EE227) used as a flying test-bed for the Rolls-Royce Trent turboprop, flying comfortably on one engine. Philip Jarrett

Clyde and Dart

A very much more ambitious turboprop was the next to be developed. Known as the Clyde, this engine was advanced enough to require the combined efforts of a team at Barnoldswick under Hooker and Adrian Lombard and another at Derby under Lionel Haworth. Designed for 3,000shp, the Clyde was a twin-spool engine with an axial compressor followed by a centrifugal compressor. The axial compressor was a nine-stage one, driven by a low-pressure, single-stage turbine, the same shaft also driving the airscrew via reduction gearing at the centre of an annular air intake. A coaxial hollow shaft, over this shaft, connected the centrifugal compressor to its driving, high-pressure, single-stage turbine. Thus air passed from the axial compressor to the inlet of the centrifugal compressor and thence to eleven Lucas combustion chambers feeding hot gas to the turbines. Connected to the rear of the centrifugal compressor was a smaller centrifugal impeller that supplied cooling air. However, cooling air for the front and rear faces of the low-pressure turbine was tapped off the axial compressor and cooling air for the turbine bearings was tapped off the centrifugal compressor. The total pressure ratio of the compressors was 6:1 and the air mass flow was 18.82kg/sec (41.5lb/sec). Most of the accessories of the engine were conveniently located in the neck formed by the reducing diameter of the axial compressor casing.

After separate component testing, the Clyde made its first run as a complete engine on 5 August 1945. It exceeded expectations and soon developed 4,000 ehp with a specific fuel consumption of 0.71. Unfortunately, because there was no suitable aircraft in prospect for it, the Clyde was shelved in 1949 after completing about 3,300 hours of test-bed and flight testing. Strangely, straight-toothed spur gears were used in the Clyde that tended to build up destructive resonances, whereas the way forward had already been shown in the Trent, which used helical gears for the high-speed train. Perhaps helical gears would have been used if the development had continued. In April 1945, while the Clyde was being developed, and just before the war in Europe ended, Lionel Haworth at Derby was at work designing another turboprop, this time with tandem centrifugal compressors. This turboprop was designated RB.53 and was to become the famous Dart. Originally it had been suggested at Barnoldswick in response to an official requirement for a 1,000hp turboprop for use in trainers such as the Boulton Paul Balliol, and was in competition with the AS Mamba (see Chapter 15) and Napier Naiad (see Chapter 17) turboprops.

The first Dart, the R.Da.1, was bench tested in late 1946 and made its first test flight in the nose of a Lancaster test-bed in October 1947. Another flying test-bed was the Wellington (LN715), which flew with two Dart engines. At that stage, the engine was overweight at 454kg (1,000lb) and delivered only 890shp. Redesigning reduced the weight to 363 kg (800 lb), the main changes being the replacement of the high-speed spur gears with helical gears and a new air intake casing that housed the oil cooler and was therefore kept free of ice.

The Dart was a compact engine using two centrifugal compressors, close together, driven by a two-stage turbine. The compressors were made of light alloy but the centre, curved part of the vanes was of steel. Air was drawn in through an annular inlet to the first-stage compressor and passed from this via a diffuser and turnover passage down to the centre of the second-stage compressor. From there it was led by elbows into seven interconnected combustion chambers that were mounted at a skewed angle, leading into the turbines. The simple exhaust cone had a fixed central bullet,

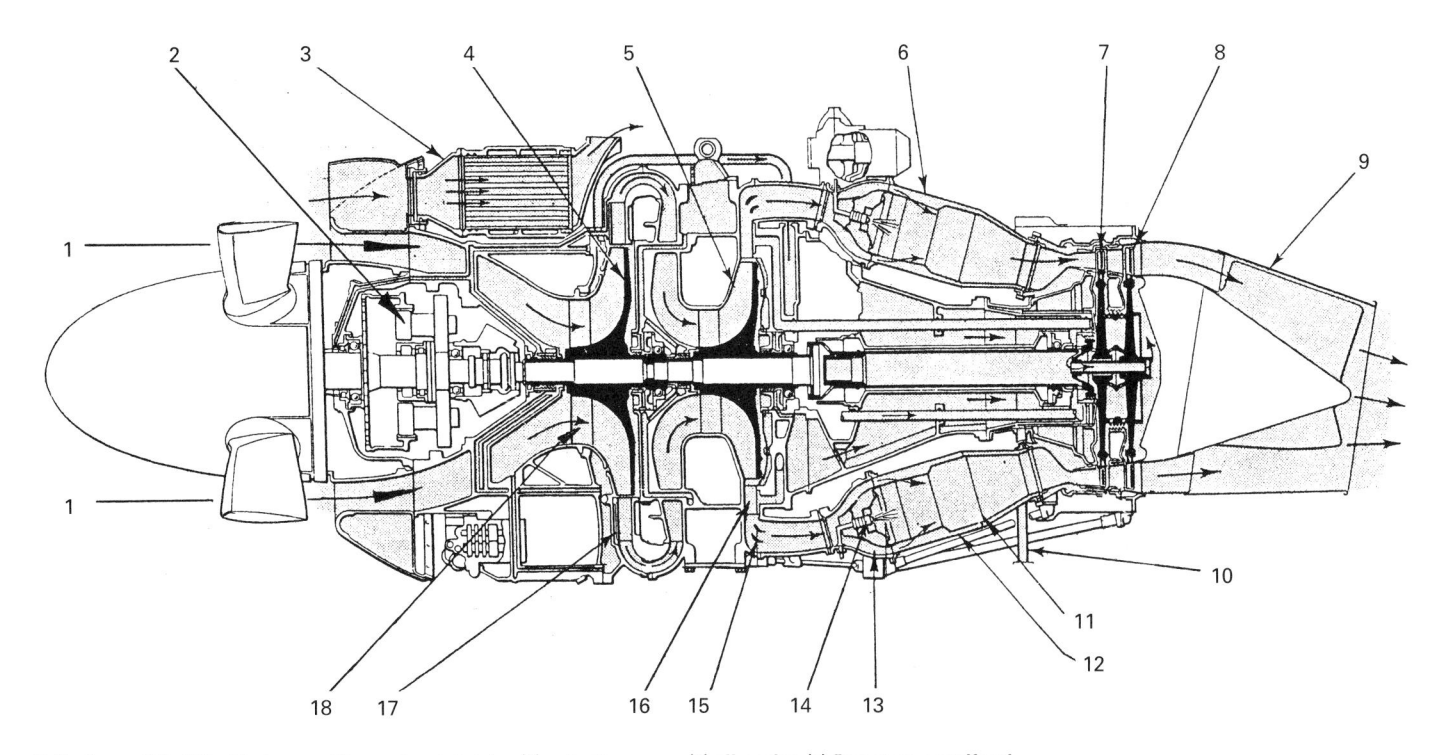

Rolls-Royce R.Da 6 Dart turboprop: (1) annular air intake; (2) reduction gears; (3) oil cooler; (4) first-stage centrifugal compressor; (5) second-stage centrifugal compressor; (6) combustion chambers; (7) HP turbine; (8) LP turbine; (9) exhaust nozzle; (10) cooling air outlet; (11) flame tube; (12) air casing; (13) expansion chamber; (14) burner; (15) cascade vanes; (16) second-stage diffuser; (17) first-stage diffuser; (18) rotating guide vanes.

Rolls-Royce Dart centrifugal turboprop. Rolls-Royce Heritage Trust

A Wellington flying test-bed (LN715) with Rolls-Royce Dart 602 turboprops. Philip Jarrett

was set at an angle and gave a small amount of residual thrust. The Dart had three bearings, there being a coupling with gear teeth on the shaft near the centre roller bearing. These gear teeth also drove the auxiliary gearbox. Various Nimonic alloys were used in the turbines, the blades being without air cooling. Air was tapped off the first-stage compressor to feed oil seals throughout the engine and was tapped off the second-stage compressor to cool the turbine discs. In the centre of the annular air intake was a streamlined egg, the rear, fixed section of which enclosed the airscrew reduction gears and the forward section of which formed the airscrew spinner and enclosed the hydraulic pitch-change mechanism.

After passing a 150-hour type test at 1,000shp, the R.Da.1, designated Dart 502, was test flown in a Wellington that flew in June 1948. The first and most successful aircraft with which the Dart was associated, and the first aircraft specifically designed for turboprop propulsion, was the Vickers Viscount, a combination that became a resounding success. (The original aircraft for which the Dart had been designed, the Balliol and Apollo, finished up with piston engines.)

Viscount, the world's first turboprop-powered passenger aircraft

The Viscount airliner was the most successful outcome of the deliberations and recommendations of the Brabazon Committee, reconvened before the end of the war. With its Type IIB, the committee conceptualized a turboprop short-haul airliner for European routes. (Type IIA was to be a piston-engined version, later realized by the Airspeed Ambassador.) The MoS ordered prototype airliners from Armstrong Whitworth and Vickers-Armstrong, these competing companies respectively offering the A.W.55 Apollo and the VC.2. The official specification of April 1946 called for a turboprop airliner with twenty-four seats, a 3,400kg (7,500lb) payload and a range of 1,600km (1,000 miles). The choice of engines fluctuated between the AS Mamba, R-R Dart and Napier Naiad. In March 1946 two Mamba-powered VC.2s and Apollos were ordered, but Chief Designer George Edwards at Vickers decided the following spring to power their prototype with R-R Darts.

The prototype Vickers V.630 airliner (G-AHRF), later named the Viscount, made its first flight on 16 July 1948, piloted by Capt J. 'Mutt' Summers. It was powered by four 1,000shp R.Da.1 turboprops. Results were outstanding but, inexplicably, British European Airways (BEA) went ahead and ordered twenty of the piston-engined Ambassador airliners. The loss of the main customer in this way caused the Viscount programme to be throttled back and could have been a terminal blow. Meanwhile Rolls-Royce had developed the Dart into the R.Da.3 of 1,400shp and, using this more powerful engine, the Viscount was stretched in wingspan and length to accommodate forty-seven passengers, the aircraft in this form being designated V.700.

At this time, the second prototype was built as the V.663, powered by two R-R Tay turbojets. As VX217 it made the world's first scheduled airline operations with a turbojet aircraft, starting on 29 July 1950 from London to Paris. This success swung BEA back in favour of the Viscount, albeit the turboprop-powered V.700.

On 19 April 1950 the V.700 prototype (G-AMAV) made its first flight, from Weybridge, and on 3 August 1950 BEA ordered twenty Viscounts of the type V.701, which could accommodate fifty-three passengers. These airliners went into BEA service on 18 April 1953 and were the first regular turboprop transport services for passengers. The first such flight was by a V.701 from London to Nicosia (Cyprus) via Rome and Athens.

The Viscount was an elegantly proportioned aircraft with a conventional fuselage, having fore and aft entrance doors and large, oval windows for the passengers. There was a crew of three and underfloor cargo space. Its main features of note were

The prototype Vickers V.630 (G-AHRF, VX211), predecessor of the very successful Viscount airliner. It was powered by four 1,000shp Rolls-Royce R.Da.1 Dart turboprops. Philip Jarrett

A Viscount airliner (N7411) of Capital Airlines, showing the installation of its No. 3 engine, a Rolls-Royce Dart turboprop. The large, hinged nacelle panels afford easy access. Philip Jarrett

Vickers Viscount V.708 with 1,600shp Dart 510 turboprops		
Span	28.56m	(93ft 8½in)
Length	24.74m	(81ft 2in)
Wing area	89.56sq m	(963sq ft)
Height	8.15m	(26ft 9in)
Empty weight	14,665kg	(32,330lb)
Maximum loaded weight	32,886kg	(72,500lb)
Passengers	65	
Cruising speed	500km/h	(311mph)
Range with maximum payload of 5,670kg (12,500lb)	1,591km	(940 miles)

the equi-tapered wings of high aspect-ratio with main and auxiliary spars, ribs and double-skinned leading edges, the flaps being of the double-slotted Fowler type. The wing spars carried through the fuselage below the passenger floor and loads from the fuselage were transmitted by circular frames. Each turboprop engine was carried by bearer struts entirely forward of the wing leading edges, leaving the wings largely clear for fuel tanks. The retractable, tricycle undercarriage had double wheels on each member.

Following the BEA orders, world orders flooded in for the Viscount, Air France and Aer Lingus being the first to follow. The Dart turboprop continued to be up-rated, development concentrating on the

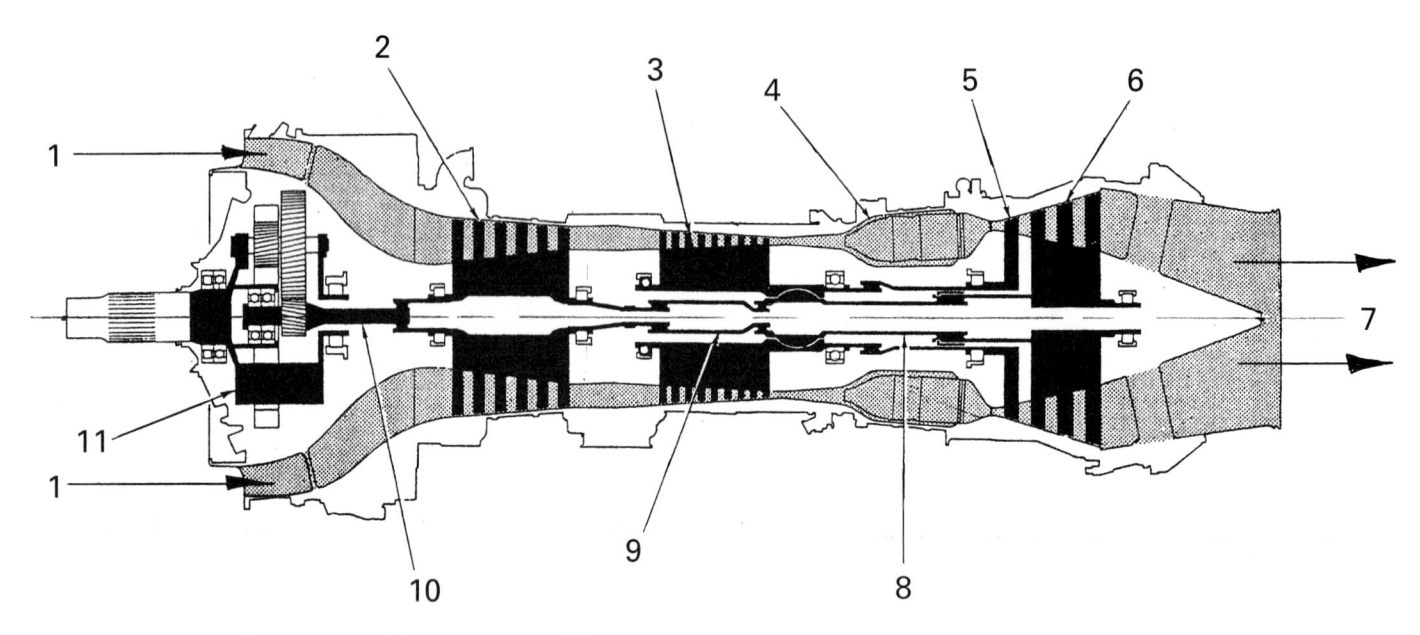

Schematic of Rolls-Royce Tyne turboprop: (1) annular air intake; (2) 6-stage LP compressor; (3) 9-stage HP compressor; (4) combustion chambers; (5) single-stage HP turbine; (6) 3-stage LP turbine; (7) exhaust efflux; (8) coupling shaft; (9) intermediate shaft; (10) high-speed pinion shaft; (11) airscrew reduction gears.

refinement of every component, the R.Da.6 (Dart 510) giving 1,600shp and the R.Da.7 (Dart 520 with three-stage turbine) of 1954 giving 1,800shp. With the Dart 506, water-methanol injection was available to boost power during take-off if required, but pilots felt that this feature made little difference. To some the Dart, with its centrifugal compressors, was of primitive layout, but it was totally reliable and delivered the power economically. Eventually the Dart was delivering some three times the power of the original 1,000shp engines and, in forty years of production, 7,100 were built.

As the orders came in from different airlines and the engines improved, so the marks of Viscount proliferated and its size and capacity increased. With the final variant, the V.810 powered by 2,100ehp R.Da.7/1 (Dart 525) engines, the passenger capacity had increased to seventy-seven and the weight had grown to 32,886kg (72,500lb) or 1.8 times that of the original aircraft. By 1964, when production ceased, some 444 Viscounts had been built, the last being ordered by China. The Viscount was flown by sixty operators around the

Rolls-Royce Tyne turboprop. Philip Jarrett

A Rolls-Royce Tyne turboprop fitted neatly into the nose of an Avro Lincoln at Hucknall airfield. The water spray rig is for icing tests. Philip Jarrett

world and the market for used Viscounts was always buoyant.

In addition to the Viscount, the Dart also powered the Fokker F27, Hs 748, Hs Argosy, HP Herald, Breguet Alizé and Convair 660 Metropolitan, amongst others.

Tyne turboprop

During the early 1950s, with the Dart turboprop nicely established, Lionel Haworth began considering a successor for it. The idea was to start with a clean sheet but benefit from experience with high-efficiency components in the Dart engine and to produce an engine in the range of 2,500 to 4,000shp. A number of layouts were considered and, finally, a 4,000shp engine was decided upon because Vickers, buoyed up with the success of the Viscount, was planning a new airliner requiring four such engines. This was the V.950 Vanguard, designed in 1953 to BEA's requirements for an airliner with Viscount-like economy and comfort but for up to 139 passengers and with greater speed. The Vanguard was first flown on 20 January 1959, piloted by G.R. Bryce. Unfortunately, excellent as the airliner was, it suffered a similar fate as the Bristol Britannia in that turboprops were mistakenly being ousted by airlines for faster turbojet aircraft. Consequently, only forty-three Vanguards were built (the last in March 1962), the customers being BEA and Trans-Canada.

Despite the lukewarm reception for the Vanguard, the Tyne, as the new turboprop was called, was a very successful engine and went on to power other aircraft. The Tyne, designated RB.109, was a two-shaft, high-pressure ratio, axial engine. It had a six-stage low-pressure compressor driven by a three-stage low-pressure turbine. In between these, coaxially, was a nine-stage high-pressure compressor driven by a single-stage high-pressure turbine. A cannular combustion system with ten combustion tubes, each having a Duplex burner, was used. Inside the annular air intake was a streamlined casing containing the spur and helical reduction gears for the de Havilland Hydromatic, four-bladed airscrew. The hydraulic pitch control unit was housed in the airscrew spinner and the reduction gears were driven from the low-pressure shaft. Ball and roller bearings were at the ends and centre of the gearbox, both ends of both compressors and at both ends of the turbines.

The power take-offs for the port and starboard auxiliaries gearboxes were taken from the centre of the engine, between the

The Airspeed Ambassador (G-37-3, later G-AKRD) used to flight test the Rolls-Royce Tyne turboprop.
Philip Jarrett

compressors. One take-off was driven by the low-pressure shaft and the other by the high-pressure shaft. Hot air was bled from the high-pressure compressor's final stages to feed in and out of the inlet guide vanes and first-stage stator blades of the low-pressure compressor. From there the air went into an annulus and thence to the bottom of three struts of the air intake annulus. The top three struts of this annulus were heated by scavenged hot oil. Air bled from the low-pressure compressor was used for seals and to cool the rear of the fourth-stage turbine disc. Air bled from the seventh stage of the high-pressure compressor was used to cool the discs of the turbine's first three stages. The exhaust outlet had a centre cone with ten streamlined arms down which cooling air was passed to the turbine disc via the rear roller bearings. All in all, the Tyne was considerably more complex than the Dart. The initial design (RB.109) was changed to give 4,500shp and designated RTy.1. In this form the pressure ratio was 13.67:1 and the air mass flow was 21.14kg/sec (46.60lb/sec).

The RTy.1 Tyne was first run in January 1956 and it passed its first 150-hour test that December. Flight testing began in June 1956 with a Tyne mounted in the nose of a Lincoln that also retained its four Merlin piston engines. The first public demonstration of this aircraft was at Farnborough that year in September. Icing flight tests were later made by mounting a spray grid around the engine. In August 1958 the first flights purely on Tyne power were made with two of the turboprops fitted to the second prototype of the Airspeed Ambassador airliner (G-AKRD), the same aircraft also being used to test Dart and Bristol Proteus turboprops. From 6 March to 16 October 1959 a Tyne-powered Ambassador was flown from Hucknall by Rolls-Royce and BEA crews to gain endurance experience with the turboprops, some 1,642 hours of flight experience being gained.

By November 1959 the Canadair CL-44 powered by Tynes had flown and this engine went on to power the Short Belfast, Breguet Atlantic and Transall C-160. Eighteen developmental engines were used in the Tyne programme, including the original three RB.109s, which were modified up to RTy.1 standards. Without radically altering the engine and, at the same time, retaining the same pressure ratio and air mass flow, the Tyne was uprated by increasing the operating temperature and by making minor internal improvements. An example of the latter is that the throat area of the first-stage, low-pressure nozzle guide vanes was altered by changing their aerofoil angle. With the introduction of the RTy.20 engine in 1959, the power had risen to 6,100ehp for take-off, the dry weight being 975kg (2,149lb) and the specific fuel consumption dropping to a mere 0.391. Some 1,467 Tynes were sold by 1960 and the engine lived on into the 1990s, a tribute to its excellence.

Conway, the first bypass turbojet

During 1946 it was decided that the time had come to look more closely at the bypass or turbofan engine. In the bypass turbojet, part of the airflow bypasses the combustion chamber and rejoins the flow

A Rolls-Royce RCo.12 Conway bypass turbojet undergoing final assembly.
Rolls-Royce Heritage Trust

of gases at the exhaust end of the engine. This mixed flow has a reduced velocity, making it cooler and quieter than in a turbojet without bypass. Because of the lower exhaust velocity, propulsive efficiency is better and so the specific fuel consumption is lower. This makes the type of engine suitable for long ranges. Dr Griffith suggested that, to obtain some early data quickly, an engine could be built using components from an Avon and from an experimental turbojet known as the Tweed. A more advanced engine of about 2,268kp (5,000lb) thrust was proposed in April 1947. Other projects were examined until, in October 1948, an engine designated RB.80 (later to become the Conway) was planned.

The RB.80 was designed as a 4,195kp (9,250lb) thrust bypass engine and was offered to the MoS to power the Vickers Pathfinder, a Mk. 2 version of the Valiant bomber that was subsequently cancelled. By January 1950 further design studies had led to the Conway RCo.2, which was a two-shaft engine having a four-stage low-pressure compressor driven by a two-stage turbine and, in between, an eight-stage high-pressure compressor driven by another two-stage turbine. Only one RCo.2 was built. It first ran in July 1952 and a year later it was producing a thrust of 4,536kp (10,000lb). According to A.A. Lombard, Director of Engineering, 'It must have known that it was going to be superseded, as it ran perfectly for the whole of its 133 hours life'.

Because, in the early Conway engines the bypass ratio was kept low (perhaps 30 per cent or less), the improvement in fuel consumption was not expected to be dramatic but this was done in order to keep the engine diameter down for the purpose of burying it in a wing. Consequently it was only necessary to have an enlarged low-pressure compressor and to divert some of its airflow for the bypass portion. Of course, the turbojet without bypass could be made to produce a lower jet velocity by lowering the combustion temperature, but this would detrimentally reduce its cycle efficiency. In any event, the Conway became the first bypass turbojet.

Official sponsorship for the Conway came about because engines of sufficient power and efficiency were needed for the Vickers V-1000, a large military transport for the RAF that could accompany V-bombers to any part of the world. Work on the V.1000 began in October 1952 in response to the MoS Specification C.123D. Its shape, at least, was reminiscent of a larger Comet. From this aircraft, Vickers also planned to develop the VC.7 airliner, initially for BOAC. In designing the V.1000, experience with the Valiant bomber was drawn upon. A low-mounted wing of 42.67m (140ft) span was designed with compound sweep on the leading edges and a wing root intake for each pair of Conway engines. The weight was to be 104,328kg (230,000lb). Construction of the V.1000 prototype was well under way at Wisley when, in summer 1955, the aircraft was cancelled. This was due to the RAF decision that perhaps there was no need for a transport that could keep up with the V-bombers, as well as the overall need for budget cuts. The cancellation of the bomber also put paid to the VC.7 airliner.

This was by no means the end of the Conway engine. Government sponsorship of its development continued in the hope that it would be used to power at least one of the V-bombers. In the USA, the Conway was in competition with Pratt & Whitney's J-75 turbojet to power the new Boeing 707 and Douglas DC-8 airliners.

The design of a more powerful Conway, the RCo.5, had been started in January 1952 for the V.1000 transport. It had a six-stage low-pressure compressor driven by a two-stage turbine and a nine-stage high-pressure compressor driven by a single-stage turbine. First run in July 1953, the RCo.5 passed an official type test in August 1955 at 5,897kp (13,000lb) thrust. Thirteen of these engines were built for development and flight tests were made beneath an Avro Ashton test-bed and then in a Vulcan. Altogether some 980 flight hours were made.

In January 1956 test runs began on a generally similar engine, the RCo.8, which differed chiefly in having a more efficient air-cooling system. Intended for the Victor V-bomber, the RCo.8 produced a thrust of 6,577kp (14,500lb) but did not go into production. Instead, the RCo.10 of 7,484kp (16,500lb) thrust, developed specifically for civil use, was built; four were tested in the Vulcan V-bomber test-bed (VX770), beginning on 9 August 1957. The RCo.11 of similar thrust was then produced to power the HP.80 Victor V-bomber, the first flight of the first production model so powered (XH668) taking place on 20 February 1959.

Considerable effort went into developing the RCo.10 for civilian use, for which longer life and more stringent requirements are necessary than in military use. The impetus for this was as a result of the interest of Trans-Canada Air Lines (TCA) in the engine and the subsequent persuasion of the Boeing and Douglas companies that they should offer Conway-engined 707 and DC-8 airliners respectively. It was anticipated that the range of these airliners

The Rolls-Royce RCo.12 Conway bypass turbojet with silencer, used to power the Boeing 707-420 and the Douglas DC-8 airliners. This type of 'silencer' was only moderately successful but, with the advent of larger, quieter turbofan engines, it was no longer needed The left thrust reverser is also shown. Philip Jarrett

Two of the four Rolls-Royce RCo.42 Conway engines on a rear-engined Vickers VC-10 airliner (G-ASGC). The VC-10 was specifically designed to operate from 'high and hot' airfields and from short runways. Author

would be increased by using Conway instead of the American J-75 engines. Studies by Boeing, for example, showed that a Conway-powered 707-420 would have more than an 8 per cent range advantage over a J-75-powered 707-320. In May 1956 TCA ordered Conway-powered DC-8s and this was followed by more orders for such aircraft from Alitalia and Canadian Pacific Air Lines, and orders for Conway-powered 707s from BOAC, Lufthansa, Varig and Air India. First flights of production versions of these aircraft with Conway engines were on 19 May 1959 for a Boeing 707-420 (G-APFB) and 23 July 1959 for a Douglas DC-8 (CF-TJA). The Conways used were RCo.12s of 7,780kp (17,150lb) thrust, first tested in November 1957 and uprated from the RCo.10.

Conways of the RCo.10 type were delivered to Boeing and Douglas in September 1958 for certification flying. The RCo.10 differed from the RCo.8 in having an additional stage added in front of the low-pressure compressor, a high-pressure compressor of 10 per cent greater capacity and low-pressure turbine blades of greater length. Also, engine accessories were designed and arranged for pod installation and, for Boeing, R-R supplied the engine complete with pod. Because development proceeded ahead of schedule, Conways went into airline service at the RCo.12 rating of 7,780kp (17,150lb). Also for these engines R-R developed a multi-lobe silencer and a thrust reverser for braking when landed. This reverser employed a pair of jack-operated doors (or 'eyelids'), fitted before the silencer, which deflected some of the exhaust gases into vane cascades that deflected the gases in a forward direction. Maximum reverse thrust was about 50 per cent of forward thrust and the jacks were operated by air tapped off from the high-pressure compressor via pressure-reducing and control valves.

The RCo.12 Conway had a seven-stage low-pressure compressor driven by a two-stage turbine and, in between, a nine-stage high-pressure compressor driven by a two-stage high-pressure turbine. Only 23 per cent of the low-pressure compressor's air was bypassed and ducted around the engine to join the hot gas flow just before the thrust reverser. Pin fixing of the compressor rotor blades was used, aluminium alloy being used for the low-pressure blades except for the last stage, which used titanium. For the high-pressure rotor blades, titanium was used except for the last two stages, which were in steel. Titanium was also used for the bypass duct casting.

The combustion system was of the cannular type, with ten interconnected Nimonic flame tubes. For the turbine rotor blades, fir-tree roots were used for attachment to their discs. Hollow, air-cooled blades were used for the high-pressure turbine. Auxiliaries were clustered around the front section of the engine, which was of a roughly parallel, cylindrical shape. The RCo.12 had a dry weight of 2,043kg (4,504lb), an air mass flow of 127.91kg/sec (282lb/sec) and a specific fuel consumption of 0.870 in the cruise. Despite the proven benefits of using the Conway, of the 1,519 707s and DC-8s built, only 69 were powered with this type of engine.

The final development of the RCo.12, the RCo.15, takes us into the 1960s but will be mentioned for completeness. It had a larger zero-stage in front of the first stage of the low-pressure compressor and detail improvements to reduce leakage losses. In the RCo.15, mass flow was increased to 134.31kg/sec (296.1lb/sec), specific fuel consumption in the cruise was improved by about 3 per cent and take-off thrust was up at 8,392kp (18,500lb). During overhaul, an RCo.12 could be rebuilt to RCo.15 standards.

Finally, an extensive redesign of the RCo.15 resulted in the RCo.42, specifically for the Vickers VC.10 airliner. Because it was designed from the outset for the rear pods of the VC.10, the RCo.42 was not restricted in its diameter. This meant that the engine could have a low-pressure compressor giving the greater mass flow of 164.65kg/sec (363lb/sec) and the bypass ratio increased to a useful 60 per cent. The low-pressure turbine was suitably increased in diameter also. First run in March 1961, the RCo.42 gave a take-off thrust of at least 9,185kp (20,250lb) for a dry weight of 2,268kg (5,001lb). Its specific fuel consumption in the cruise was 0.785. Development continued well beyond the timescale of this history in such bypass engines as the Medway, Spey, Spey Junior and Tay (the second use of that name, since it was also used to name R-R's last centrifugal turbojet).

RB.108, the first lift engine

Before leaving Rolls-Royce, something must be said about the first lift-jet that the company developed in the mid-1950s for VSTOL research. Following the success of the R-R Flying Bedstead test rig (*see above*), which proved the feasibility of using a puff-pipe system for controlling a jet-supported vehicle, it was decided to develop a special lift engine. Studies by A.A. Griffith indicated that the most efficient VSTOL aircraft would be produced if the propulsion engine were optimized

An assembly of two Rolls-Royce RB.108 turbojets, specifically developed for VTOL aircraft experiments. Philip Jarrett

The second prototype (XG905) of the Short SC.1 experimental VTOL aircraft in the hover. It was powered by five Rolls-Royce RB.108 lift engines, four for vertical lift and one for forward flight. Philip Jarrett

for forward flight and separate jet thrust engines were provided for vertical take-off and landing. Accordingly, the RB.108 was designed as the simplest and lightest turbojet possible having a high thrust-to-weight ratio. An axial engine was produced having an eight-stage compressor with a pressure ratio of 5.5:1. There was an annular combustion chamber with twelve Duplex burners, followed by a two-stage turbine and no exhaust nozzle. A pressurized tank supplied the engine with oil, which was then drained into the jet exhaust without being recirculated. For starting, the turbojet was spooled up by an external air supply impinging upon the turbine blades, combustion then being initiated by a high-energy ignition system. One main propulsion engine in the 4,536kg (10,000lb) thrust class could provide enough bleed air to start up eight RB.108s simultaneously within 30 seconds. Air bled from the RB.108s could then be supplied to puff-pipes for hovering aircraft control.

The RB.108 was first run in July 1955 and success was rapid. Its thrust-to-weight ratio of almost 8:1 was revolutionary, thrust being 912kp (2,010lb), later developed to 1,002kp (2,210lb), with 11 per cent available for bleeding off. Dry weight was 129kg (285lb), diameter 0.508m (1ft 8in) and length 1.06m (3ft 6in). The necessary very rapid throttle response was achieved by using a low-inertia rotor, idling to maximum thrust taking only one second. To prove that the RB.108 could be restarted during flight, a single engine was mounted vertically in a special bay in the centre fuselage of a Meteor PR.9 (VZ608) and flown from Hucknall. Ram air to the engine was admitted through a door on top of the fuselage.

Short SC.1 V/STOL aircraft

Following a design competition, the Short Brothers company was chosen to build an experimental aircraft to test the lift engine system and the MoS sponsored two prototypes. Designated the Short SC.1, this was an all-metal, small, delta wing aircraft with a glazed nose section, a small fin and a tricycle undercarriage. It was powered by five RB.108 engines, four for vertical lift and one for forward propulsion. The forward propulsion engine was installed in the rear fuselage and had an air intake on top of the fuselage, below the fin. Air reached the four lift engines via louvres on top of the fuselage, which only opened on demand of the lift engines. Further air could be supplied via pilot-controlled gills. In the hover, the SC.1 was controlled fore and aft by the gimbal-mounted lift engines and by puffer-pipes in the wingtips, nose and tail. A comprehensive, full three-axis, auto-stabilization system, which could be overridden by the pilot, was fitted.

The first SC.1 prototype (XG900) had no lift engines and was used for horizontal flight only, its maiden flight being on 2 April 1957, piloted by Tom Brooke-Smith. The second prototype (XG905) had all engines fitted and made its first tethered flight on 26 May 1958, and its first free vertical take-off and landing on 25 October 1958. Tom Brooke-Smith made the first transition from hover to forward flight, and vice versa, in XG905 on 6 April 1960.

There were many problems needing attention, not least being the high pilot workload during a vertical landing and the high fuel consumption during hovering. Tragically, on 2 October 1963 the auto-stabilization system failed and the pilot was killed in the ensuing crash. The programme continued after this aircraft was rebuilt, and RB.108s were used all over the world in VSTOL experiments during the 1960s and later. As many as eight RB.108 lift engines were used in the Dassault Balzac research aircraft. The RB.108 proved to be very reliable and research aircraft of the SC.1 type of layout proliferated in the 1960s and beyond. Rolls-Royce continued as one of the world's greatest aero-engine companies and, in due course, all the surviving British competitors came under its control. This process was largely facilitated by the merging of Rolls-Royce and the Bristol-Siddeley organization in January 1968.

With its high power-to-weight ratio, the Rolls-Royce RB.108 lift engine was used all over the world in V/STOL experiments and such research aircraft proliferated in the 1960s and beyond.
Rolls-Royce Heritage Trust

CHAPTER FIFTEEN

Armstrong Siddeley Motors Ltd

Armstrong Siddeley Motors (AS) of Coventry began building aero-engines in 1917, during the First World War, its most prolific product of those days being the Puma in-line six-cylinder piston engine. Unfortunately none of its subsequent piston engines really excelled in either production or performance. The company entered the gas turbine field in 1939. Initially, this was only in a small way when it became involved in assisting the RAE with axial compressor construction. Such constructions included the contraflow compressor devised by A.A. Griffith and the eight-stage 'Anne' and the fourteen-stage 'Sarah' compressors.

Armstrong Siddeley's gas turbine work came under the Director of Engine Development, Bill Saxton, with A. Thomas as his Chief Designer and W.H. Lindsey as Deputy Chief Engineer. Previous to this, in 1936, the Chief Engineer of the company, Col Fell, had made a study of a German combined steam and gas turbine aero-engine (presumably after Vorkauf's Drehkessel scheme) but nothing came of it.

ASX turbojet and ASP/Python turboprop

On 7 November 1942 the company was given its first gas turbine contract. After making many studies it centred on the ASX turbojet, which in a turboprop form was designated the ASP. The ASX consisted of the RAE-designed fourteen-stage 'Sarah' axial compressor feeding reverse-flow air through 90-degree elbows to eleven combustion chambers, which exhausted into an annulus before entering the two-stage turbine. Air entered the compressor at the rear of the engine via ports between the combustion chambers and turbine. The compressor was built onto a drum-type rotor that tapered at its rear to connect with the turbine discs. There were only two bearing points, one in front of the compressor and one in front of the turbine, the latter being overhung. A shaft extended forward of the compressor bearing for connection to airscrew gearing if the engine was to be used in its ASP form, which was probably the reason for the unusual rear configuration of the air intakes. Another unusual feature was that, in order to relieve the thrust load on the bearings in front of the compressor, an equalizer piston on the shaft in front of the bearing was fed with high-pressure air tapped from the compressor.

By April 1943 the ASX was running but producing a static thrust of only 363kp (800lb); by September 1943, however, it was giving its calculated thrust of 1,157kp (2,550lb), the specific fuel consumption being 1.0. Flight tests in a Lancaster followed and by 1944 the ASX was delivering 1,270kp (2,800lb) of thrust at 8,000rpm for a weight of 862kg (1,900lb). Since this was an unexciting performance, the ASP version was tried next and run with an airscrew in April 1945, delivering some 3,600shp plus 499kp (1,100lb) of thrust. Unfortunately, the gearbox and airscrew pushed the weight of the ASP up to 1,565kg (3,450lb). However, a start had been made and, despite the ASP's indifferent performance, the company was awarded a contract to develop it as the Python. Development then continued with the Python fitted with a Rotol eight-bladed contra-rotating airscrew, test flights being made with Pythons in the outer positions of a Lancaster (TW911) and a Lincoln (RF403).

Out of the work on the ASX turbojet and ASP/Python turboprop engines emerged an important development regarding the combustion system, which was to have a considerable influence on subsequent engines. The original ASX combustion chambers were designed and made by Joseph Lucas Ltd and were of typical British type using fuel atomization. Armstrong Siddeley, however, decided to make its own investigations into combustion and focused on the idea of a vaporizing fuel

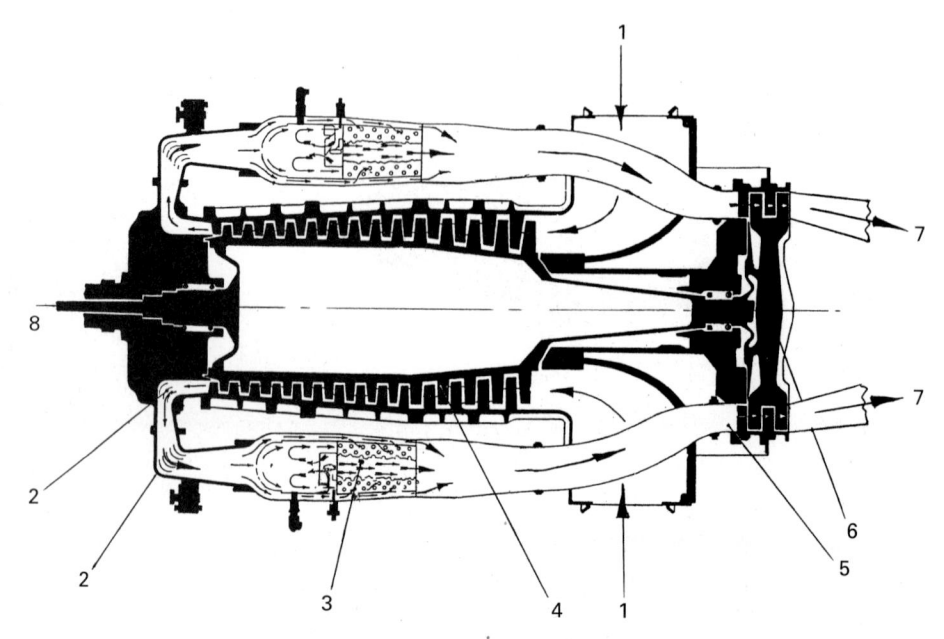

Schematic of the Armstrong Siddeley ASX turbojet: (1) air inlets; (2) guide vanes and elbows; (3) eleven combustion chambers; (4) 14-stage axial compressor; (5) annulus; (6) 2-stage turbine; (7) exhaust nozzle; (8) shaft for connection to airscrew reduction gears, if used in ASP turboprop form.

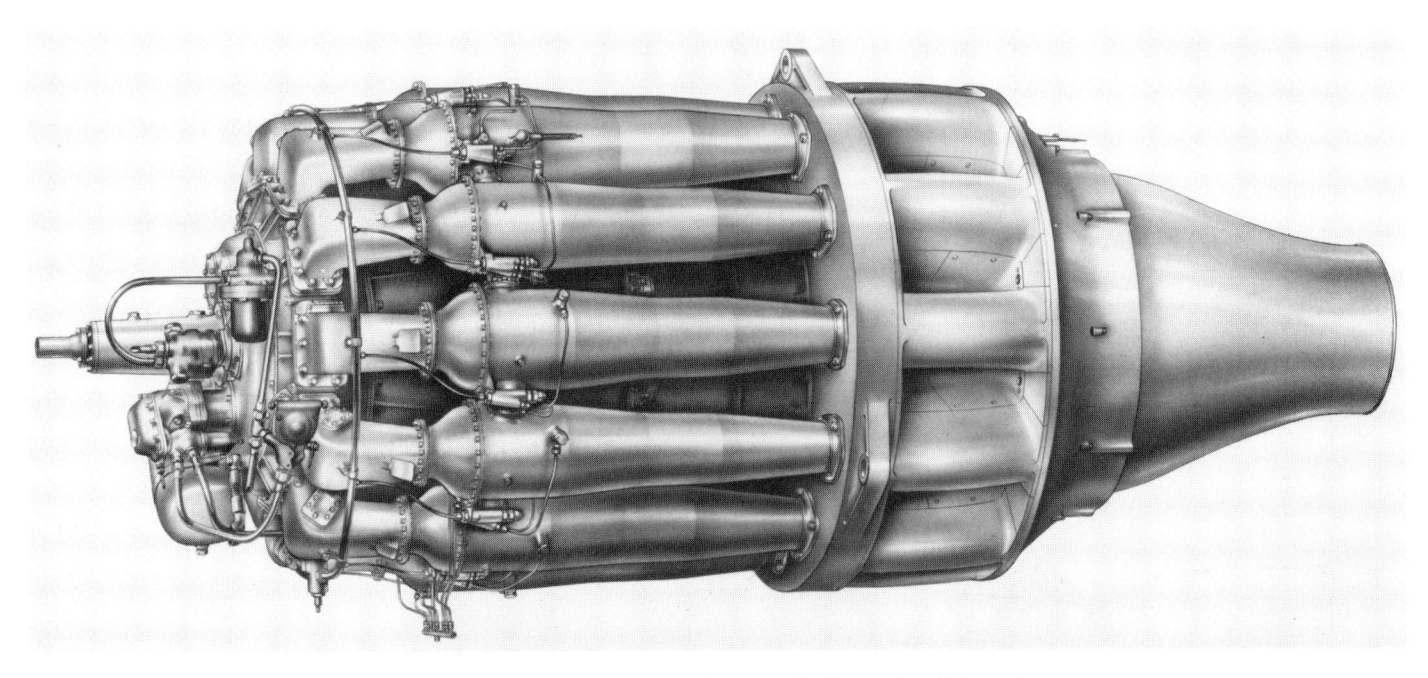

Armstrong Siddeley ASX turbojet. The forward shaft could be connected to an airscrew reduction gearbox, if the engine was to be used in its ASP turboprop form. Rolls-Royce Heritage Trust

An Armstrong Siddeley Python turboprop installed in a special Wing Test Rig designed to simulate flight conditions. The oil coolers below the wing were not required when the engine was installed in an aircraft. Philip Jarrett

Lancaster (TW911) flying test-bed with Armstrong Siddeley Python turboprops and contra-rotating airscrews in the outboard positions. Philip Jarrett

system. By 1945 the company had its own combustion test rigs installed and had fitted a vaporizing fuel system to an ASX turbojet. This engine was tested in a Lancaster up to the aircraft's ceiling of 10,700m (35,000ft) and the slow running and re-starting performance of the ASX was most encouraging. Advantages claimed for the vaporizing fuel system were that the combustion was easier to control under widely varying conditions, the use of specialized, troublesome atomizing jets was avoided and a high-pressure fuel system was unnecessary, thereby rendering fuel pumps more reliable. Most important was that the system gave near-perfect combustion under all flight conditions. (The reader may recall Whittle's lack of success with a vaporizing fuel system.)

Westland Wyvern fighter

Considerable improvement in performance was given by using the vaporizing fuel system employed in the ASP or Python turboprop. By 1949 the Python turboprop had been chosen as the powerplant for the Westland Wyvern T.F.2 torpedo strike fighter after the withdrawal of the Rolls-Royce Clyde turboprop from this project.

The original Wyvern, the T.F.1, first flew on 12 December 1946 but this was powered, as were a few other prototypes and pre-production aircraft, by a 2,690hp R-R Eagle 22 piston engine with de Havilland three-bladed contra-rotating airscrews in the absence of the intended turboprop. The first turboprop Wyvern (VP120) with a Clyde engine of 4,030ehp made its first flight on 18 January 1949, piloted by Harald Penrose; this was followed by the Python-powered prototype (VP109), making its maiden flight on 22 March 1949, piloted by Mike Graves.

The Wyvern was of all-metal, stressed-skin construction and very striking in appearance. The huge spinner, with its contra-rotating blades, led back to a neat annular air intake. The fuselage, in side profile, had an almost flat bottom but a pronounced curve on the top with a bubble canopy over the cockpit positioned at its highest point. A very large but gracefully curved fin and rudder, necessary to counteract the huge airscrew, had at its base a large tailplane. The semi-elliptical wing could be folded at its dihedral points and tips for carrier stowage. Two large exhausts for the turboprop emerged from each side of the fuselage at about wing trailing edge level. The undercarriage was of the tailwheel type and there was a hook, large flaps and air brakes. All controls were hydraulically operated. Armament on the production machines consisted of four 20mm cannon in the wings plus the means to carry rocket projectiles, bombs or a single 20in torpedo. At 10 tons, the Wyvern was a very large fighter and had a wing loading, at about 293kg/sq m (60lb/sq ft), three times that of a Spitfire.

Test flying of the Wyvern was largely carried out by Westland's Experimental Flight from Merryfield airfield, some 20km (12 miles) from the company's Yeovil factory. Much of the test flying was performed by the redoubtable Harald Penrose. During its development there were many forced landings and crashes, which cost the lives of four test pilots. Virtually all of these incidents were caused by engine or airscrew failures. Early Wyverns had no means of feathering the airscrew blades and so, in the event of an engine failure, it was necessary, because of the high drag, to dive the Wyvern very steeply in order to have sufficient speed in hand for the final landing flare.

With the fitting of the Python engine, considerable modification of the Wyvern had to be made including the raising of the fuselage hump and cockpit and the

The first turboprop version of the Westland Wyvern torpedo strike fighter (VP120) was powered by a Rolls-Royce Clyde turboprop. Philip Jarrett

addition of a dihedralled tailplane and larger fin and rudder. Urgently required was a means of changing the airscrew pitch and cutting the engine swiftly enough for deck landings, but not too sensitive for diving. In the Python 2 turboprop, a PCU (propeller control unit) and an 'anticipator' replaced the combined operation of the fuel control unit and CSU (constant speed unit). By October 1951 Armstrong Siddeley and Rotol had together developed the mechanically complex propeller inertia governor or ICU (inertia control unit), which was sensitive to the amount of deceleration and acceleration rather than simply a change in speed. By 1952 both the ICU and 'anticipator' were incorporated into the Python 3 and tested in the Wyvern VP109, the original Python-powered prototype. Unfortunately this machine was later written off in a crash at Farnborough, due to engine failure.

Further intensive flight testing of the ICU included 1,000 dives with pull-outs at between 3 and 6g. By 1953 the Wyvern S.4 was developed and entered shore-based service with 813 Squadron at Ford on 20 May and then carrier-based service, embarking on HMS Eagle in May 1955. No. 830 Squadron flew its Wyverns operationally in ground-attack sorties in November 1956 during the brief Suez war. Wyverns performed eighty-two sorties in the Canal Zone, two aircraft being lost and three others damaged. Naval pilots liked the big Wyvern and found it roomy, enjoyable to fly and an excellent and stable weapons platform.

It was straightforward to take off and land, with or without the use of its large landing flaps. However, once it had used its attack load, it was not much use as a fighter.

More troublesome was its Python turboprop. Several Wyverns were lost following engine failure when being catapulted, one aircraft crashing in front of HMS *Albion* which passed over it. (The pilot, Lt Bruce McFarlane, made the first underwater ejection and was picked up astern of the carrier.) Unlike the Double Mamba

Following the cancellation of the Rolls-Royce Clyde turboprop, the Wyvern was powered by the Armstrong Siddeley Python turboprop, this being the first prototype (VP109). Philip Jarrett

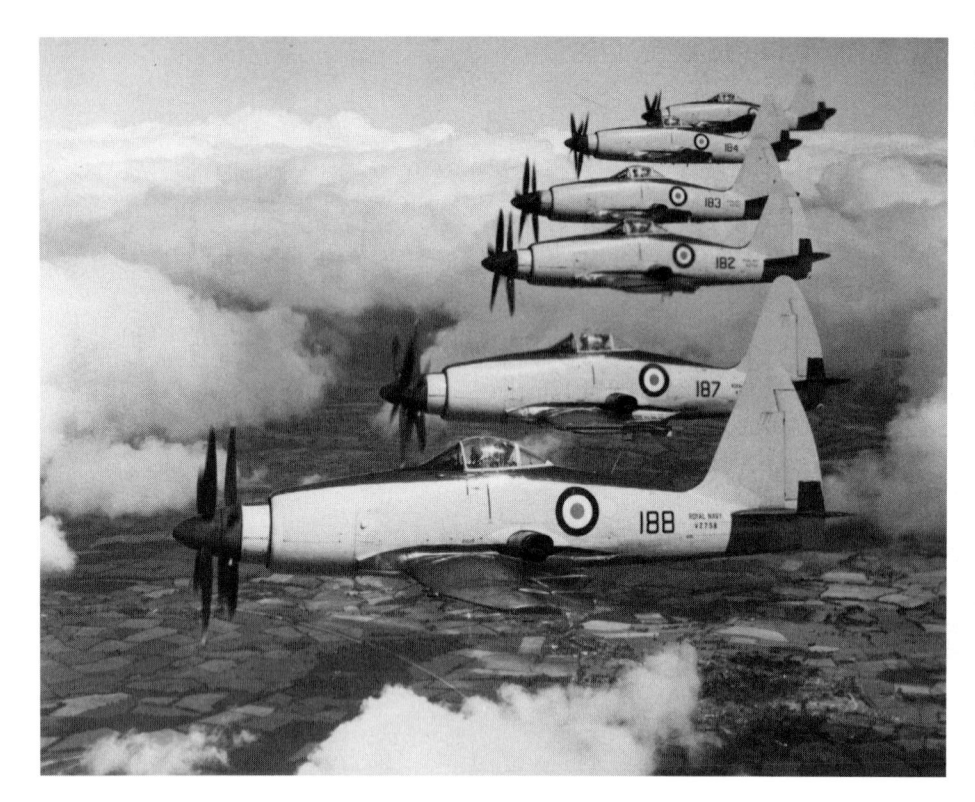

These 813 Naval Air Squadron Wyvern S.4s from Ford in 1954, each powered by the Python 3 turboprop, make a fine sight. VZ758 is nearest. Philip Jarrett

turboprop in the Fairey Gannet, the Python did not have a recuperator and a special booster pump to supply fuel during catapult launching. AS made improvements to the Python in service but the problem was never fully solved. Much less common was a failure of the airscrew gears, which happened to VZ756 from HMS *Eagle* on 24 September 1957 while on exercise north of the Arctic Circle. This aircraft crashed into Harstadtfjord but the pilot, Lt D.J. Doust, ejected at low level and was saved.

Mamba turboprop

At about the time that Armstrong Siddeley was working on the ASP or Python turboprop project, Bill Saxton directed his team to look into the potential of a turboprop to do the same job as an 850hp radial piston engine that the company was developing for civil and trainer aircraft application. Because such a turboprop was expected to have a higher fuel consumption than the piston engine, special emphasis was to be placed on light weight and a small diameter as compensation. By autumn 1945 the project work was completed and the proposal looked attractive. It was estimated that the turboprop would weigh about 295kg (650lb) and have a diameter of only 0.69m (2ft 3in), this comparing with 445kg (980lb) and 1.27m (4ft 2in) for the radial engine.

It was therefore decided to build a prototype of the turboprop and this was ready to run on the test-bed by April 1946. The weight of the engine actually came out at 321kg (707lb), but this was simply to speed up its manufacture and to take advantage of simplifications in construction. The first runs of this turboprop were disappointing and the cause was largely due to the compressor blades fouling the casing. This was caused, upon shut down, by the rise of hot gases making the rotor bow and distort. The cure was to fit a centre bearing.

This early turboprop was redesigned into what became known as the Mamba engine. In its original form, the Mamba had a ten-stage axial compressor, six can-type combustion chambers with a vaporizing fuel system and a two-stage turbine. The airflow was straight through (unlike the Python's reverse flow) and the rate of only 6.12kg/sec (13.5lb/sec) compared with the 23.81kg/sec (52.5lb/sec) of the Python. Power was taken from the front of the compressor rotor at 14,500rpm and this

Westland Wyvern S.4 with Python 3 engine		
Span	13.41m	(44ft 0in)
Length	12.88m	(42ft 3in)
Wing area	33sq m	(355sq ft)
Height	4.80m	(15ft 9in)
Empty weight	7,086kg	(15,608lb)
Maximum loaded weight	11,123kg	(24,500lb)
Maximum speed at sea level	617km/h	(383mph)
Climb rate	716m/min	(2,350ft/min)
Approach speed	222–241km/h	(138–150mph)
Touchdown speed	180km/h	(112mph)
Service ceiling	8,534m	(28,000ft)
Range	1,455km	(904 miles)

Armstrong Siddeley Mamba turboprop. Philip Jarrett

speed was reduced by compound epicyclic gears down to 1,640rpm for a 3.04m (10ft) diameter airscrew. The estimated power was 1,010shp plus 145kp (320lb) of jet thrust for a fuel consumption of 472ltr/hour (104gal/hour).

Originally the Mamba had a double ball bearing at the front of the compressor and a single roller bearing just forward of the first turbine. The compressor rotor was a steel drum onto which were fitted steel discs riveted together to sandwich the T-roots of the rotor blades. This gave, in section, a triangular structure that was both light and stiff. The compressor stator blades were held in a forged aluminium alloy casing. Extensive air cooling of the turbines was used. This comprised high pressure air from the end of the compressor led to the front face of the first turbine and then out into the main stream, air taken through holes in the compressor drum rotor and led between the turbine discs, and air leaking past the compressor sealing glands and exhausted via the back of the second turbine disc. A separate supply of intermediate compressor air was used to cool the turbine bearing.

With the addition of the third, centre bearing already mentioned, the early Mamba was approaching its design value by December 1946 and was giving 1,013shp. The next step was to improve the reduction gearing to eliminate tooth meshing frequency vibrations and give it a longer life. Help was obtained from experience gained with the Python engine at that time. Vibration was eliminated by careful design of the gear teeth and by very high accuracy of manufacture. This followed on from correct analysis of the problems and the principles of high-speed gear design were thereby laid down and, it was felt, fully understood.

By May 1947, with improved reduction gearing, Mamba development accelerated and the first test flights, in a neat installation in the nose of a Lancaster, took place from 14 October that year. In February 1948 a Mamba engine completed a 150-hour military and civil type test.

Development and redesign led to the Mamba 2. Without increasing the overall dimensions, the compressor and turbines were redesigned for an increase of 26 per cent in the air mass flow, which increased the power to 1,260shp. Its weight was 635kg (1,400lb), including 181kg (400lb) for the duralumin airscrew. During the redesign, the diffuser casing, combustion chamber manifold and mounting ring were all combined into a single casting, a difficult design job but rewarded by a reduction in weight. Blading was produced by stamping accurately to form, leaving only the root attachments requiring machining. Completely cowled, the Mamba 2 was only 0.76m (2ft 6in) in diameter. The successful development of the Mamba is all the more creditable since it was carried out with completely inadequate test equipment: full-scale tests of a design or modification could usually be carried out only by running the engine itself.

Mamba engines were flown in Boulton Paul Balliol and Avro Athena trainers and, in civil versions, in the twin-engined Miles Marathon and the four-engined Armstrong Whitworth Apollo airliners. A Mamba 6 was also test flown in a Douglas DC-3 Dakota in 1954, but unfortunately the Mamba was not taken up in any big way. The Short Seamew, a light, two-seat, anti-submarine search and strike aircraft, used the Mamba engine and at one time it was hoped the Mamba would be chosen to power the Vickers Viscount airliner, but it lost out to the R-R Dart turboprop.

Adder and Viper turbojets

In November 1948 the Adder turbojet or ASA.1 was created from the Mamba 2 (ASM.3) by removal of the airscrew gearing, this resulting in a very slender turbojet with a long intake spinner housing the generator. The Adder was only 0.71m (2ft 4in) in diameter. It gave a static thrust of 476kp (1,050lb) for a weight of 249kg (550lb). An Adder was used to power the Swedish SAAB 210, a small experimental

LEFT: The Armstrong Siddeley ASA.1 Adder turbojet was created from the company's Mamba 2 turboprop by removal of the airscrew reduction gearing. Philip Jarrett

BELOW: The small Viper turbojet represented Armstrong Siddeley's first use of the annular combustion chamber. Philip Jarrett

aircraft built to test the double delta wing configuration later used for the SAAB J35 Draken fighter. The Helsinki University of Technology also used the Adder to power its PIK-A experimental jet aircraft. Thus, although the Adder found no commercial application, it proved to be a valuable point on the learning curve and the designers went on to develop the Viper turbojet. Also, from 1 January 1950 the Adder was used for Armstrong Siddeley's first afterburning experiments.

The intention in the design of the Viper turbojet was to produce a small, expendable engine. For the first time the company used an annular combustion chamber, this having a vaporizing fuel system with twenty-four burners. The compressor was a new eleven-stage axial type passing 14.51kg/sec (32lb/sec) of air and with a pressure ratio of only 4:1. With the Viper turbojet, which was to power the Jindivik target drone, Armstrong Siddeley had an outstanding success, the engine producing an amazing 744kp (1,640lb) of thrust for a weight of only 165kg (365lb). Not surprisingly, therefore, Bill Saxton was encouraged to develop a long-life version of the Viper and this powered the Folland Midge fighter and the Hunting Jet Provost trainer. The Viper was also built, as the M.D.30, by Marcel Dassault in France.

The Midge was conceived in 1952 by W.E.W. Petter as a small, lightweight fighter of comparatively simple design. Being far cheaper than other fighters then extant, it was thought it would be of particular interest to emerging countries: it was, for example, estimated that three of these small fighters could be bought for the price of one Hunter. The Viper-powered Midge or Fo.139 prototype (G-39-1) made its first flight on 11 August 1954, piloted by E.A. Tennant, from Boscombe Down. Subsequent aircraft used the Bristol Orpheus engine. Its success lay in India and as a two-seat advanced trainer for the RAF. Subsequent development of the Viper by Bristol Siddeley saw its employment in the BAC Strikemaster, the experimental Handley Page HP.115 and the Hs 125 executive jet. The small HP.115 had a wing leading edge sweepback of 74° 42' and provided much data that was used towards the development of the supersonic Concorde airliner. Another experimental aircraft employing the Viper turbojet was the Saunders Roe SR.53 mixed-powerplant, delta-wing interceptor. The idea of the SR.53 was that it would cruise on a small turbojet but use a liquid-fuelled rocket motor for rapid take-off, climb and interception of the enemy. It was designed for speeds of up to Mach 2.2, altitudes in excess of 13,700m (45,000ft) and be armed with two Firestreak missiles. The first of three prototypes (XD145) made its first flight on 16 May 1957, piloted by J.S. Booth. Subsequent tests proceeded smoothly, but Booth was later killed in an SR.53 take-off accident and the project was killed off from the time of the 1957 government decision that missiles and not manned aircraft were the future.

Another interesting use was found for the Viper. Two were fitted to the Shackleton MR.3 long-range reconnaissance aircraft to boost take-off and climb performance. A Viper 203 was fitted beneath

The Armstrong Siddeley Double Mamba turboprop, illustrating its two coaxial shafts for driving contra-rotating, four-bladed airscrews. Each of the two coupled engines could be operated together, independently or, for economy and range, with one shut down. Philip Jarrett

each of the outboard Griffon engines, each nacelle of which had a ventral air inlet door that was closed once the turbojets were closed down. This arrangement was first successfully tested in 1963 on the Shackleton XF711 at Woodford. Subsequent MR.3s so fitted were then designated Shackleton MR.3 Phase 3.

Double Mamba turboprop

To continue the story of the company's Mamba, the most important and successful development of this turboprop was the Double Mamba. This consisted of two Mamba engines driving two, coaxial, four-bladed airscrews via a coupling gearbox. The Double Mamba was developed to power the Fairey Type Q (later known as the Gannet) carrier-borne anti-submarine aircraft, which was ordered by the MoS on 12 August 1946. It was designed under Hollis Williams.

The Gannet was the first aircraft in the world to fly with a double turboprop engine, which conferred characteristics highly desirable for naval operations. Firstly, the safety of twin engines was combined with the qualities of a single-engine configuration. Range was extended by the ability to shut down one engine and feather its airscrew during patrol, while the power of both engines could be called upon immediately for combat. The contra-rotating airscrews eliminated any problems associated with torque or asymmetric thrust.

Finally, there was no need for the navy to carry petrol since the Double Mamba could run on kerosene, diesel fuel or whatever was available on the aircraft carrier.

The Double Mamba engine, designated ASMD.1, consisted of two Mamba 2s, without integral gears, mounted side-by-side and connected through a coupling gearbox to two independent, coaxial, contra-rotating airscrew shafts. The air

Fairey Gannet A.S.4		
Span	16.56m	(54ft 4in)
Length	13.11m	(43ft 0in)
Height	4.18m	(13ft 8½in)
Empty weight	6,835kg	(15,069lb)
Loaded weight	8,890kg	(19,600lb)
Maximum speed	481km/h	(299mph)
Initial climb rate	610m/min	(2,000ft/min)
Service ceiling	7,620m	(25,000ft)
Range	1,185km	(736 miles)
Endurance	cruising at 241km/h (150mph) 4.9 hours	

ABOVE: The prototype (VR546) of the Fairey Gannet anti-submarine patrol aircraft, powered by the Double Mamba. In June 1950 this aircraft made the first landing by a turboprop-powered aircraft on an aircraft carrier. This unique aircraft was a great achievement and was without vices. *Philip Jarrett*

The contra-rotating airscrews on a Fairey Gannet AS.6 anti-submarine aircraft, in the RAF Cosford Museum. Once airborne, the aircraft could cruise on one of its Mamba gas turbines, with one airscrew feathered, for maximum range and loiter time. *Author*

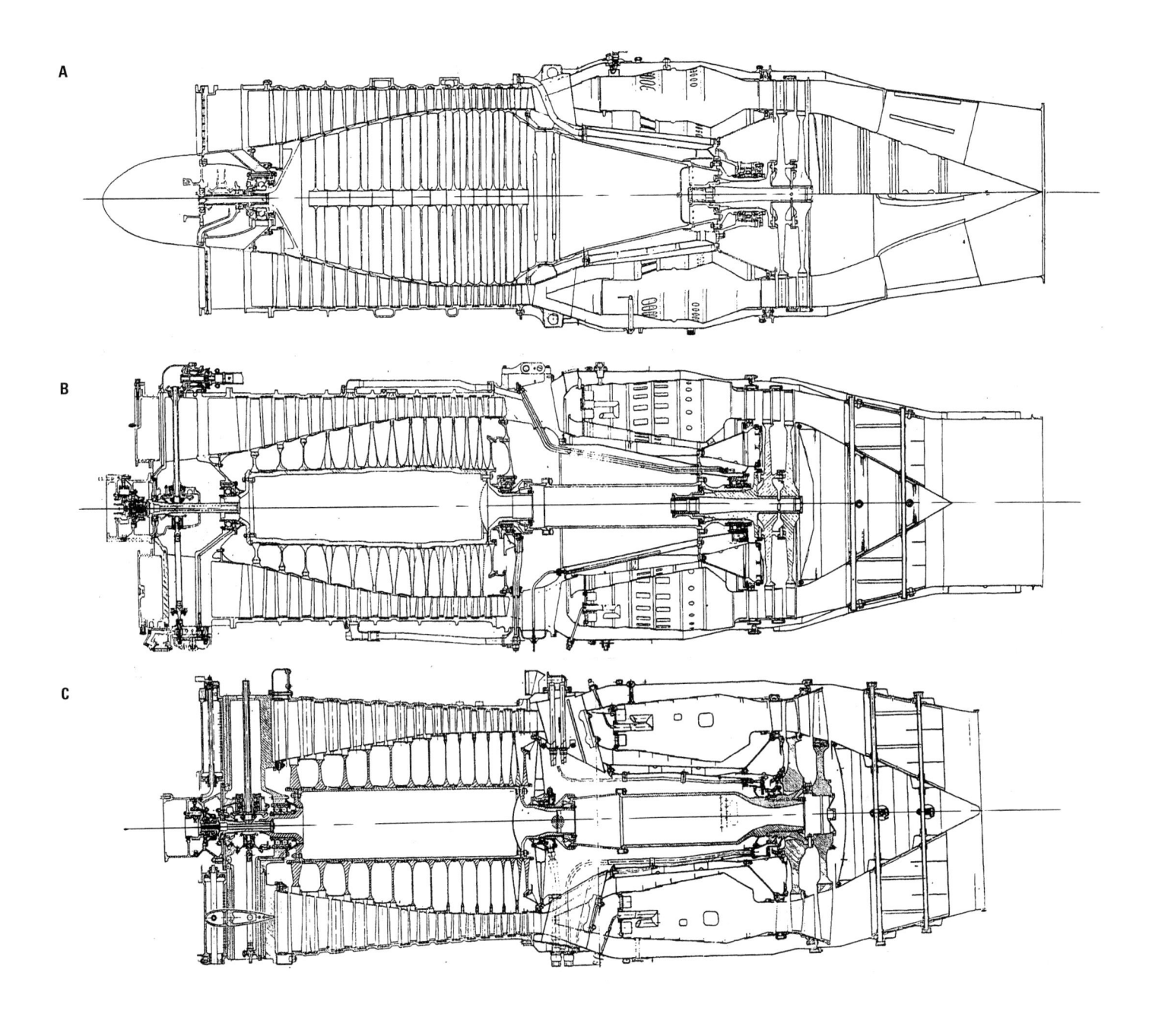

Sapphire turbojet development:

(A) Original Metropolitan-Vickers F.9 with its two-bearing arrangement, drum-type compressor rotor and upstream fuel atomizing combustion system.

(B) Armstrong Siddeley Sa.6 Sapphire with three-bearing arrangement, combined drum and disc rotor and a fuel vaporization combustion system.

(C) Armstrong Siddeley Sa.7 Sapphire with a flared annulus for the unshrouded turbine. Simplifications and a shortening of the engine were made.

intakes were through ducts in the coupling gearbox casing, which was a three-piece magnesium alloy casting. Between each engine output shaft and its airscrew shaft were compound helical and spur reduction gears of total ratio 10.36:1. The axis of the airscrew shafts was 0.279m (11in) above the engine axes. The well-developed AS single-lever master control, with automated fuel and airscrew controls and Rotol CSU unit, was used. The take-off or static power of the Double Mamba Mk.100 was 2,540shp plus 365kp (810lb) of thrust at 15,000rpm, equating to 2,950eshp.

Gannet anti-submarine aircraft

The first test flight of the Double Mamba was made in the prototype Fairey Gannet (VR546) on 19 September 1949 from Aldermaston, piloted by Gp Capt Slade. This and the second prototype (VR557) were two-seaters, but from the third prototype (WE488) the aircraft were built with three seats in tandem. They were of all-metal, stressed-skin construction and the fuselage had a somewhat bulky appearance due to the cockpits being positioned above

An Armstrong Siddeley F.9 Sapphire turbojet in the RAE cold chamber.

the engine and exhaust installation with the large internal weapons bay below it. The engine exhaust ducts slightly protruded from the fuselage sides just aft of the wing trailing edges. A large, rounded fin carried at its base a large tailplane that, later, had small auxiliary fins. Attached above the weapons bay line were the wings, which had a slightly inverted-gull form and an equi-tapered planform. Each wing was foldable at two points. The tricycle undercarriage was retractable and there was a long arrester hook.

The first occasion that a turboprop-powered aircraft landed on an aircraft carrier was on 19 June 1950, when VR546 landed on HMS Illustrious at sea. The extensive carrier trials with the Gannet prototypes that followed resulted in the eventual elimination of the competing Blackburn Y.B.1, which was also powered by the Double Mamba. The Gannet A.S.1 was ordered into production at Fairey's Hayes and Stockport facilities on 14 March 1951. Peter Twiss flew the first production Gannet (WN339) on 9 June 1953.

Various marks of Gannet were produced, including the A.S.5 and the A.E.W.3 early warning reconnaissance aircraft with a completely new fuselage. For the Gannet A.S.4 an uprated Double Mamba, the Mk.101, was fitted, which gave 3,035eshp. Eventually the Gannet equipped more than twenty squadrons of the Royal Navy and RNVR and was a complete success. From 1958 Whirlwind helicopters began replacing the Gannet, the last operations of which were in January 1960 from Cyprus.

Sapphire turbojets

When, under government pressure, Metropolitan-Vickers pulled out of turbojet development in 1947, its very promising F.9 axial turbojet was handed over to Armstrong Siddeley. This turned out to be most fortunate for AS, since the F.9 went on to be developed into the very successful Sapphire engine. At the time it was MoS policy to sponsor the development of a given category of engine in two companies, as an insurance against failure, and so the Sapphire became a competitor of the R-R Avon.

The work of Metropolitan-Vickers on the Sapphire (or MV Sa.1) was taken over by AS in January 1948, at which stage the engine was capable of reliably producing a thrust of about 3,175kp (7,000lb). The Sa.1 was intended as just a static research engine but the second MV engine, the Sa.2, was intended for flight and had a design rating of 3,348kp (7,380lb). Armstrong Siddeley therefore began some redesign of the MV Sa.2.

On 1 October 1948 AS began bench testing of the Sa.1, which then passed its first official acceptance test in the same month at 2,948kp (6,500lb) thrust. The Sa.1 first took to the air on 19 January 1950 with two of them fitted into the outboard nacelles of the Avro 691 Lancastrian flying test-bed (VM733). The Sa.1 was short-lived, however, because the Sa.2 passed its official acceptance test in December 1949 at a thrust of 3,348kp (7,380lb).

Meanwhile a third, more complete redesign of the engine was undertaken, leading to the Sa.3. This retained the very successful MV aerodynamic design of the compressor and turbine but was considerably changed mechanically in line with AS thinking. The main mechanical changes were the incorporation of a centre bearing and the replacement of the heavy, welded drum with a combined drum and disc design reminiscent of the successful Mamba. The centre bearing rendered the engine more suitable for a lively fighter aircraft. Even more important was the substitution of a fuel vaporizing combustion system in place of the MV injection system. The Sa.3 passed its official acceptance test in April 1950 and three months later it passed its first 150-hour type test at 3,275kp (7,220lb) thrust with a specific fuel consumption of 0.916.

During the early part of 1951 bench running of the Sa.3 had shown that, by raising the maximum speed by 300rpm (to 8,600rpm) and by reducing the swirl angle of the compressor inlet guide vanes, an increased air mass flow could be obtained. By this means, the engine passed a type test in March 1951 at 3,630kp (8,000lb) of thrust. By January 1952 the Sa.3 had passed a further type test at its full design rating of 3,443kp (7,950lb) thrust and its specific fuel consumption had been lowered to 0.91. Further development of the Sa.3 led to the Sa.6, which passed a type test in March 1952 at 3,765kp (8,300lb) thrust and a specific fuel consumption of 0.90.

Earlier, test flying was carried out with

The first Armstrong Siddeley Sapphire turbojets were test flown with two Sa.1s in the outboard positions of this Avro 691 Lancastrian (VM733). Here the airscrews are feathered. Philip Jarrett

This, the third prototype of the Hawker P.1067 Hunter fighter (WB202), was powered by an Armstrong Siddeley Sa.6 Sapphire turbojet. Philip Jarrett

the Sa.2 version. Two were installed in a Gloster G.41K Meteor F.8 (WA820) and first flew on 14 August 1950, the thrust of each being 3,275kp (7,220lb). On 31 August 1951 this Meteor established a new time-to-altitude world record when it attained 12,000m (39,370ft) in 3 minutes 7 seconds from a standing start. Another flying test-bed was the HP Hastings transport aircraft (TE583), which was fitted with two Sa.2s in the outer positions to give Handley Page turbojet experience prior to their use in the Victor V-bomber. The Hastings first flew in this form on 13 November 1950. A much more spectacular test-bed was the English Electric (E.E.) Canberra B.2 (WD933), which began flying with two Sa.6 Sapphires in August 1951 and made many flights up to extreme altitudes.

Beginning on 30 November 1952, an Sa.6 with 3,720kp (8,200lb) of thrust began test flying in the third prototype of the Hawker P.1067 (Hunter, WB202) and from 12 October 1953 four such Sa.6s were flown in the Avro 698 Vulcan B.1 flying test-bed (VX770).

By summer 1951 it was obvious that the Sapphire was not only a great success but that it also had a great development potential and many would be needed. Accordingly, a special production facility was set up for the engine, warranted by the urgent expansion brought about by the Korean War. A new company, Brockworth Engineering Co. Ltd, was formed by the Hawker Siddeley Group, which took over most of the 9,300sq m (1,000,000sq ft) of floor space at the Hawksley Constructions factory, which had been built at Hucclecote airfield in Gloucestershire during the Second World War to produce Albermarle bombers. This extensive engine factory, the most modern of its kind in Europe, began Sapphire production in summer 1953. Its departments included machine and assembly shops, a mechanized precision casting foundry, a metrology department, a continuous X-ray department, plating and heating shops, drawing offices and so forth. On the far side of Hucclecote airfield were engine test-beds with silencers good enough to prevent the sound of the engines reaching the nearest built-up area. Licence production of the Sapphire as the J-65 by the Curtiss-Wright company is discussed in the USA section of Volume 2.

Gloster Javelin fighter

The first aircraft into RAF service specifically designed for the Sapphire engine was

The world's first delta aircraft to enter service, the Gloster GA.5 all-weather fighter was powered by two Sapphire turbojets. These two fine studies (this page and p. 132) show XA644 performing a loop and XH966 armed with Firestreak missiles. Russell Adams Collection

F(AW).7 Javelin with Sa.7 Sapphire engines			
Span		15.80m	(52ft 0in)
Length		17.10m	(56ft 3in)
Wing area		86sq m	(927sq ft)
Height		4.80m	(16ft 0in)
Normal take-off weight		16,188kg	(35,690lb)
Maximum overload weight		18,266kg	(40,270lb)
Maximum speed	at sea level	1,141km/h	(709mph)
	at 13,716m (45,000ft)	1,000km/h Mach 0.945	(621mph) or
Climb to 13,716m (45,000ft)		6.6 minutes	
Service ceiling		16,039m	(52,800ft)
Range		1,545km	(960 miles)

the Gloster GA.5 Javelin delta-wing, all-weather fighter. It was both the world's first delta aircraft to enter service and the largest fighter to serve with the RAF. Originally designed in 1948 to specification F.4/48, in competition with the de Havilland DH.110, it was intended to be powered by two MV F.9s. The official specification included a maximum speed at altitude of 973km/h (604mph), a maximum diving speed of Mach 0.95, a service ceiling of 13,700m (45,000ft) and an endurance of 2 hours. The aircraft was to reach its maximum altitude within ten minutes of the starting buttons being pressed.

Glosters evolved the design of the Javelin using results from the work of the German scientist Alexander Lippisch. It was also backed up by the building and flying of various versions of the Avro 707 delta-wing research aircraft, the first of which began

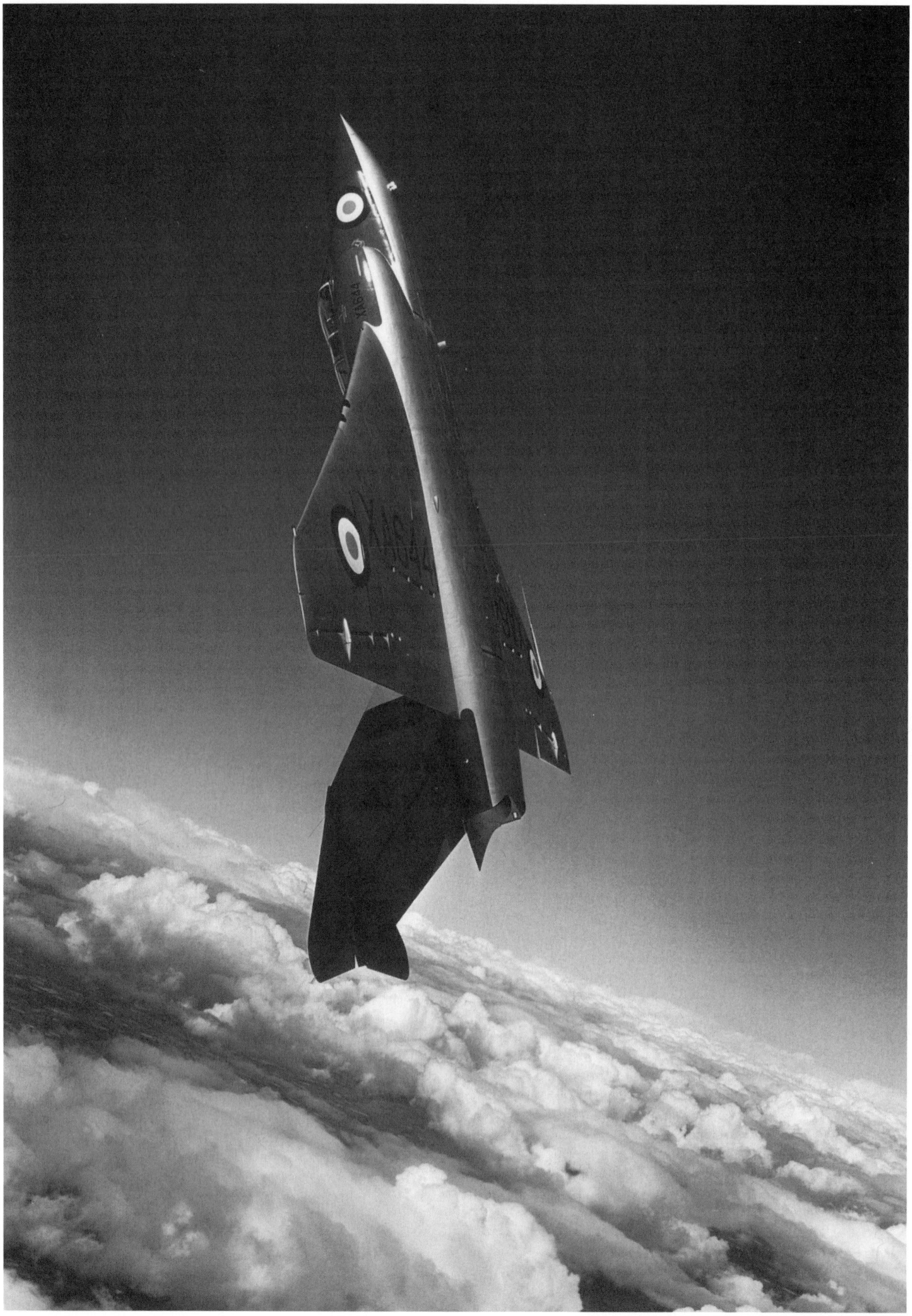
XA644

flying in September 1949. The MoS ordered the construction of four prototypes of the Javelin on 13 April 1949. Vacillation by the MoS concerning how many prototypes were actually wanted led to disruption of the Gloster programme with consequent delays.

A large aircraft for a fighter, the GA.5 Javelin was dominated by its 86sq m (927sq ft) delta wing, an adjustable tailplane of similar planform being mounted on top of a large, swept-back fin. Large circular intakes with ducts leading to the two engines were built separately and attached to the fuselage sides, the engine bays being formed by large frames attached to the sides of the box beam structure forming the central fuselage. Although a complex aircraft, its airframe used conventional construction, chiefly in aluminium alloy. The engines could be fitted and removed on rails through the open rear end of the fuselage. For the wing, a torsion box was formed by the main, front spar and the covered leading-edge ribs. Fuel was carried in the inboard wing sections, which were built as units complete with the main landing gear, guns, ammunition storage, flaps and air-brakes. The complete wing was attached to the fuselage by pin joints on the front and rear spars. A crew of two was housed in a pressurized cockpit and the radar, housed in the nose, was initially British wartime AI type. All controls were powered.

The maiden flight of the first Javelin prototype (WD804) took place on 26 November 1951 from Morton Valence, the pilot being W.A. Waterton. This flight revealed severe rudder flutter, which was readily seen to be caused by interference of the jet exhausts and airflow around the rear fuselage. The evidence for this was shown by oil streaks across the lower surfaces of the rudder, the oil coming from the centre and rear bearings of the Sapphire engines and venting below the rudder. At the time, those bearings were lubricated on a total loss system. The aerodynamic remedy was to lengthen the fairing around and between the jet pipes. On 29 June 1952 this prototype lost both elevators following violent fluttering during a high-speed run. Waterton exhibited great skill by landing the aircraft at Boscombe Down by using the slowly responding, electrically operated, variable incidence tailplane. The aircraft was damaged upon landing but, by August that year, the second prototype (WD808) was flying.

On 7 July 1952 the MoS ordered the Javelin into series production under a super-priority classification following the wake-up call brought about by the Korean War. Soon, from summer 1953, the Sa.6 Sapphire was being built in Mk. 101, 102 and 103 versions, rated at a thrust of 3,629kp (8,000lb). Tragically, the second Javelin prototype (WD808) crashed on 11 June 1953 and the pilot Peter Lawrence lost his life. This resulted from the aircraft super stalling at a high angle of attack, the elevator control being lost at the same time due to blanketing of the airflow by the wing.

The first production Javelin, an F(AW).1 (XA544) made its first flight on 22 July 1954 piloted by Dicky Martin. The type entered RAF service initially with No. 46 Squadron at Odiham. A total of 302 Javelins were produced by Gloster Aircraft at Hucclecote, plus another 133 by W.G. Armstrong Whitworth Aircraft at Baginton. A number of versions were made, up to Mk.6 being powered by Sa.6 Sapphire engines (except for the Mk.2, which was powered by R-R Conways). The Mk.2 used US AI.22 radar, the Mk.3 was a trainer, the Mk.4 had an all-moving tailplane, the Mk.5 had extra internal fuel tankage and the Mk.6 was similar to the Mk.5 but with AI.22 radar.

More of the F(AW).7 Javelin were built than any other. It carried as standard armament Firestreak air-to-air missiles and two 30mm Aden cannon. It could also carry four underwing drop tanks and two ventral

Armstrong Siddeley Sapphire 202 turbojet. Rolls-Royce Heritage Trust

tanks for increasing its range. This and the subsequent two marks were powered by two Sa.7 Sapphire engines each of 4,990kp (11,000lb) thrust. Javelins flew all over the world with the RAF, the final disbandment of the type being at Tengah in June 1967 with No. 64 Squadron.

200-series Sapphire turbojets

The Sa.7 Sapphire engine had its origin in the Sa.4, which launched a complete redesign from the 100-series of engines that stemmed from the original Metropolitan-Vickers engines (AS Sa.3, Sa.6, Sa.5 and Sa.12). This redesign was known as the 200-series and its chief aims were to provide a much more powerful engine suitable for high subsonic and supersonic speeds, while at the same time keeping within the dimensions of the Sa.6 and using the same trunnion mountings. It was also intended to develop a similar afterburner to that then being developed for the Sa.6. Designated Sa.4, the new engine was aimed at a thrust increase of 25 per cent over the Sa.6 (bringing it up to 10,000lb thrust) plus 25 per cent increase over the take-off rating in afterburner mode. More afterburning was possible but was constrained by keeping the tailpipe temperature down and minimizing installation changes.

The Sa.4 retained the same basic layout as previous Sapphires with an axial compressor, fully annular fuel vaporizing combustion system and a two-stage turbine. Its increased thrust was largely achieved by an increase in air mass flow. It first ran in July 1952 and was immediately found to be suitable for a slight increase in rotational speed to gain more power. This engine was designated Sa.4/7. It passed its 25-hour special category acceptance test in February 1953 and ultimately achieved a rating of 4,763kp (10,500lb) thrust.

Despite this great performance, the aerodynamics of the turbine were not satisfactory at the higher speed and so corrective action was taken by redesigning the turbine to have a flared annulus instead of the constant-diameter annulus of all previous engines. In this form, which included general modifications and a shortening of the tailpipe, the engine was designated the Sa.7 and made its first run in March 1954. Work also went ahead to manufacture components such as the compressor in steel rather than aluminium alloy because the 200-series Sapphires were to be suitable for supersonic flight. By June 1954 the Sa.7 had passed a 25-hour special category flight approval test and two of them were fitted into a Canberra flying testbed (WD933), which first flew on 13 August 1954. The following month this aircraft gave a most impressive take-off and climb at the Farnborough Air Show.

In March 1955 the Sa.7 passed a 150-hour type test at 4,763kp (10,500lb) and went into production at Brockworth in April. By March 1956 production engines were being passed out at 4,990kp (11,000lb) thrust, this being achieved by improvement of the combustion system, unshrouding the turbine rotor blades and improving turbine sealing, thereby allowing a higher operating temperature and greater efficiency. This was the Mk.202 Sa.7 engine, which had a diameter of 0.950m (3ft 1.4in), a length of 3.353m (11ft 0in), a dry weight of 1,395kg (3,075lb) and a specific fuel consumption of 0.89. A simple afterburner, developed for the Javelin, could boost the thrust to 6,075kp (13,390lb). Test flying was supplemented by fitting two Sa.7s in a Javelin F(AW).7 (XA560), which began flights on 30 September 1955.

Apart from the Javelin, the other main British aircraft to be powered by the Sapphire were the Hawker Hunter fighter and the Handley Page Victor V-bomber (American aircraft powered by licence-built Sapphires are discussed in the USA section) but, before discussing these aircraft, a description of the construction of one of the Sa.7 Sapphire engines, the Mk.202, will be given.

The backbone of the engine was the centre section (between the compressor and combustion section), which was of fabricated, welded steel construction and carried the centre roller bearing and to which were attached the side-mounting trunnions and a top steadying trunnion. These side trunnions transmitted the thrust loads to the airframe. A matched pair of ball races comprised the front bearing for the rotor and this took the main thrust. Ten aerofoil-section radial struts connected the inner and outer portions of the centre section. The rear bearings, from which the turbine was overhung, were of the roller type.

A conical extension for the rear bearing and also the combustion system outer casing were bolted to the centre section. The outer casing for the compressor was also bolted to the centre section. The compressor was a thirteen-stage axial type and its intake area was much enlarged over previous engines to match the greater air mass flow. This compressor necked down towards the centre of the engine (previous types were parallel) and this fortuitously provided space for some of the accessories, thus avoiding an increase in installed diameter. Four hollow struts in the intake supported the cone for the front bearings and also supplied hot air for de-icing purposes.

For the compressor casing, work done to develop this in steel was found to be unnecessary since the rigours of supersonic flight were not as great as had been envisaged. The casing was therefore made from aluminium alloy with carrier rings for the steel compressor blades. The first seven rows of stator blades, because of their length, were supported at their inner ends by shroud rings, the remaining six rows being unsupported at their inner ends. The steel drum for the rotor blades was untapered and fitted with discs having interlocking rings at their bases. To these rings were fitted steel rotor blades with serrated root fixings. Despite having no variable-incidence inlet guide vanes or blow-off valves for the compressor, the Sapphire was easy to start, easy to handle and surge free. At one time work was carried out on a variable-area intake, but was later found to be unnecessary.

The combustion section was connected at its front end to the ten streamlined support struts of the centre section. Airflow from the compressor was divided into three separate annuli, the middle one of which handled primary air for the combustion process. Primary air led into twenty-four fuel vaporizing tubes fitted downstream along with six primer jets. Initially two of the primer jets were lit by igniter plugs. Secondary air flowed into the combustion section from the surrounding inner and outer annuli. The Sapphire was one of the few British turbojets in which Nimonic alloy metal was not used for the flame tubes.

The combustion gases entered the first turbine via nozzle blades made from Jessops G.39 heat-resisting steel and entered the second turbine via nozzle blades in Nimonic 80 alloy. Nimonic alloys were used for the turbine rotor blades, which were attached to forged discs by fir-tree roots. The two turbine discs were individually mounted directly onto the rear flange of a hollow stub shaft. This shaft led through the rear bearing, forward to a semi-flexible coupling to connect to the compressor drum. The exhaust bullet and cone assembly was manufactured from stainless steel and fitted with two sets of cruciform stay tubes. These stay tubes were inside floating streamlined struts that were only attached at their inner ends to the exhaust bullet. This was to avoid failures at the fixing points and care was taken to spread loads out over large areas.

Among the various simplifications introduced with the Sa.7 was the cooling of the bearings with oil flow only, thus avoiding the complications of air cooling. Cooling air was only used between the turbine discs. Cooling of the oil was by means of passing it through a fuel-cooled cooler. Starting of the engine was by means of an iso-propyl nitrate turbine starter and gearbox (for fighters) or an electric motor and gearing (for bombers).

Handley Page Victor B.1 with As.7 engines		
Span	33.83m	(111ft 0in)
Length	30.45m	(99ft 11in)
Wing area	223.76sq m	(2,406sq ft)
Normal loaded weight	92,990kg	(205,000lb)
Maximum speed at 12,200m (40,000ft)	1,030km/h Mach 0.92	(640mph) or
Service ceiling	17,070m	(56,000ft)
Range	6,440km	(4,000 miles)

Although it was initially superior to the R-R Avon, the Sapphire was built in smaller numbers in Britain, but very large numbers under licence in the USA as the J-65. Thus, while most Hawker Hunter fighters were powered by Avons, the F.2 and F.5 were Sapphire powered. It was in June 1948 that Hawker received an official order for three prototypes of the P.1067, later named the Hunter, two to be Avon powered and the third (WB202) to be Sa.6 Sapphire powered. The silver-painted Sapphire Hunter first flew on 30 November 1952, at Dunsfold, its performance being better than Avon-powered machines. The Korean War brought forth super-priority orders, including one for 200 Sapphire-powered Hunters to be built by Armstrong Whitworth.

The Sa.7 engine was first flown in a Hunter Mk.2 (WN888) on 12 October 1953 but the Hunter Mk.5 was fitted with the Sa.6 engine because the Sa.7 was not in production at that time. The first Hunter Mk.5 (WN954) made its maiden flight with the Sa.6 engine on 19 October 1954. The F.5, like the F.4, had small, extra fuel tanks inside the wing leading edges and a

Powered by four Armstrong Siddeley As.7 Sapphire turbojets, this crescent-winged Handley Page Victor B.1 bomber (XH615) is from 232 Operational Conversion Unit based at RAF Gaydon, Warwickshire. This large bomber could exceed Mach 1.0 in a shallow dive and was the largest aircraft to do so in 1957. Philip Jarrett

pair of underwing pylons for drop tanks. Out of a total of about 2,320 Hunters built, inexplicably only 45 F.2 and 105 F.5 were built with the Sapphire engine. For unknown reasons, further development of the Sapphire after the Sa.7 version was officially curtailed.

Handley Page HP.80 Victor V-bomber

The main use of the Sapphire in Britain was to power the Handley Page HP.80 Victor V-bomber, which was designed and developed in parallel with Avro's Vulcan V-bomber to Specification B.35/46. The Victor emerged as a totally different solution to this specification. Designed under the leadership of R.S.Stafford, advantage was taken of German wartime research into the crescent wing, in which the angle of sweepback of the wing was progressively reduced towards the wing tips. Such a wing shape promised the best performance at high subsonic speeds.

For the Victor, the shoulder-mounted crescent wing had the leading-edge sweepback reduced in three stages (48.5, 37.5 and 26.75 degrees) towards the tips, the trailing edge being at a constant sweptback angle. It had Fowler flaps on the inboard trailing edges and retractable leading edge flaps for the outer wing panels (later replaced with fixed, cambered leading edges) to ensure aileron authority close to the stall. Since the wing section was relatively thin, the housing of the four Sapphire engines produced a root section bulged downwards and with large, elongated air intakes. The sharply swept-back and dihedralled tailplane, with compound sweepback on its leading edge, was mounted on top of a swept-back fin. Apart from its flying surfaces, a distinctive feature of the Victor was its sharply pointed fuselage nose, which had a ventral bulge, back to about where the air intakes ended, containing radar equipment and batteries.

Pioneering methods of construction were used in the Victor, including honeycomb and corrugated aluminium alloy sandwich, which was both light and very rigid and was used particularly on the wings and tailplane and to carry fin loads. The centre section of the wing began just behind the cockpit pressure bulkhead and ended before the bulkhead of the bomb bay. Attached to the centre section were the three main spars of the wing, which had openings for the intake ducts. Behind these main spars began the engines, mounted inside spectacle frames that connected to the fuselage multiple spars of the wing. Thus, because the main structure of the wing was connected well forward into the fuselage, a large, unobstructed bomb bay resulted (actually twice as large as the Vulcan's bomb bay). Fuel was contained in the wing centre section and its outer panels, and later in large, underwing tanks. All control surfaces were power operated. Large air-brakes extended from either side of the fuselage tailcone. Like the Vulcan, the Victor had a crew of five.

The Victor's large bomb bay could house a 4,500kg (10,000lb) nuclear weapon or 15,900kg (35,000lb) of conventional bombs. The bomb bay doors retracted into the fuselage, not into the airflow. Altogether, the airframe construction was remarkable, innovative and also long-lived, as evinced by flying tanker versions serving into the late 1980s.

The all-silver Victor prototype (WB771) made its maiden flight on 24 December 1952 from Boscombe Down, the pilot being Sqn Ldr H.G. Hazelden, Handley Page's Chief Test Pilot. The landing from this first flight immediately showed the benefit of the high-mounted tailplane and crescent wing. The wing tended to produce a nose-up attitude, while the tailplane was unaffected by ground effect. Thus, once trimmed for landing, the aircraft always stayed in the correct attitude until touchdown and virtually landed itself.

Flight trials produced relatively few problems and the MoS issued an initial order for twenty-five of the bombers. Then, after 100 hours of testing, the prototype crashed on 14 July 1954, after the tailplane broke away due to fatigue failure of the three connecting bolts. On 11 September 1954 the second prototype (WB775) made its first flight, from Radlett, and the programme continued. Sa.6 Sapphires of 3,720kp (8,200lb) thrust each were used for the prototypes, but on the first production Victor B.1 (XA917), which made its maiden flight on 1 February 1956, Sa.7 engines of 4,990kp (11,000lb) thrust were fitted. For take-off under overload conditions, it was possible to fit a de Havilland Spectre rocket of 3,630kp (8,000lb) thrust beneath and between each pair of turbojets.

On 1 June 1957 XA917 inadvertently exceeded Mach 1.0 in a shallow dive and became the largest aircraft at that time to become supersonic. The transition was so smooth that only the Machmeter informed the crew of the event.

The Victor entered RAF service on 28 November 1957 with No. 232 Operational Conversion Unit. Following the first fifty production Victor B.1s, subsequent aircraft were powered by more powerful R-R Conway engines because Sapphire development had been officially cancelled.

Because the further development of the Sapphire was cancelled, the chance to power further aircraft was lost to Rolls-Royce, chiefly to their Avon. A notable exclusion for the Sapphire was the English Electric Lightning interceptor. Sapphires were only used to power the P.1A prototypes of the Lightning, beginning with the first prototype (WG760), which first flew on 4 August 1954, and the second prototype (WG763), which first flew on 18 July 1955. Each of these had two 100-series engines of 3,265kp (7,200lb) thrust each, the engines being designated Sa.5 for the vertical installation. The Sa.5s of WG760 were later fitted with simple afterburners, with fixed-area nozzles, which increased the maximum thrust of each to about 4,173kp (9,200lb); in that form it first flew on 31 January 1956. These P.1A prototypes, which were capable of Mach 1.0, were a far cry from the completely redesigned P.1B and Lightning aircraft that followed, capable of Mach 2.0. Most striking was the replacement of the simple, sharp-lipped, oval air intake for a conical centrebody in a circular intake. In April 1959 Armstrong Siddeley merged with the Bristol company to become Bristol Siddeley; this, in turn, was merged with Rolls-Royce in January 1968.

CHAPTER SIXTEEN

Bristol

The Bristol Aeroplane Company Ltd traced its origins back to February 1910 at Filton. Its earliest product was the Boxkite, which was followed by the Scout and F.2B Fighter during the First World War and the Bulldog fighter between the wars. The nucleus of its Aero-Engine Division was formed in January 1920 when it acquired the Cosmos engine company; thus it became, like de Havilland, a manufacturer of both airframes and engines. Under the leadership of Roy Fedden, the company produced many successful air-cooled radial piston engines, notably the Pegasus, Hercules and Centaurus, the latter two types with sleeve valves. Under Fedden, the best aero-engine metallurgy laboratory in the world was established, since he it was who foresaw the paramount importance of this discipline.

As early as 1924 Bristol was flying its exhaust turbo-supercharger on a modified Jupiter piston engine, which reached altitudes of more than 9,150m (30,000ft). However, from 1926, this promising development was abandoned to concentrate on gear-driven superchargers. This work culminated on 30 June 1937 when a Bristol 138A, powered by a Bristol Pegasus P.E.VI S with two gear-driven superchargers and an intercooler, reached a new record altitude of 18,269m (59,937ft). From 1937, however, work on turbo-superchargers was renewed since it was considered that they were ideal for sleeve valve engines, such as the Hercules.

Because the sleeve valve engine could withstand very high back pressure, it was thought that continued development might raise the exhaust turbine power to the point where it could drive the airscrew alone and leave the blower to be driven by all the piston power. Thus, the turbocharged piston engine would become a gas generator. However, calculations indicated that a boost pressure of some 4 atmospheres would be needed, which made the project a long-term one and an early casualty of more urgent war work. These studies were carried out under Frank M. Owner, who had invented a constant-volume or explosion gas turbine in 1920 during his time at Manchester University. His university professor at the time pointed out that such a gas turbine, although apparently sound, had already been invented and built by Hans Holzwarth in 1908 in Germany, but had not yet become viable.

In 1931 Frank Whittle visited Bristol to discuss his ideas. Owner's subsequent report on these considered that it would take ten years to bring them to fruition. In this he was correct. One of Whittle's ideas, for fused silica turbine blades, was considered impractical. Apart from study, very little else was done about the gas turbine at Bristol for ten years due to the company's concentration on developing its sleeve valve engines. In any case, it was thought that the lack of suitable heat-resisting steels would prevent the development of a turbine. With the outbreak of war in 1939, the company soon had about 50,000 employees, the majority working on piston engine production, and a new 18,600sq m (200,000sq ft) engine factory was built. Bristol's wartime production of piston engines reached over 100,000, just a little behind Rolls-Royce; between them, these two great companies produced 82 per cent of the aero-engines manufactured in the country. With production of the famous Blenheim and later Beaufighter aircraft, Bristol was one of the largest aircraft and engine manufacturers in the world.

Theseus turboprop

On 7 July 1941 Owner again met Whittle, when he and Roy Fedden went to Brownsover Hall, Rugby, where Whittle had his office. There they discovered the amazing advances made in high-temperature alloys. This put the gas turbine in an entirely new light. Owner then attended the second meeting of the Gas Turbine Collaboration Committee in November 1941 and became aware of the great number of turbojets being designed and developed in the country. There seemed little point in Bristol attempting to add to this number and so it was decided to enter the less populated field of the turboprop, which, in any case, was more suited to Bristol's background of bombers and civil aircraft.

Following the visit to Whittle's offices, Owner had requested Bristol's Project Technical Office to offer suggestions for a 4,000hp turboprop with a fuel consumption and weight comparable to a piston engine at 483km/h (300mph) at an altitude of 6,100m (20,000ft). As a side issue, in December 1941 the engine department produced a scheme to add Whittle W.2B/700 turbojets to the rear of the two Centaurus engines in the Bristol Buckingham bomber (which first flew on 4 February 1943), but the scheme was not approved because the MAP wanted to 'kill off' the W.2/700 engine.

By May 1942 F.G. Evans and P. Fortescue had produced a scheme, requested by Owner, for a turboprop that embodied an axial compressor followed by a centrifugal compressor, a heat exchanger and a separate, free turbine to drive an airscrew. This was the origin of the engine, later named the Theseus, which was ready in a project brochure form and submitted to the MAP for support in April 1943. By mid-1943 Bristol had received instructions from the MAP to carry out heat-exchanger research, this device being considered important for efficiency and hence range. Soon after this, it was decided to scale the engine down to 2,000hp as more practical for aircraft and test facilities.

By early 1944 Bristol had received an order from the MAP to build five Theseus engines plus three more complete sets of parts for developmental purposes. Frank Owner set up a design office at a dispersed site at Tockington Manor, although, due to war work, his staff consisted of only six apprentices and trainees from Bristol's main drawing office at Somerdale. Some guidance for these young people, who did a remarkable job, was forthcoming from more senior designers such as Stanley Mansell,

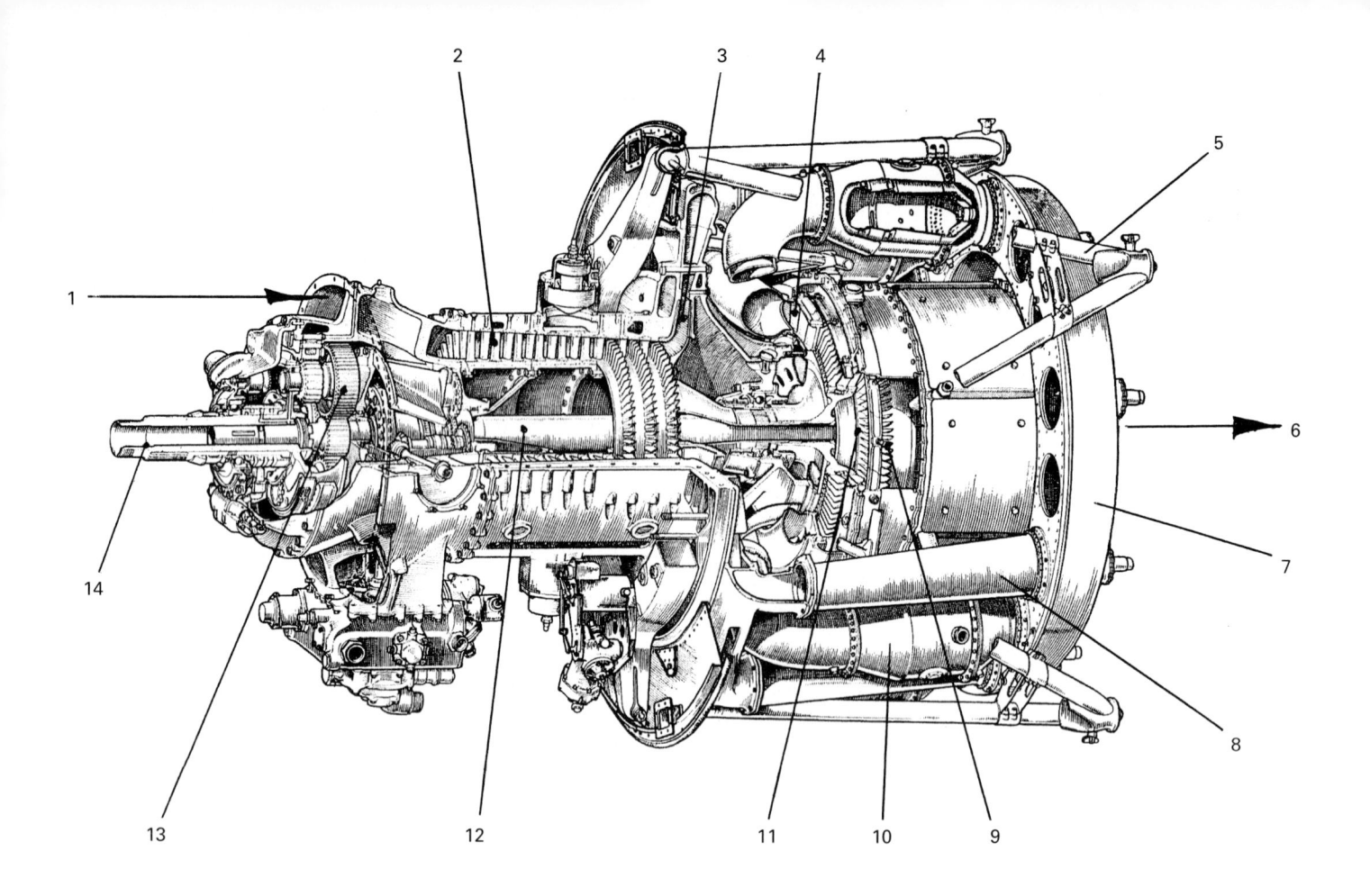

ABOVE: **Bristol Theseus II turboprop:** (1) annular air intake casting; (2) 9-stage axial compressor on drum rotor; (3) centrifugal compressor; (4) turbine inlet nozzles; (5) tubular support structure; (6) exhaust efflux; (7) annulus feeding air from diffusers to combustion chambers; (8) eight diffusers; (9) single-stage turbine driving airscrew gears; (10) eight reverse-flow, interconnected combustion chambers; (11) 2-stage turbine driving compressor; (12) drive shaft to reduction gears; (13) epicyclic reduction gears; (14) airscrew shaft.

Bristol Theseus II turboprop.
Rolls-Royce Heritage Trust

This Bristol Hermes V airliner was fitted with Bristol Theseus turboprop engines. These engines were notably quiet and heralded turboprop airliners to come. Note the fine airframe construction, something the company was famous for.
Philip Jarrett

Chief Designer (Turbines), and Cameron. There was also continuous collaboration with the engineering company of Whitehead & Luby. By about July 1946 the design of the Theseus turboprop was largely complete. It was dominated by the heat exchanger.

The Theseus had a nine-stage axial compressor followed by a centrifugal compressor to feed air via eight diffuser pipes to the heat exchanger at the rear of the engine. From the heat exchanger, the air was reversed through 180 degrees to enter eight interconnected combustion chambers, each having a single fuel burner and downstream injection. From the combustion chambers, the hot gases (800°C) were reversed 180 degrees to enter the turbines and thence through the stainless-steel tubes of the heat-exchanger, entering at 550°C before leaving the exhaust cone. Thus, the air entering the combustion chambers was already heated to 300°C, thereby, it was estimated, saving some 68kg (150lb) of fuel per hour.

For the turbines, there were two stages to drive the compressors and a third, independent or free, stage to drive an airscrew via a shaft passing forward through the hollow compressor rotor to reduction gearing. The airscrew was a Rotol five-bladed type, the reduction gearing for which consisted of compound epicyclic spur gears giving a reduction of 8.41:1. A torque dynamometer was incorporated in the reduction gearing and, fortuitously, its oil cushion was thought to absorb harmful high-frequency vibrations. The maximum speed of the airscrew turbine was 9,000rpm. Surrounding the reduction gearbox was an annular air inlet comprising a casting with eight support vanes.

When the Theseus I was first run on the test-bed, in May 1945, it did so without the heat exchanger, which was not ready. The engine would not run in a self-sustaining manner, although the highest temperatures were not used at this stage. After the turbine stator blades were reduced in width (to increase the throat area) and blow-off valves were fitted to the compressor, the engine ran in a sustained manner on 18 July 1945. It gave plenty of power but the specific fuel consumption was poor due to the low efficiency of the extemporized turbine stators. Fom then on, however, progress was rapid.

The heat exchanger was a complex piece of equipment containing some 1,700 oval tubes. Research on and manufacture of the heat exchanger was carried out by the Coventry Radiator & Presswork Company Ltd. (At one time, the Ritz rotary heat exchanger, details of which became available from a defeated Germany, was considered but rejected.) The Theseus was first run with its heat exchanger in December 1945 but, straightaway, this was found to be too small. At full throttle the fuel saving was only 8 per cent (20 per cent had been hoped for) and the power loss was considerable. The heat exchanger was therefore abandoned.

After the deletion of the heat exchanger the Theseus II became the first turboprop to be type tested, passing an MoS 100-hour test at 1,950ehp in December 1946. However, it was a dauntingly complex engine: according to Owner, 'I remember two US colonels who, after being shown the Theseus in 1946, said "Gee, you boys certainly have got your nerve!" Privately, we agreed with them, but did not allow our nervousness to affect our enthusiasms.' The Theseus was also the first turboprop to pass a 500-hour endurance test, which was monitored by the MoS in July 1948. The first flight of the Theseus took place in a Lincoln in February 1947; two Lincolns subsequently carried out almost 1,000 hours of test flying. In August 1948 the first flight was made with a Hermes V civil airliner, the four Theseus engines being notably quiet and heralding turboprop airliners to come. The final type test was carried out on a Theseus in August 1949, when the engine gave 2,450ehp.

Projected Janus turboprop

Early in 1945, when the Theseus was still being designed, the MAP requested design studies for a 1,000hp turboprop engine, later scaled down to 500hp to avoid duplication of effort. Owner's team evolved a two-stage centrifugal compressor for this engine and, because their discs were positioned back-to-back, the engine was given the name of Janus (the two-faced Roman deity). Four curved pipes connected the first compressor to the second and, in between these, there were four highly skewed combustion chambers. One turbine drove the compressors and another the airscrew. This neat little engine appeared to have a future, especially for helicopters, but was not built due to lack of an official order.

To speed development of the Proteus turboprop, the engine was first built as the Phoebus turbojet without a gearbox. Here a mock-up of the Phoebus is being used to create the installation beneath an Avro Lincoln flying test-bed. *Philip Jarrett*

Proteus turboprops

Even before the Theseus was built, Owner charged Charles Marchant to design a much larger turboprop of 3,500eshp. Work began in September 1944 and the engine was named, again from mythology, the Proteus. It was destined to be one of the most troublesome engines in history to develop. The purpose of developing such a large engine was to power a very large airliner for use after the war. Five types of aircraft for post-war development had been recommended by the committee chaired by Lord Brabazon in 1943. The first of these, designated the Brabazon Type I, was given priority. The Bristol company was selected to design this huge airliner on the condition that its war production did not suffer.

Only a small part of the Bristol Engine Division work-force was employed on gas turbine work since the Division's Development Engineer responsible for all piston and turbine engines, Roche Swinchatt, believed that the future of aero-engines lay with the likes of the excellent Centaurus piston engine and not gas turbines. In this view he was supported by the management. Discouraging as this atmosphere was, Chief Engineer Frank Owner pressed on with the Proteus turboprop project with his small team, which included the brothers Stanley and Harvey Mansell, who were, respectively, Design Engineer and Research Engineer. Jimmie Fell was responsible for the construction and modification of engines.

Although the free turbine and reduction gear was based on the Theseus, the Proteus differed considerably in its layout. The compressor consisted of a twelve-stage aluminium alloy axial unit followed by two successive aluminium alloy centrifugal stages and was centrally located with eight interconnected, conical combustion chambers clustered around it. Air entered the compressor at the rear of the engine, passing first through the axial stages and then the centrifugal stages to enter the combustion chambers via a steel diffuser with eight tangential elbows to reverse the flow. This reverse flow arrangement was chosen in order to keep the engine length down. The turbines consisted of a two-stage set to drive the compressors and a single-stage free turbine to drive the airscrew, the shaft for this leading from the turbine, through the hollow main rotor to the reduction gearbox for the airscrews at the front of the engine. The turbine blades were of Nimonic 80 alloy, manufactured by a pioneering precision casting process. There were 48 stator blades per row and 127 blades per disc, the latter being air cooled and made from Jessop G.18B steel. Apart from the gearbox, bearings comprised ball bearings at the front of the compressor, roller bearings either side of a splined coupling at the centre of the engine, ball bearings in front of the two-stage turbine and ball bearings in front of the free turbine. An atomizing fuel system with Lucas Simplex burners was used. Because the airscrew gearing was driven by its own free turbine and there were four compressor blow-off valves, starting was easy and required relatively little power.

To speed development, the engine was first built as a turbojet without the gearbox. Named Phoebus, this engine first ran in June 1946 and was then tested in the bomb bay of a Lincoln. Unfortunately, its performance was very poor and this was found to be caused by the poor aerodynamic characteristics of the first centrifugal compressor, which was actually causing a reduction in pressure. The engine ran red hot in places. Therefore, the first centrifugal compressor was removed and the flow-reversal diffuser passages leading to the combustion chambers were redesigned.

On 25 January 1947 the first run with a Proteus turboprop was made in a newly converted hangar at Bristol's Gipsy Patch site. This was followed by test flights with two engines fitted to a Lincoln, but unfortunately results were very poor. By 1950 the engine had been redesigned as the Proteus II. This was designed to give 3,200shp plus 363kp (800lb) of thrust for a weight of 1,383kg (3,050lb), but produced only about 2,500shp for a weight of 1,724kg (3,800lb) and was riddled with faults. Even running at half power, the Proteus II's combustion chambers were, in places, glowing red and virtually every other part had problems: the compressor blades, turbine blades and bearings, for example, were subject to frequent failure. Longer and wider turbine blades on a thicker disc were fitted. During the latter part of 1947 Bristol's compressor and turbine research station began to come on stream (and was completed in 1951), and this greatly facilitated problem solving.

On 3 January 1949 Stanley Hooker, having left Rolls-Royce, joined the Bristol Engine Division and by mid-1950 was appointed Chief Engineer in place of Frank Owner, who was 'retired' to a small office

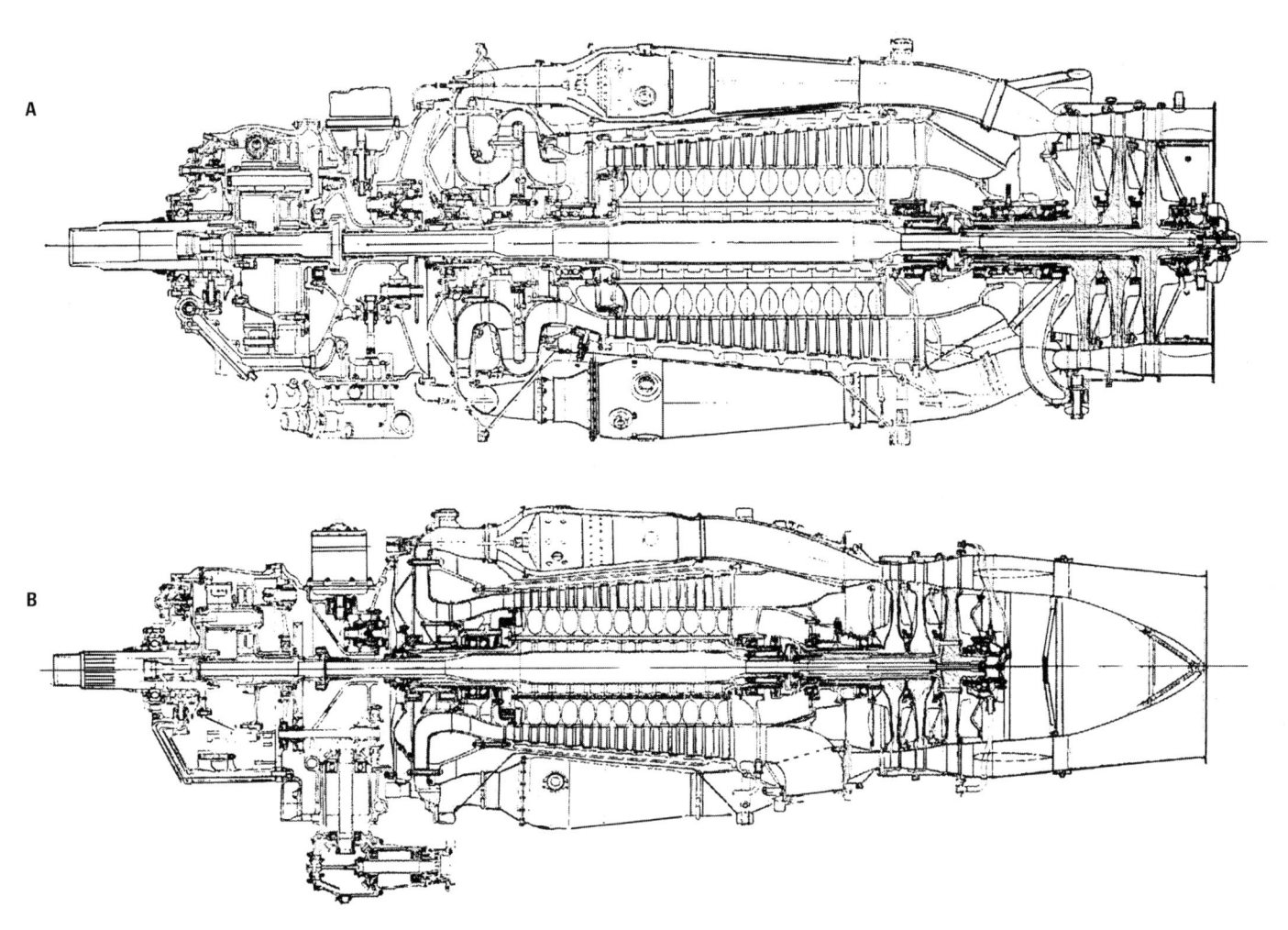

Bristol turboprops: (A) Proteus I; (B) Proteus II.

The Bristol Proteus III (700-series) turboprop, which gave 3,780shp plus 535kp (1,180lb) of residual thrust. Rolls-Royce Heritage Trust

and soon left to join de Havilland Engines. Hooker's unenviable brief was to sort out the Proteus engine urgently. He lost no time in energetically reorganizing the whole of the Bristol engine departments and he made many new appointments. The young Charles Marchant was appointed as Chief Engineer on turbine engines, while Gordon Lewis was brought from Rolls-Royce to head gas dynamic calculations and compressor design and development. Other engineers came from outside, including Bob Plumb, one of the ablest mechanical engineers in the world.

Much was done to modernize the test equipment and Hooker produced a series of formulae (known as Hooker's cookers) to help analyse Proteus test results. Hooker's possibly harsh dictum for his team was 'Work until it hurts, or get out', but he thought the biggest obstacle to progress was the superb, two-hour lunch that Bristol's top men had! With the advent of the Korean War in 1950, Bristol received much modern equipment to help, along with other companies, boost production of the R-R Avon turbojet.

Meanwhile, the Proteus made its first test flights, beginning on 12 December 1950. For this, two Proteus 2 (later known as the Proteus 600-series) engines were fitted to a Lincoln. Power was still down and specific fuel consumption was up, but the engine was at least reliable. It passed a 150-hour type test in August 1951. Hooker's team then set about completely redesigning the Proteus. The idea was to scrap it and start from scratch, most especially abandoning the reverse-flow layout. However, the engine was already designed into the Brabazon and Princess aircraft. To begin with, all combustion chamber design and production was turned over to the Lucas company, where Drs Watson and Clarke soon surmounted the problems with better, larger chambers. A new twelve-stage axial compressor, followed by a single centrifugal compressor, was developed with a total pressure ratio of 7:1. New two-stage turbines were developed and almost everthing except the original airscrew gearbox was redesigned. Even the gearbox had a problem of blowing out oil, but Bob Plumb quickly solved this hitherto intractable problem. The new engine, known as the Proteus 3 (later called the 700-series), first ran in May 1952 and was soon giving 3,475shp plus 454kp (1,000lb) of thrust for almost 454kg (1,000lb) less weight than the Proteus 2. Soon this was upgraded to 3,780shp plus 535kp (1,180lb) of thrust. The Proteus 3 had to be built within the dimensions of the previous Proteus 2 and came out as 0.978m (3ft 2½in) diameter by 2.880m (9ft 5⁵⁄₁₆in) long, the weight being 1,315kg (2,900lb) without the airscrew. It therefore emerged as shorter and more efficient than before.

Coupled Proteus turboprop

Once the Proteus 3 was licked into shape, the business of developing a coupled unit could begin. Two engines, in turbojet form, were mounted side-by-side with their centre lines about 1.07m (3ft 6in) apart and were attached to the coupling gearbox. This comprised a large, three-piece cast aluminium alloy casing housing the first stage of the transmission gearing and various drives for ancillaries. Its gear train included a double helical pinion, direct driven by each free or power turbine, each pinion meshing with an idler gear, which in turn meshed with a single output gear. The reduction gear ratio was 3.2:1. Each power turbine shaft and its pinion was connected via a dog clutch, which could be disengaged hydraulically in the event of one engine failing. The system was designed to prevent re-engagement of the clutch while in flight.

Bristol Proteus III (705-series) turboprop. Philip Jarrett

This Airspeed Ambassador airliner (G-AKRD) was fitted with two Bristol Proteus turboprops for trials.
Philip Jarrett

The single output shaft from the coupling gearbox extended forward via a flexible coupling to the gearbox for the two contra-rotating airscrews. This gearbox was a neat, self-contained unit (apart from sharing the lubrication system with the coupling gearbox) providing coaxial output shafts for the two 4.88m (16ft) diameter Rotol four-bladed, feathering and reversing airscrews. Total reduction through all the gears from the turbine (maximum speed 10,000rpm) was 11.11:1. The purpose of the flexible coupling was to accommodate small amounts of misalignment caused by the flexure of the mountings during running and flight. Incorporated in the flexible coupling, on its input side, was an electrically actuated drum brake for 'parking' the system when shut down. Drives were taken from the rear of the contra gearbox to drive ancilliaries such as the CSU and the synchronizing system (*see below*).

The complete Coupled Proteus formed a neat installation having low frontal area. Air intakes for it were to be in the wing leading edge on both sides of the engine, spaced to take maximum advantage of the ram effect from the airscrews. Intake air

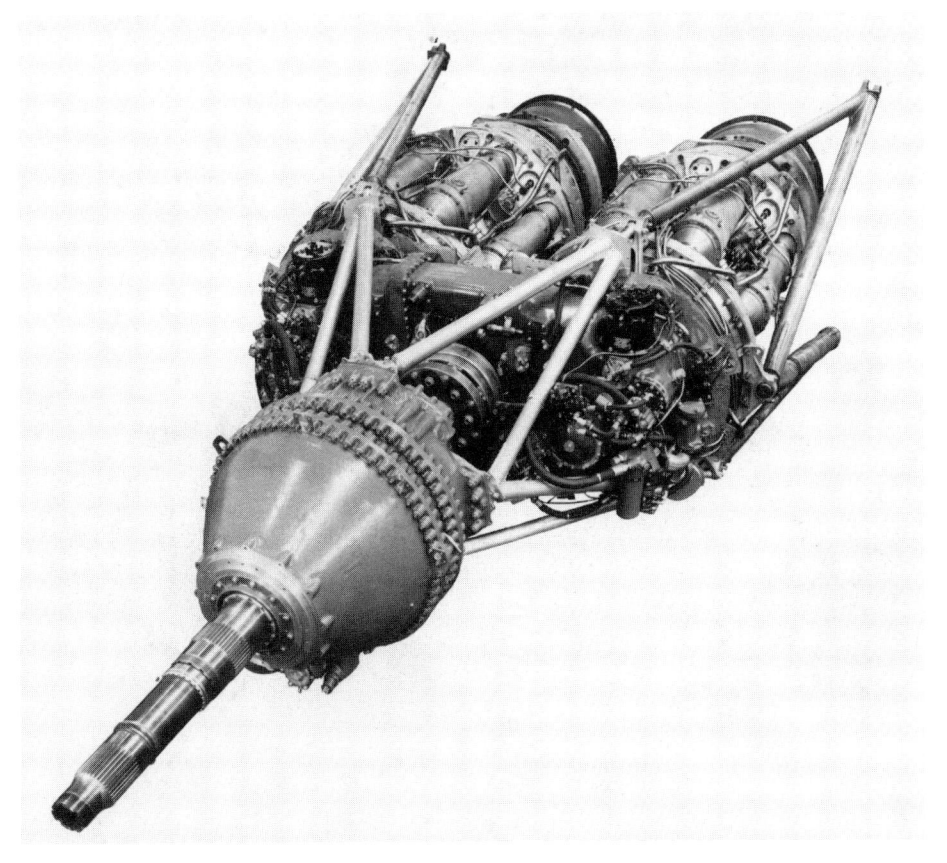

The Bristol Coupled Proteus turboprop, showing the coaxial shafts for driving the 4.88m (16ft) diameter contra-rotating airscrews. This engine could deliver some 7,000 equivalent shaft horsepower.
Rolls-Royce Heritage Trust

Another view of the Bristol Coupled Proteus turboprop. Philip Jarrett

was then ducted back to the rear of the engines where a plenum chamber enclosed the intakes of both engines. The main support points of the coupled Proteus were on the coupling gearbox joining the two engines. From this, a triangulated tubular structure extended forward to support the contra gearbox of the airscrews. Attachments to the aircraft structure were designed to accommodate thermal expansion of the coupling gearbox. Synchronization of the two Proteus engines was performed electrically by an alternator on one engine trimming the setting of the other's blade angles via the CSU.

Bristol Type 167 Brabazon

By November 1944 the giant Brabazon airliner was in the preliminary layout stage and designated as the Bristol Type 167. This was a fifty-passenger aircraft of some 113,400kg (250,000lb) loaded weight, planned to cruise at 483km/h (300mph) at 10,700m (35,000ft). Pending the development of the Proteus turboprop, it was planned to power the first prototype with four pairs of coupled Bristol Centaurus piston engines driving four contra-rotating, reversing airscrews. These engines were to be buried in the huge 70.1m (230ft) span wing.

Following mock-up construction, manufacturing drawings were issued to the workshops in April 1945, a month before the end of war in Europe. Although the Brabazon, as the Type 167 became known, was extremely large, it was of a conventional configuration. The circular-sectioned fuselage was curved over its whole length for streamlining and carried a large tail empennage well above the line of the wings. The wing had a straight trailing edge and most of its leading edge sweepback was on the panels outboard of the engines. A multi-wheeled tricycle undercarriage was used. The engines were mounted forward of the wing's front spar, the engine section of the wing being built integral with the fuselage structure, and there were air intakes either side of each airscrew nacelle. Fuel was carried in flexible tanks in the outer sections of the wing. Great care was taken to control the weight of the all-metal fuselage structure, all metal and fittings being produced to very close tolerances. Controls and undercarriage were hydraulically actuated. Of special importance for such a large aircraft was the development of a gust alleviation device in order to permit a wing built relatively lightly to survive at the speeds required in the event of being

A Coupled Proteus turboprop under test at Bristol. It is fitted with two 5m (16ft 6in) diameter four-bladed, contra-rotating de Havilland airscrews.
Philip Jarrett

The prototype (G-ALUN) of the mighty Saunders Roe SR.45 Princess transatlantic flying boat about to land. It was designed to carry up to 220 passengers and was powered by four Coupled Proteus and two single Proteus turboprop engines. Philip Jarrett

hit by a gust of wind. A gust detector was fitted to the nose some 24.4m (80ft) ahead of the wing and this authorized symmetrical movement of the ailerons to counteract the effect of a gust.

Construction of the first prototype (c/n: 12759, VX206, G-AGPW) began in October 1945 and its maiden flight was made on 4 September 1949, with A.J. 'Bill' Pegg at the controls, from the specially extended runway at Filton. This first Brabazon was fitted out with instruments but, later, some thirty seats and a bar were installed to demonstrate the comfort, quietness and smoothness of the aircraft to officials. Few problems were experienced during the test flights apart from fatigue cracks in the airscrew mounting structures.

The second Brabazon prototype (c/n: 12870) was to be fully equipped for 100 passengers and powered by four coupled Proteus engines. Bristol submitted a tender design for this, the Brabazon I Mk.II, in June 1947. Various other changes were made compared with the first prototype, but problems with the gust alleviation device could not be solved and this, together with political and economic reasons, led to the abandonment of the whole project. Both prototypes were scrapped in 1953 when the second one was still incomplete. Thus, the giant airliner had been attempted some ten years before its time had come.

Saunders Roe SR.45 Princess flying boat

Another giant, designed to be powered by the Proteus, was the SR.45 Princess long-range flying boat built by Saunders Roe in the Isle of Wight. This company had already built the world's first turbojet-powered flying boat fighter, the SR.A/1 (see Chapter 6). The Princess was designed to carry up to 220 passengers across the Atlantic, cruising at 579km/h (360mph). Power was derived from ten Proteus engines arranged in four coupled units and two single units inside the 66.90m (219ft 6in) span wing, which was shoulder-mounted on the pressurized, double-deck hull. The maiden flight of the prototype (G-ALUN) took place on 20 August 1952 and the flying boat was found to be excellent. During the mid-1960s, however, this and two other prototypes, which never flew, were scrapped due to lack of customers. The government had ordered the Princess on the assumption that BOAC or the RAF would operate them but they were not interested. Among the various projects for the Princess was one to power it by nuclear energy.

Bristol Type 175 Britannia airliner

Thus sadly ended another project for the Proteus, but a more successful aircraft for it was about to come on the scene. This was the Bristol Type 175 long-range airliner, later known as the Britannia. Although destined for BOAC, it was the MoS that ordered three prototypes on 5 July 1948. It was envisaged that the aircraft might be powered initially by four Bristol Centaurus radials, but the powerplants would later be either Napier compound-diesel Nomads or Bristol Proteus engines.

Designed by a team under Dr A.E. Russell, the Britannia benefited from lessons, systems and structural data gleaned from work on the ill-fated Brabazon. This enabled the idea of combination skin and

Final construction of the prototype Bristol Britannia airliner (c/n: 12873, G-ALBO) taking place at Bristol's Filton plant. This view shows the installation of the No.1 engine, a Bristol Proteus 625 turboprop. The nacelle hinged access doors incorporate the annular air intake. Philip Jarrett

stringers to be used, thereby reducing fuselage formers and wing ribs. Finalizing of the aircraft size came about after much vacillation by BOAC and it was decided to design for four Proteus engines at the outset. A highly pressurized fuselage was of uniform 3.66m (12ft) diameter for most of its length and layouts for up to sixty-four passengers were made. The pressurized section went back as far as the base of the fin. The wing embodied a box spar, tapered at the tips. Each Proteus engine was mounted on the forward flange of this box spar and drove a four-bladed de Havilland Hydromatic reversible airscrew. Each slim engine cowling had an annular intake around the spinner to duct air to the rear of the engine, the exhaust being led back via a long internal duct over the wing to an exhaust nozzle. Fuel was contained in the wings. Control surfaces were operated by servo tabs and double-slotted flaps were used.

The first prototype, a Britannia 101 (c/n: 12873, G-ALBO) with Proteus 2 or 625 engines, made its first flight on 16 August 1952 from Filton, the pilot being Bill Pegg. Very few modifications were subsequently required and the aircraft was of such promise that, as a world-beater, most of the world's airlines became interested. In December 1953 the second prototype (G-ALRX), powered by Proteus 3 engines, made its maiden flight, the pilot again being Bill Pegg. Unfortunately the main input gearwheel from the turbine of one engine stripped its teeth and the overspeeding turbine flew apart. Fragments from it pierced an oil tank and started a fire. Pegg made an emergency, wheels-up landing on mudflats by the Severn estuary, the aircraft later being written off by salt water and recovery damage.

As stated, the only original item on the Proteus 3 was the airscrew gearbox. This used straight-tooth spur gears and the failure was caused by resonant vibration that came about when the frequency of spur gear teeth engagement equalled the natural frequency of vibration of other

Bristol Britannia Series 310 with Proteus 755 turboprops		
Span	43.36m	(142ft 3in)
Length	37.87m	(124ft 3in)
Wing area	192.98sq m	(2,075sq ft)
Height	11.43m	(37ft 6in)
Empty weight	37,439kg	(82,537lb)
Loaded weight	83,916kg	(185,000lb)
Maximum speed	639km/h	(397mph)
Cruising speed	575km/h	(357mph)
Service ceiling	7,315m	(24,000ft)
Range with maximum payload of 15,830kg (34,900lb)	6,870km	(4,268 miles)
The CL-44D had a maximum payload of 29,995kg (66,128lb)		

A fine shot of the Bristol Britannia prototype, G-ALBO. Its aerodynamically fine airframe and quiet Proteus turboprop engines later led to the appellation of 'Whispering Giant'. Technically this airliner was an unqualified success, but its long development time damaged its export prospects. Philip Jarrett

parts of the gear train. In a kindly gesture, Hives at Rolls-Royce immediately sent five of his top engineers over to Bristol to see if they could help with the emergency. Fortunately, Bob Plumb had already designed a new gearbox with smoothly engaging helical gears and this was entirely free from resonant vibration problems. Earlier work by Armstrong Siddeley in mastering gear problems with their Python turboprop seems to have not been called upon. To insure against an overspeeding power turbine in the future, should a breakage occur in the gear train, an automatic fuel cut-off was developed that instantly shut the engine down if torque in the drive shaft fell to zero. This device was never triggered in the millions of hours of subsequent Proteus 3 running.

By 1953 the Proteus 3 was deemed to be an excellent engine and was type tested for civilian operation in August 1954. This was the Proteus 705 for use in the Britannia 102. Following a flap failure on G-ALBO, Bristol were without a flyable Britannia for test work until the first production aircraft (c/n: 12902, G-ANBA) flew on 15 September 1954, delaying delivery to BOAC. By 1955 the aircraft had been developed into the heavier, longer-range Series 300, but the engines grew to more than match the weight, the Proteus 755 developing 4,455ehp.

From December 1955 BOAC began route-proving with two Britannia 102s. In March 1956, however, all four engines stopped on one of them while flying over Africa. Using TV cameras inside the engine cowlings, Bristol found the problem to be caused by ice accretion on the inner walls of the cowlings. Glow plugs were fitted into some of the combustion chambers, these continuing to glow if the flames were extinguished by large amounts of ice and then re-igniting the fuel when the ice cleared. This system never failed and only a momentary bump was noticed before re-ignition. However, despite the unqualified success of this modification and the fact that icing could be avoided by flying at a slightly different altitude, BOAC was not satisfied and so two further years were wasted flying around looking for icing conditions. In due course Bristol devised a new cowling design that prevented a build-up of ice inside.

These unnecessary delays damaged the export prospects for the Britannia and came close to bankrupting the Bristol company. Eventually about eighty aircraft were sold, just enough to save Bristol, but the prospects for this fine airliner were further reduced when BEA ordered a similar airliner, the Vanguard, with R-R Tyne turboprops, from the Vickers company. The result was that neither Bristol nor

Close-up of the Proteus No. 3 engine of a Britannia of Monarch Airways. Author

Vickers did very well out of these aircraft. In Canada, developments of the Britannia under licence with Canadair led to the CL-28 Argus (four Wright Turbo-Cyclones) maritime patrol aircraft and the CL-44D (four R-R Tynes) swing-tail freighter. Though small in numbers, the Britannia and its derivatives have been very long-lived. As an airliner, it was dubbed the Whispering Giant and it was acknowledged as the fastest, smoothest, quietest and safest turboprop airliner in the world.

BE.25 Orion turboprop

Once the problems with the Proteus were sorted out, Hooker took the opportunity to abandon the reverse-flow layout and start afresh on a completely new, larger, axial turboprop engine. Known as the BE.25 Orion, the new turboprop was a marked advance on the Proteus. For approximately the same weight and size, the Orion had greater altitude cruising power and lower fuel consumption. It was a two-spool engine whereby a seven-stage low-pressure axial compressor was driven by a three-stage low-pressure turbine and, arranged coaxially and in between, a five-stage high-pressure compressor was driven by a single-stage high-pressure turbine. The low-pressure system was, in effect, supercharging the high-pressure system. The total compression ratio was 11.0:1. The combustion system was of the cannular type in which there were ten flame tubes within inner and outer annular air casing walls. Each flame tube had a fuel injector at its upstream end, so Hooker did not use the vaporizing system in this case.

The Orion was designed to give a cruising power of 3,500ehp at 9,150m (30,000ft) altitude and a take-off power of 4,400ehp, which remained constant up to 4,600m (15,000ft). The maximum power available was 5,150ehp at sea level. With a weight of only 1,452kg (3,200lb), the Orion had a specific fuel consumption of only 0.39. Its diameter was 1.06m (3ft 5¾in) and length 2.85m (9ft 4⅛in). This superb new turboprop offered outstanding high-speed cruise performance at altitude and Bristol designed the Britannia Series 400 to be powered by four Orions. Flight tests began in August 1956 with the port outer Proteus 755 engine replaced with an Orion on the first Britannia prototype, G-ALBO. Sadly, the Orion was cancelled in 1958 due to government support favouring the rival R-R Tyne turboprop instead.

Olympus turbojet

The most successful of Bristol's gas turbine engines was the Olympus turbojet, which had its origins in October 1945 when, with his project office free, Owner began discussions with Charles Marchant and Stanley Mansell about entering the turbojet field. The engine was to power a bomber at 805km/h (500mph) at 15,250m (50,000 ft) altitude. Once again, Owner decided that they needed to attempt something different from the general trend, as had been done when turboprop work was started. It was decided that a compression ratio of about 9:1 should be aimed at for the compressor, but it was realized that such a turbojet would have starting and acceleration difficulties. The solution to this problem was the first two-spool turbojet, in which a low-pressure turbine drove a low-pressure compressor outside of a high-pressure turbine driving a high-pressure compressor. The combined, or compounded, systems were coaxial and in this, of course, Bristol had the benefit of its earlier work on coaxial shafts for turboprops.

Bristol BE.25 Orion two-spool turboprop. Rolls-Royce Heritage Trust

Owner was keen to have an axial compressor (LP) followed by a centrifugal (HP) compressor, but these could not be kept within the desired maximum diameter of 1.02m (40in) and the flow path would have been far from straight. The objective was a high-speed bomber engine and so an all-axial layout was chosen. Pressure ratios of 2.7:1 and 3.3:1 were selected for a six-stage LP and eight-stage HP compressor respectively, in order that each could be driven by a single-stage turbine. An initial thrust of 4,080kp (9,000lb) was decided upon, with an ultimate thrust of 6,810kp (15,000lb) to be developed in the future. Little did the engineers know how much more powerful than that the engine would be in the future.

The company issued a brochure for this project to the MAP in March 1946 and this resulted, later in July, in engine specification TE1/46 being issued. Bristol received an order for six experimental engines in 1947, but it was not until January 1949 that the first manufacturing drawings were ready. Another seven months were then needed for detail design work, the engine being designated BE.10 and

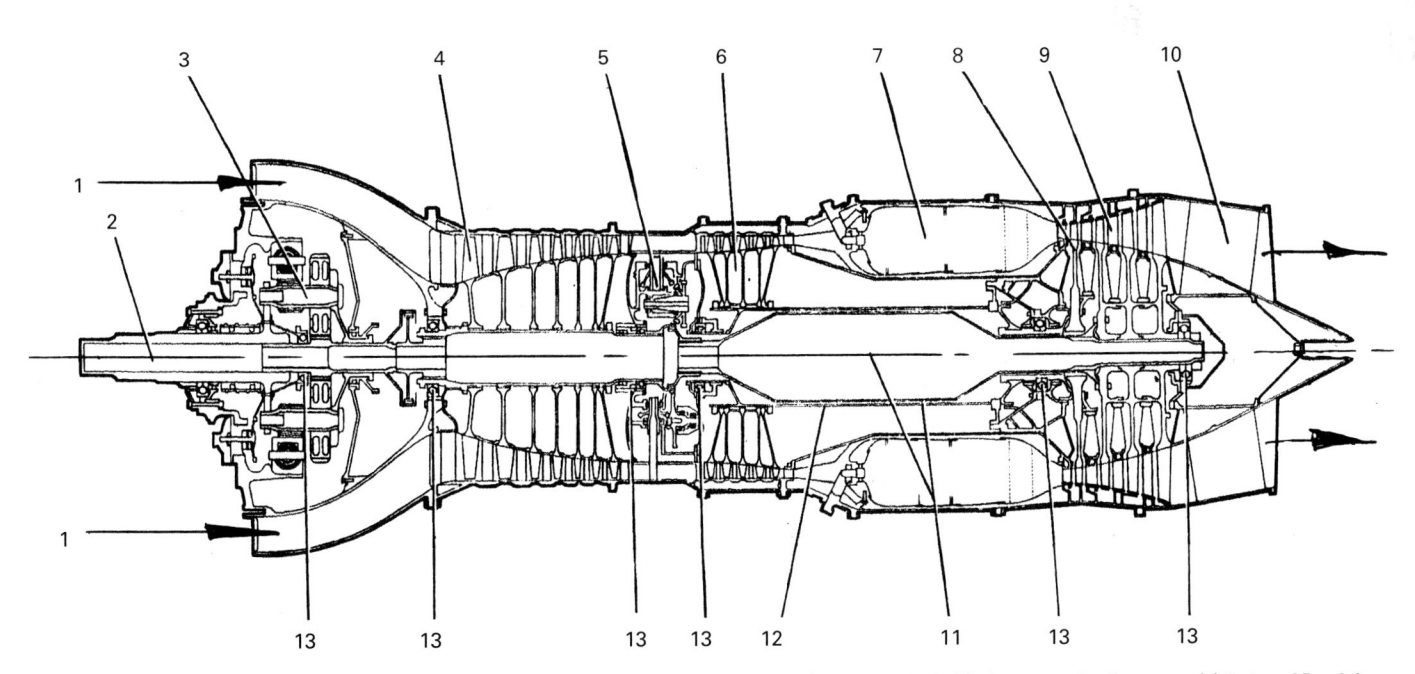

Schematic of the Bristol BE.25 Orion 5,150eshp two-spool turboprop: (1) annular air intake; (2) airscrew shaft; (3) airscrew reduction gears; (4) 7-stage LP axial compressor; (5) accessories drive gears; (6) 5-stage HP axial compressor; (7) ten flame tubes, in annular casing; (8) single-stage HP turbine; (9) 3-stage LP turbine; (10) exhaust duct; (11) LP turbine rotor, connecting to airscrew reduction gear shaft; (12) HP turbine rotor; (13) anti-friction bearings.

Bristol Olympus 201 turbojet. Rolls-Royce Heritage Trust

named the Olympus. Despite the name, there was nothing mythological about the engine that first ran on 16 May 1950 at an impressive full throttle thrust of 4,540kp (10,000lb), making it the most powerful British turbojet at that time. From the beginning Owner knew they had a winner, not only with the thrust but with good fuel consumption and very impressive acceleration.

Of course, there were developmental problems to be solved but, unlike in the case of the Proteus, these were mechanical rather than aerodynamic. The most dramatic incident during testing was when, due to too much deflection, a turbine diaphragm cut through a shaft and the released turbine disc flew out of the engine and spun around the test-bed floor, walls and roof! Fortunately nobody was hurt.

Flight testing began on 18 December 1951 with an Olympus slung beneath an Avro Ashton Mk3 (WB493), also powered by four R-R Nene turbojets. On 5 August 1952, however, a Canberra B.2 (WD952) made the first flight with an Olympus 2B of 4,425kp (9,750lb) thrust and this became the test-bed of choice. The twin-jet Canberra, designed by W.E.W. Petter and built by English Electric, had made its maiden flight on 13 May 1949. It was Britain's first jet bomber and was destined to be long-lived and one of the world's greatest aircraft. Despite Dr A.A. Griffith telling the Air Ministry that a two-spool turbojet such as the Olympus would not operate above 11,000m (36,000ft) without stalling, the Olympus-powered Canberra was soon reaching record altitudes of 19,410m (63,668ft) and then 20,080m (65,876ft). Nevertheless, the Olympus was not to be the chosen powerplant for the Canberra.

Avro 698 Vulcan V-bomber

Following research with the small Avro 707 delta aircraft, Avro received in July 1948 a contract to build two prototypes of a large, delta-wing bomber, later known as the Type 698 Vulcan. The prototype, VX770, made its maiden flight on 30 August 1952, piloted by Roly Falk. It was powered by four R-R Avon engines because the Olympus, for which the Vulcan was designed, was not ready. Even though the Avons gave only 2,950kp (6,500lb) of thrust each, VX770 gave an excellent fighter-like performance at the Farnborough Air Show in September 1952, despite having flown for only one hour previously. Roly Falk was again the pilot. By 3 October 1952 the Olympus 101 was giving a static thrust of 4,990kp (11,000lb); on 3 September 1953 it was tested in the second prototype of the Vulcan (VX777). On 27 July 1954 the Olympus-powered VX777 was badly damaged following a heavy landing at Farnborough.

By 1955 the Olympus 101 was in production for the Vulcan B.1, the first production example of which (XA889) made its first flight on 4 February 1955. However, the Air Ministry switched its support to the R-R Conway and instructed Avro to redesign the Vulcan for that engine. Undaunted, Bristol continued Olympus development at its own expense: it was too good for aircraft companies to ignore and, besides, it turned out to be cheaper than the Conway. Therefore, all production Vulcans were powered by Olympus engines.

The Vulcan V-bomber was conceived by Avro's Technical Director Roy Chadwick to meet the Air Ministry Specification B.35/46. The large delta wing of 30.18m (99ft) span was considered the best solution to the specification's requirement for carrying a heavy load, such as a nuclear bomb, at high altitude and at a high subsonic cruising speed while still being manoeuvrable. The delta shape evolved by virtue of the fact that a gross weight of not more than 45,360kg (100,000lb) had been officially specified. A short nose section and a large single fin and rudder were added to the wing and the four turbojets were completely buried in the wing roots with a single air intake for each pair. Two elevators and two ailerons were carried on each wing trailing edge. A large amount of fuel was carried in the wings and part of the fuselage and there was a large, central bomb bay. High-pressure hydraulics operated the control surfaces and undercarriage. For production aircraft, the crew consisted of five members, including two pilots. The first Vulcans went into RAF service in May 1957. Nuclear weaponry was the Blue Steel rocket-propelled stand-off missile; a proposal to replace this with the American Skybolt missile foundered when the US government cancelled the project. Soon, however, the British government decided upon Polaris submarines for the country's nuclear deterrent and the Vulcan then carried conventional bombs

Once production was under way, development of the Olympus showed just how much 'stretch' there was in it. In the Mk.102, a 'zero-stage' in front of the first stage was added to the low-pressure compressor to increase the overall compression ratio to 12.0:1 and this raised the thrust to 5,443kp (12,000lb). Improved alloy with higher heat resistance and better creep properties allowed temperature and rotational speed to be increased in the Mk.104 and so the thrust rose to 6,124kp (13,500lb). When Vulcan engines went in for overhaul, they were retrofitted to the Mk.104 specifications.

During 1954 Dr Stanley Hooker headed a redesign of the Olympus within its original diameter and this was designated the 200-series. Keeping the overall pressure ratio to 12.0:1, both compressors were redesigned to give a third more airflow at about 108.86kg/sec (240lb/sec), and yet the stages were reduced to only five on the low-pressure compressor and only seven on the high-pressure compressor. First run on 27 September 1954, the Olympus 200 gave a thrust of 7,260kp (16,000lb) – considerably higher than designed for. Later, with further modification, including the addition of a zero-stage in front of the first stage to the low-pressure compressor to increase the air mass flow, the Olympus 300-series was created and that gave 9,075kp (20,000lb) of thrust. The Olympus went on to further development and greatness, beyond the remit of this history.

Flight testing of the Bristol Olympus turbojet began with two such engines outboard of the four Rolls-Royce Nene turbojets on this Avro Ashton Mk 3 (WB493). Here the aircraft is seen landing at Bristol's Filton airfield. Philip Jarrett

Avro Vulcan B Mk.2 V-bomber with four Olympus 201 turbojets			
Span		33.83m	(111ft 0in)
Length		30.46m	(99ft 11in)
Wing area		368.65sq m	(3,964sq ft)
Height		8.28m	(27ft 2in)
Loaded weight		102,060kg	(225,000lb)
Maximum speed	at 16,760m (55,000ft)	998km/h (Mach 0.939)	(620mph)
Service ceiling		19,812m	(65,000ft)
Range		7,400km	(4,600 miles)

Suffice it to say that some of the eighty-nine Vulcans built went on to make some of the longest bombing missions in history, covering almost 12,900km (8,000miles), using air-to-air refuelling, during the Falklands War.

BE.17 expendable turbojet

By late 1959 the Bristol Engine Division, which had been renamed Bristol Aero Engines in 1956, had merged with Armstrong Siddeley Motors to become Bristol Siddeley Engines Ltd (BSEL), under the overall technical direction of Stanley Hooker. In 1961 BSEL then absorbed de Havilland Engines and Blackburn Engines and the respective airframe companies were also merged.

With the advent of the stand-off missile and projects for cruise missiles, a need was identified for a small, inexpensive, expendable turbojet. It seemed, in the 1950s, that subsonic speeds would be adequate for such missiles, provided they were launched from very high altitude. The engine was expected to last for about an hour and a range of 805km (500 miles). Bristol was given an MoS contract to work up a design within the parameters of 1,360kp (3,000lb) of thrust for a weight of 227kg (500lb) and a cost of £500, if produced in thousands.

The job was taken on by Owner and Whitehead, who also worked out a factory layout to mass-produce the engine, designated the BE.17, as cheaply as possible. Its axial compressor had only six stages, which were driven by a single-stage turbine. In the whole of the engine there were only two or three screw threads, nearly all other assembly being permanent and accomplished by welding. By these means, weight and cost were kept to a minimum.

BE.26 Orpheus turbojet and the Folland Gnat lightweight fighter

Although the BE.17 was cancelled, a serviceable version of it, the BE.22 Saturn, was created for Folland's Gnat lightweight jet fighter. This comparatively simple

The second prototype (VX777) of the Avro 698 Vulcan bomber, powered by Olympus 101 turbojets. The huge delta wing had a straight leading edge at that time. Philip Jarrett

An operational Avro Vulcan B.2A V-bomber (XL330). It was powered by four Bristol Olympus turbojets of up to 9,075kp (20,000lb) thrust each. Philip Jarrett

fighter was the brainchild of W.E.W. Petter, who hoped it would be of interest to emerging nations. By 1952 he had planned two designs of the Gnat, the Fo.140/1 to be powered by the 1,700kp (3,750lb) thrust Saturn engine and the Fo.140/2 to be powered by a R-R Derwent engine, but when the Saturn engine was cancelled, it was also decided that the Derwent engine was not suitable. In the meantime, however, Bristol was working on another small axial turbojet, the BE.26, later named the Orpheus. This was designed under the leadership of Bernard Massey and was without official support.

The Orpheus turbojet had a seven-stage compressor (derived from the low-pressure compressor of the Orion) driven by a single-stage turbine. The connection between the two was a thin-walled drum, the large diameter of which obviated the need for a centre bearing. A cannular combustion system incorporated seven flame tubes.

During 1953 Folland replanned the Gnat around the awaited Orpheus engine but, in the interim, decided to build a prototype, designated Fo.139 Midge, using an AS Viper engine for power. It was considered that three Midges could be bought for the cost of a typical fighter of the day, such as a Hunter. The Midge had shoulder-mounted, swept-back wings beneath which were the side air intakes for the engine. Its tailplane was mounted low, out of the wing's wake, and there was a swept-back fin. The completed prototype Midge (G-39-I) made its first flight on 11 August 1954 from Boscombe Down, the pilot being E.A. Tennant. The considerable interest the machine aroused at Farnborough that year was followed up by a number of foreign pilots visiting the company's Chilbolton test unit to fly it: sadly the Midge crashed on take-off on 26 September 1955 when being flown by a Swiss pilot.

Britol's Orpheus engine was making good progress meanwhile, the first run having been made on 17 December 1954 with a thrust of 1,360kp (3,000lb). Subsequent flight testing was made under the Avro 706 Ashton WB493 engine test-bed (powered by four R-R Nenes). On 18 July 1955 the prototype of the Gnat (G-39-2) made its maiden flight, powered by an Orpheus engine. Its success was such that the MoS placed an order for six examples for evaluation, each armed with two 30mm Aden cannon. Although, following this evaluation, it was decided not to order it

153

Bristol Orpheus 803 turbojet.
Rolls-Royce Heritage Trust

for the RAF as a fighter, the Fo.144 two-seat, advanced trainer version, with an Orpheus 100 engine of 1,920kp (4,230lb) thrust, was ordered in 1957 for the RAF's Central Flying School. Subsequently, the Gnat became legendary as the mount for the first Red Arrows aerobatic team. Foreign orders for the Gnat came from Finland, Yugoslavia and India, including licence production in the latter country. The Gnat saw action in the India/Pakistan conflict.

Following the first small successes of the Orpheus with the Gnat, the engine was built in thousands by Bristol Aero Engines under licence in Italy and India. It powered, for example, the Fiat G91, the NATO lightweight fighter. Final versions of the engine were the Orpheus Mk.703 of 2,200kp (4,850lb) thrust and the Mk.803 of 2,268kp (5,000lb) thrust.

BE.53 Pegasus vectored-thrust VTOL turbojet

The final story of the Orpheus can be mentioned without going too far beyond this history's timeline. It involves the first vectored-thrust turbojet for vertical take-off. Hooker received a letter from Sydney Camm, Hawker Aircraft's Chief Designer, asking what he was doing about vertical take-off. From 1958 Hooker's team was working on the BE.53 engine, later known as the Pegasus, but because Duncan Sandy's government White Paper the previous year had concluded that there would be no manned military aircraft in the future (so entranced was the government with missiles), official sponsorship for a new fighter was not forthcoming. Therefore, both the Pegasus engine and the projected Hawker P.1127 VTOL fighter were privately financed.

The original concept of the Pegasus engine was that it would have two side nozzles, for cold air, which could be rotated through 90 degrees to provide forward thrust or vertical lift, as required. An Orpheus engine was used with the addition of a shaft driving a fan (actually the first three low-pressure stages of an Olympus engine) to provide cold air for the rotating nozzles. The Orpheus and the fan were later arranged to rotate in opposite directions to minimize gyroscopic forces. By the time

A Bristol Orpheus 803 turbojet being prepared for testing, with a static bell-mouth air intake and duct fitted. Rolls-Royce Heritage Trust

Folland's Gnat lightweight fighter prototype (G-39-2), powered by a Bristol Orpheus turbojet. It was estimated that three Gnats could be built for the price of a typical fighter of the day, such as a Hawker Hunter. Philip Jarrett

the Pegasus 1 engine was developed, a new, two-stage cold air fan was fitted, ahead of the front bearing, and the hot exhaust was led via a bifurcated duct to a second pair of rotating side nozzles. Thus there were available two cold and two hot nozzles for forward thrust or lift, as required. Compressor air was bled off to cool the bearings of the hot exhaust nozzles and also to the nose, tail and wing tips to balance and control the aircraft during hovering and low air speeds.

Although the initial aim of the engine was a thrust of 3,630kp (8,000lb), the first run of the Pegasus 1 in August 1959 gave 4,085kp (9,000lb). During the maiden VTOL hover (tethered) of the Hawker P.1127 (XP831) on 21 October 1960, the Pegasus 2 engine was delivering a take-off thrust of 5,220kp (11,500lb). This first hover was carried out at Dunsfold over a grid, to deflect the hot gases, the pilot being Bill Bedford. By now the project had the interest and assistance of the USA, the NATO Mutual Weapons Development programme and, belatedly, the MoS. Eventually the programme led to the unique Harrier (under the design direction of J.W. Fozzard) and AV-8 VTOL fighters. By 1960 Bristol Siddeley Engines was providing competition on an equal footing with Rolls-Royce and the companies soon joined forces.

Folland Gnat T.1		
Span	7.31m	(24ft 0in)
Length	9.68m	(31ft 9in)
Wing area	16.27sq m	(175sq ft)
Height	2.93m	(9ft 7½in)
Loaded weight	3,915kg	(8,630lb)
Maximum speed	Mach 0.97	
Service ceiling	15,240m	(50,000ft)
Range	1,900km	(1,180 miles)

Bristol Pegasus 2 vectored-thrust turbojet.
Rolls-Royce Heritage Trust

LEFT: **The prototype (XP831) of Hawker's P.1127 uses the downturned nozzles of its Pegasus 2 engine to hover.**
Philip Jarrett

FACING PAGE: **A formation of Hawker P.1127 V/STOL aircraft, predecessor of the Harrier. Centre is XP831, the first of two prototypes. The other two are XP980 (nearest) and XP976, two of the development batch of four aircraft (XP836 crashed in December 1961). The development aircraft were powered by the Pegasus 3 engine of 6,125kp (13,500lb) static thrust.**

This display engine shows the major components of the Bristol Pegasus 5 vectored-thrust turbojet. The rotating nozzles for cold air thrust are on the left and those for hot gases are on the right. This unique engine at first powered the Hawker Kestrel V/STOL aircraft, predecessor of the famous Harrier. Philip Jarrett

CHAPTER SEVENTEEN

D. Napier & Son Ltd

The company's origins can be traced back to its establishment by David Napier in 1808 in Soho and Lambeth, London, and its move to Acton in 1903. Napier (b. 1785) came from a family of high mechanical and mathematical ability and his first factory was set up to manufacture printing machinery. With a reputation for the highest quality, the company entered into automobile manufacture around the turn of the twentieth century and then diversified into other types of vehicle. Napier's first aero-engine of note was the excellent Lion water-cooled, W-12 piston engine of 450hp, which went into service in the latter stages of the First World War. This engine won races after the war and one powered a de Havilland aircraft to an official record altitude of 9,297m (30,500ft) in 68 minutes in January 1919. Napier equipped a Lion with an exhaust-driven turbo supercharger in 1924. At that time Napier was making engines for almost half of the types of British aircraft and, in 1928, engaged Frank Halford (see Chapter 7) to design small aero-engines. They also built Junkers Diesel engines under licence in the 1930s. By 1932 the Lion engine had become outmoded and the 555hp Lion XV was the last of that line.

In the years leading up to the Second World War, Napier was far behind Rolls-Royce and Bristol in engine production. However, in 1935 they began the development of the Sabre, a powerful, 24-cylinder, horizontal-H, liquid-cooled, sleeve-valve engine that was to become the powerplant of the Hawker Tempest and Typhoon fighters in the second half of the war. For test flights the Napier Flight Development Establishment was created at Northolt; this was moved to a safer location at Luton by 1940. This flight establishment grew until, by 1958, it had a staff of 1,100 and covered more than 28.3ha (70 acres). By 1944 Tempests were catching and destroying German V1 flying bombs over southern England and rocket-firing Typhoons were raining wholesale destruction onto German armour on the Continent. Although never fully developed, Napier's Sabre engine was a big success and it transformed the company's fortunes. From type testing at 2,200hp in 1940, it was giving up to 3,055hp by the war's end, by which time the company had been taken over by the English Electric company.

Although the Sabre was one of the world's most powerful engines, it was the end of the line for piston aero-engines. Even so, owing to the intense pressure during the war to produce Sabre engines, it had not been possible for the small company of Napier to plan beyond it. By late 1944, however, when English Electric production facilities, especially in Liverpool, took some of the pressure off, a Napier team under Chief Designer Herbert Sammons began looking to the future,

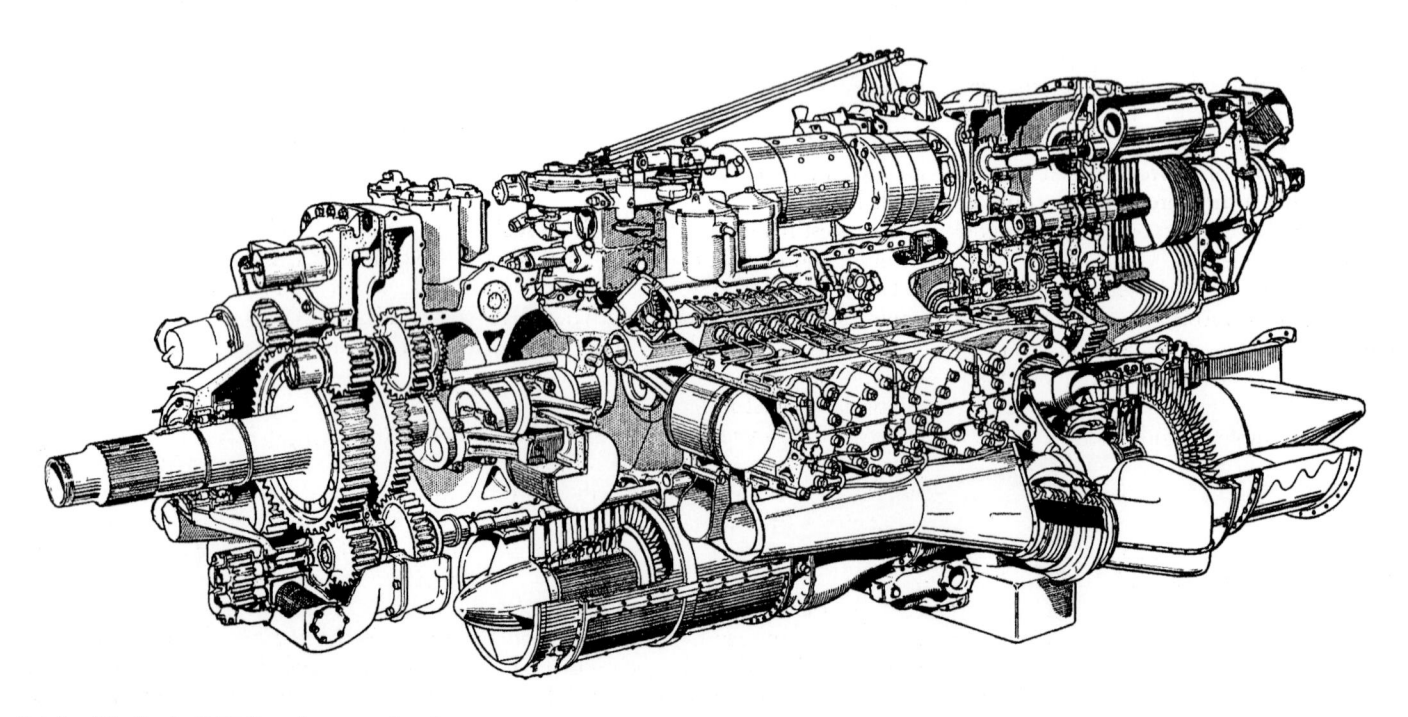

Details of the Napier E.125 Nomad compound engine.

A Napier E.125 Nomad compound engine at East Fortune. Napier Power Heritage Trust

which obviously lay with the gas turbine in one form or another. Included in his team were the engineers B. Bayne and A.J. Penn. They concluded that their efforts should be directed towards powerplants for post-war airliners and that, for these, a gas turbine driving an airscrew was going to be most suitable. Furthermore, such powerplants had been largely neglected during the war in favour of the turbojet, which was more suited to fast military aircraft of shorter range. They also realized that a turboprop engine could be readily fitted to airliners previously powered by piston engines.

The first task for Napier engineers was to design a gas turbine research establishment, which was to be built at Liverpool and sponsored by the MoS. This establishment had the most modern test plant, research equipment and machinery: what was not bought was designed and built by the company.

E.125 Nomad compound engine

By 1949 Herbert Sammons had been elevated to become Napier's Managing Director and his place as Chief Engineer was taken by Ernest Chatterton. The first project for his team was to satisfy an MoS requirement for a low fuel consumption engine for long-range flights. Originating in 1944, this requirement was for a 6,000hp engine, although by 1946 the MoS had changed this to 3,000hp. The engine was to be of the compound type combining a two-stroke diesel engine and a gas turbine. This concept doubtless followed on from the work of Harry R. Ricardo, who headed advanced diesel and sleeve-valve work at Rolls-Royce.

It was estimated that the weight of such a compound engine would be somewhere between that of the equivalent piston engine and turbojet, but that it would give a much reduced fuel consumption and hence lower the aircraft loaded weight, owing to the need to carry less fuel. This benefit would increase with range. Moreover, the compound engine would operate on any type of fuel, be almost insensitive to changes in ambient temperature and capable of constant performance at any altitude.

This was the origin of Napier's first design of compound engine, designated the E.125 Nomad. A complex engine, the E.125 consisted of a liquid-cooled, flat-12 diesel engine of 41.1 litres capacity and with valveless, two-stroke operation. It was supercharged by an axial and a centrifugal compressor connected in series with an intercooler in between them. The diesel engine had piston-controlled air inlet and exhaust ports. The turbine driving the compressors was run on the exhaust gases from the diesel engine, these being cooler (less than 700°C) than from an equivalent petrol engine. Two coaxial shafts drove a contra-rotating airscrew, one being driven by the diesel's crankshaft and the other by the exhaust turbine. To boost power, there was a reheat chamber and auxiliary turbine. The first Nomad was running by 1949 and produced 3,125ehp (3,000shp) for a weight of 1,905kg (4,200lb). It was exhibited and flown at Farnborough.

E.145 Nomad II

By 1951 this engine had been completely redesigned and simplified as the E.145 Nomad II. Gone were the centrifugal compressor, reheat chamber and auxiliary turbine, and a single airscrew was now driven by the crankshaft via gearing. The compressor supplying air to the engine was a twelve-stage axial type with a pressure ratio of 8.25:1 and a mass flow of 5.89kg/sec

Napier E.145 Nomad II compound engine. Following the Nomad I, this engine was a completely redesigned and simplified engine. It could develop more than 3,500eshp and set a new low in specific fuel consumption. Napier Power Heritage Trust

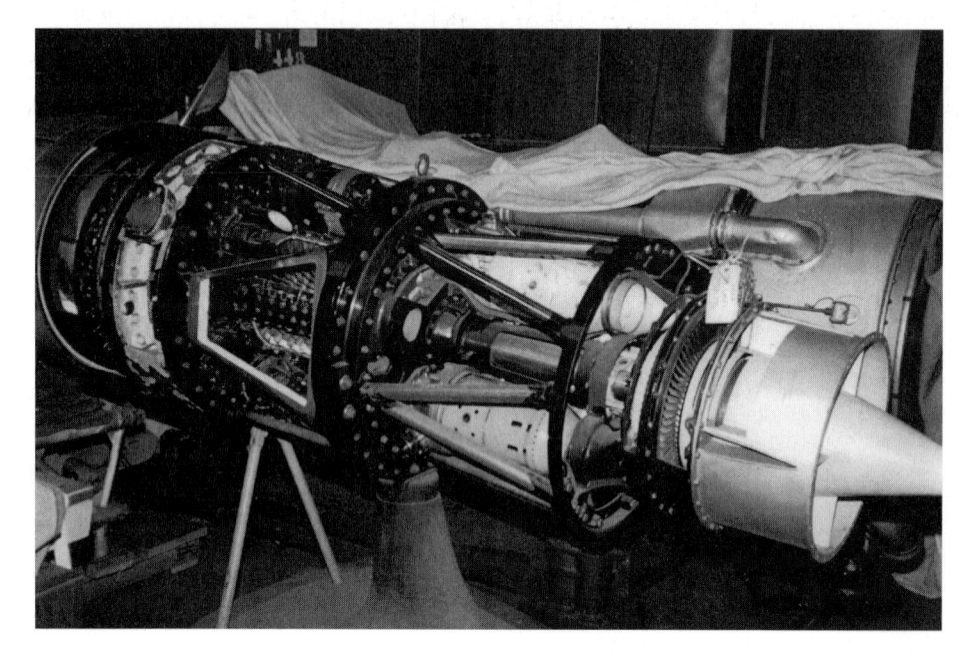

The Napier NNa.1 Naiad, one of the world's first axial-flow turboprops. Napier Power Heritage Trust

A Napier NNaC.1 Coupled Naiad to drive contra-rotating airscrews in a very neat installation.

(13lb/sec). The incidence of the air inlet guide vanes was automatically adjustable to extend the operating range at low speeds. Steel was used for the first five stages of the rotor blades and aluminium-bronze for the rest. The compressor was driven by a coaxially mounted, exhaust-driven, three-stage, reaction turbine. Surplus power from this turbine was added to the airscrew via Beier-type gearing, housed in the rear casing, and a quill shaft to the airscrew gearing. The purpose of the variable gearing, which worked on the principle of sliding cones, was that, at speeds below optimum cruising, when there was insufficient exhaust energy to operate the turbine/compressor as required, additional energy was supplied from the diesel engine to the compressor via the variable gearing. The turbine rotor blades were mounted onto discs on a hollow shaft, which had a

Napier Eland axial-flow turboprop. Philip Jarrett

front roller bearing and a rear ball bearing, the latter having a balance piston to absorb axial thrust. Air tapped from the compressor was used to cool the fir-tree roots of the rotor blades. Only a small amount of jet thrust remained after the energy extracted by the turbine. A single-lever control system was developed for the engine.

The Nomad II was developed to give 3,135ehp and even 3,580ehp with water injection into the engine manifolds. An additional increase in take-off power was possible by burning a small amount of extra fuel in the exhaust manifold before the turbine, there being enough oxygen left in the exhaust gases to support this. By this procedure and the water injection, the take-off power could be increased to 4,095ehp. The engine's specific fuel consumption set a new low of 0.345, but the long-range freighter for which it was most suitable did not materialize. It was then planned to use the engine in the Avro Shackleton MR.Mk 4, but by 1955 the Nomad II was cancelled, since its development was still not complete and the Shackleton carried on with its four R-R Griffon piston engines.

NNa.1 Naiad axial-flow turboprop

Even while work was proceeding on the Nomad, Napiers had begun work on one of the world's first axial-flow turboprops, having decided that a single-spool turboprop of this type would be best to power airliners. By 1947 a turboprop designated the NNa.1 Naiad was running. It featured a twelve-stage axial compressor, five combustion chambers and a two-stage turbine driving a single airscrew. The compressor rotor discs were mounted on a hollow shaft that was connected at its front end by an annular coupling to a reduction gearbox. Its compression ratio was 5.5:1 and air mass flow 7.80kg/sec (17.2lb/sec). The five combustion chambers were of the straight-through type with downstream injection and were interconnected and made of Nimonic 75 alloy. Nimonic 80 was generally used for the turbine, to deal with an inlet temperature of 875°C, and its discs were air cooled. The Naiad produced a take-off power of 1,590eshp including a thrust of 109kp (240lb) at a speed of 18,250rpm. Its weight was 497kg (1,095lb) without the airscrew.

A coupled Naiad, the NNaC.1, was produced to drive contra-rotating airscrews. In this, two Naiads were taken, without internal gears, and mounted side-by-side. They were connected via a coupling

Napier Eland turboprops were first test-flown in 1954, installed in this Vickers Varsity and employing over-wing jet efflux. Philip Jarrett

Two Napier Eland turboprops were fitted to this Airspeed Ambassador airliner (G-ALFR, *Golden Hind*), which later served with Dan Air. Philip Jarrett

gearbox having two coaxial shafts, each Naiad driving both contra-rotating airscrews. The coupled Naiad gave a take-off power of 3,150eshp including a thrust of 218kp (480lb). Its total weight, without airscrews, was 1,157kg (2,550lb) and its specific fuel consumption was 0.72.

Work on the Naiad engines remained in the experimental stage, as did that on a small 535hp turboprop known as the Nymph.

Eland turboprop

By 1950 Napiers had turned its attention to another turboprop project known as the NE.1 Eland. The Eland featured a ten-stage axial compressor, six can-type combustion chambers and a three-stage turbine. It first ran in 1952 and gave 3,000eshp (2,690shp) driving a single airscrew. No trouble was spared to make the Eland safe and it had automatic safety systems. It first flew in

1954, installed in a Vickers Varsity flying test-bed. This was followed by adapting the third prototype of the Airspeed Ambassador airliner (G-ALFR, *Golden Hind*) to take two Elands. (This aircraft was later converted to airline standards and sold to Dan Air). A similar conversion was later carried out on a Convair 340 airliner.

In a conversion from piston engine power, two Napier Eland turboprops were supplied as a kit and fitted to this Convair 340 airliner in the USA. Philip Jarrett

The Napier N.El.3 Eland turboprop with auxiliary compressor to supply air to the rotor tip jets of the Fairey Rotodyne convertiplane. Philip Jarrett

Fairey Rotodyne convertiplane

The Eland's finest hour came in powering the unique Fairey Rotodyne convertiplane. The concept, tested in the small Fairey Gyrodyne, was that the aircraft would use its rotor to take off and land but use conventional airscrews for forward flight. The idea of the Rotodyne was conceived by Dr J.A.J. Bennett and Capt A.G. Forsyth in 1947. For take-off and landing the Rotodyne's rotor was powered by small tip jets that operated on kerosene and compressed air. Such rotor jets were tested in the Fairey Jet Gyrodyne.

The Rotodyne had a squarish-sectioned fuselage with shoulder-mounted, rectangular, short wings to carry the two N.E1.7 Eland turboprops and to share the lift with the freely rotating rotor in forward flight. The tail empennage comprised a rectangular tailplane and three fins. Above the fuselage was a large, streamlined pylon carrying the 27.43m (90ft) diameter, four-bladed, stainless-steel rotor. Initially there was capacity for forty passengers and the fuselage had oval windows and, at the rear, full width cargo doors. To supply air to the rotor jets, each Eland engine was modified to drive an auxiliary compressor. Each compressor supplied two opposing blade jets so that, in the event of an engine failure, enough air would be supplied to the rotor in a balanced manner.

With the interest of BEA, Fairey received an MoS research contract in 1953 to build a prototype. This prototype (XE521) first flew in 1957; on 5 January 1959 it set a record for a convertiplane by flying at an average speed of 307.2km/h (190.9mph) over a 100km (62.1 mile) closed circuit. The fine-looking Rotodyne, decked out in white and blue, caused much excitement and seemed to herald a new era in transport from city centres on short and medium-haul routes. There was much interest worldwide, from both civilian and military quarters, in buying it and also in manufacturing it under licence. Sadly, this rosy future was not to be. Fairey was taken over by Westland in 1960 and this company had its competing, conventional but very large Westminster helicopter to promote. The Westminster was also powered by two Eland engines, but in the manner familiar today. Orders for the Rotodyne evaporated because of the long delivery times and the development, in any case, began to put on weight. To cater for its increased weight, a more powerful Eland was developed by 1955 to give 4,200eshp. By air cooling the turbine blades, hotter operating temperatures were possible. There may well, however, have been considerable opposition to the very noisy rotor jets and Fairey's attempts at reducing the noise did not meet with much success. Paradoxically, Westland

The prototype Fairey Rotodyne convertiplane (XE521), powered by two Napier Eland turboprops with auxiliary compressors. The rotor was powered by its tip jets only for take-off and landing. Designed to carry forty passengers, the Rotodyne received worldwide interest. Philip Jarrett

Fairey Rotodyne		
Wingspan	14.7m	(46ft 5¾in)
Rotor diameter	27.43m	(90ft 0in)
Rotor disc area	591sq m	(6,359sq ft)
Length	17.88m	(58ft 7¾in)
Height	6.76m	(22ft 2in)
Maximum take-off weight	14,969kg	(33,007lb)
Cruising speed	298km/h	(185mph)
Range	724km	(450 miles)

had to scrap its promising Westminster helicopter to concentrate on the government-sponsored Rotodyne; in the end both projects foundered.

Oryx gas generator

Because of the simplicity of driving a helicopter rotor by tip jets, Napier was encouraged to develop a gas turbine engine specifically tailored as a gas generator for such use. This gas generator was known as the NOr.1 Oryx and was designed to generate a take-off power of 754ghp (gas horsepower, assumed to be the power obtained from the flow of gas through a turbine of 100 per cent efficiency) for a dry weight of 225kg (495lb). It measured 0.49m (1ft 7¼in) in diameter by 2.12m (6ft 11½in) long and was designed as a self-contained unit that could be fitted inside a streamlined pod for external mounting on a helicopter.

The Oryx had a twelve-stage axial compressor, five combustion chambers with upstream fuel injection and a two-stage turbine. The turbine also drove an auxiliary four-stage axial compressor with its air intake at the opposite end to the main compressor. Air from the auxiliary compressor plus the exhaust gases were passed into separate chambers and then into the rotor jets. On its way there, the auxiliary air was passed around the exhaust volute for cooling purposes. The air intakes at both ends had bell mouths but the main compressor was preceded by variable inlet guide vanes.

Two Oryx gas generators were fitted to the Hunting Percival P.74 tip-drive helicopter, which had a three-bladed stainless-steel rotor. Its fuselage was a neat, streamlined pear shape with four under-

Also powered by two Napier Elands, but in turboshaft form, was the Westland Westminster, at that time the largest single rotor helicopter in the Western world. This is the crane version (G-APLE), which could lift a payload of 6 tons or more. The military version was designed to carry, for example, 51 fully equipped troops. Philip Jarrett

Napier Gazelle 13 turboshaft engine. The Gazelle was the first turboshaft in the world to pass a type-test for helicopters. Napier Power Heritage Trust

carriage members and a conventional fin and tailplane. Seating was for ten. The P.74 first flew in 1952 but, despite the simple elegance of its concept, £3 million and five years later it was abandoned as a failure. Accordingly the larger P.105, which was projected in 1954, was not proceeded with. The P.105 was of generally similar layout to the P.74 but had a tail rotor and its Oryx gas generators were at the ends of stubby wings extending from the main rotor pylon. For more power, Napier developed or projected the Oryx NOr.5 of 950ghp and the NOr.10 of 1,250ghp, but by 1957 this line of helicopter exploration had come to an end.

Gazelle free turbine engine

Regardless of that, it had been decided that Napier should develop the gas turbine to drive a helicopter rotor mechanically in the manner that is favoured today, as a so-called free-turbine engine in which a separate turbine (not connected to the compressor/turbine shaft) is connected by a shaft to a gearbox near to the rotor hub. Napier's most successful design for this type of engine was the Gazelle, which was developed under an MoS order for a Bristol twin-rotor helicopter. The Gazelle had an eleven-stage axial compressor, six flame tubes and a two-stage turbine that was followed by a single-stage free turbine connecting to the mechanical drive shaft. Four exhaust nozzles turned the exhaust gases through 90 degrees before the drive shaft coupling.

The Gazelle was first run in December 1955, only seventeen months after the first drawings were made, and produced 1,260 shp. It later passed an MoS type test for helicopters, the first in the world to do so, at 1,650shp.

Westland Wessex helicopter

While Bristol was developing its Type 191 twin-rotor helicopter, to be powered by two Gazelle gas turbines, Westland obtained a licence in 1956 to build Sikorsky's very successful S-58 with the idea of converting

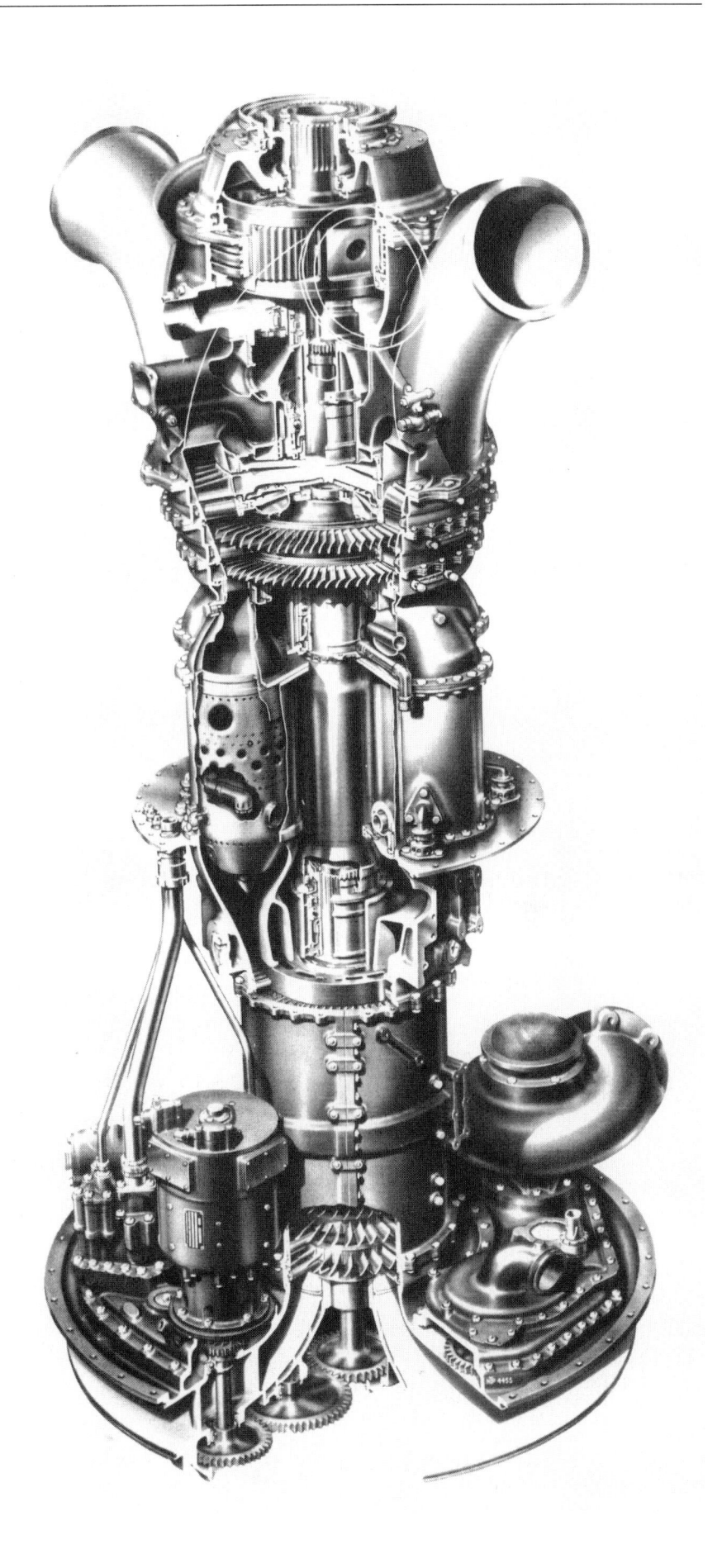

The first Westland Wessex helicopter (XL722) modified by replacing its piston engine with a Napier Gazelle turboshaft engine. Here the helicopter is tethered to the concrete apron during ground running. This helicopter was the first in the world to fly on turboshaft power. Philip Jarrett

it to gas turbine power instead of the piston power of a Wright Cyclone engine. It was aimed at satisfying the Royal Navy's requirement for an anti-submarine helicopter that could use a ship's fuel.

Napier's Gazelle engine had already been marinized and so was chosen as the powerplant. Unfortunately the Gazelle could not be fitted near to the rotor hub, as was desirable, due to the large diameter of its radial air intake and the time needed for the consequent design work. Therefore, because of the urgency of the job, the Gazelle was mounted in the nose where the former piston engine was and drove the angled shaft up to the rotor head. The nose of the S-58, or Wessex as it became named, was lengthened to preserve the centre of gravity and there were two exhaust nozzles at each side, below the cockpit windscreen. The air intake was beneath the extended nose section.

The Wessex had a raised section comprising the cockpit and a compartment for the rotor gearbox, brake and servo controls. Beneath this raised section was the cargo and personnel compartment, the floor of which contained crash-proof fuel cells. Two main wheels were braced from the forward fuselage and there was a castoring tailwheel. The aft fuselage led to an angled fin carrying the tail rotor and a tailplane. Armament comprised two homing torpedoes, four missiles and 2in rocket launchers or gun pods. Other equipment included a dipping sonar, Doppler and auto-stabilization.

The first Wessex (XL722) was a modified S-58 with an early NGa.11 Gazelle gas turbine of 1,100shp. It began ground running on 22 March 1957 and the maiden flight was made on 17 May 1957. Few problems were encountered but many minor modifications had to be made to the Gazelle installation before production was authorized. Altogether some 140 HAS.1 Wessex helicopters were built, with Gazelle 161 engines, followed by three HAS.3s with Gazelle 165 engines. Other versions were built for commando use and for export but with coupled Gnome engines. The Gazelle-powered Wessex saw service in the South Atlantic in 1982 during the Falklands War. One HAS.3 (XP142, 'Humphrey') flew from the destroyer HMS *Antrim* in April 1982, helped retake South Georgia, carried out the rescue of downed helicopter crews and attacked and crippled the Argentinian submarine *Santa Fe*. Following the attack on the submarine, on 25 April, it survived being shot up during the landing of marines.

Bristol Type 192 Belvedere twin-rotor helicopter

Less successful were Bristol's efforts in developing the twin-rotor Type 192 helicopter. This began as the piston-powered Type 191 for the Royal Navy but the project was cancelled in 1956. In April of that year, however, twenty-two Bristol Type 192s, powered by two Gazelle gas turbines, were ordered for the RAF. The Type 192 had a long, parallel-sectioned fuselage with a rotor above the front end and another on a pylon at the rear. It had four landing wheels. Each Gazelle engine was mounted on top of the fuselage, near to each rotor gearbox. The first flight of the 192 was on 5 July 1958 from Weston-super-Mare and was followed by nine pre-production machines, which were later modified to production standards for the RAF. During production of the Belvedere, as the Type 192 became known, the Bristol Helicopter Division was taken over by Westland. The Belvedere went on to give good service in theatres around the world until the end of the 1960s. Napier was also active in the fields of ramjets and rockets, such as the Scorpion, but it was absorbed by Rolls-Royce in 1962.

PART TWO

Germany

British turbojet development ran parallel with that in Germany. It too had its early pioneers who often dreamed up turbine schemes that were before their time since the technology and appropriate materials did not exist to allow their construction.

Joseph Wertheim of Frankfurt was granted a gas turbine patent (2023) in 1877 for an Atmospheric Gas Engine in which a mixture of gas and air was to be exploded in a turbine. The first patent (3944) covering an open combustion chamber and an explosion turbine to operate on gasoline was granted to Tecklenburg of Darmstadt in 1878. The earliest German patent (53322) for a constant-volume explosion turbine, with a closed combustion chamber in which a gas and air mixture was to be exploded, was granted to Christian Broeker in 1890. Soon August Rohrbach of Erfurt devised a constant-pressure type of gas turbine in which the turbine was to be driven by expanding air and he was granted a patent (8927) in 1896.

Many other patents, both in Germany and from other countries, formed a pool of ideas for others to study. Then, in 1908, Hans Holzwarth built the first German gas turbine. It was of the constant-volume or explosion type and was complicated. Although a few of this type were installed in industrial works, it was unsuitable for aeronautical use and never seriously competed with the constant-pressure type of gas turbine.

This and other early work did not seem particularly to influence Hans von Ohain, Germany's first successful pioneer of the turbojet engine. He began his studies in the early 1930s and his work led, as we shall see, to the flight of the world's first turbojet-powered aircraft, the Heinkel He 178, in 1939.

Less successful, but in several ways more advanced, was Herbert Wagner's so-called RT0 turbojet (it is thought the '0' stood for zero) at Junkers, which was worked on at the same time as von Ohain's engine. It was unfortunate that the geometry of this engine, with its advanced reaction, axial compressor and annular combustion chamber, was not correct enough to allow a successful run early in its short life. Wagner's engine was, no doubt, a couple of steps too far ahead and, as in Whittle's case, von Ohain took the surer course of using a centrifugal compressor to get early success.

Turbojet and turboprop development in Germany mostly took place just before and during the Second World War (1 September 1939 to 7 May 1945). Since 1932 the country had been dominated by Hitler's Nazi Party, a dictatorship under which few people were truly safe. Thanks to Hitler's fanaticism, one could be persecuted, incarcerated and even murdered for one's politics, religion or anything else not to the regime's liking. Consequently, many physicists and engineers of talent, particularly Jews, escaped from Germany in the early days of Nazi rule and their talents were thereby denied to the country. Bad as the situation was, though, it was arguably even worse in Stalin's Communist Soviet Union, another dictatorship that employed concentration camps and slave labour.

Following secretive beginnings in 1936 at the Heinkel and Junkers companies, the Reichsluftfahrtministerium (RLM; Reich Aviation Ministry) was soon giving its wholehearted support to turbojet development in Germany. Unlike the early days of Whittle's pioneering turbojet work in England, funding and facilities were not a problem. Soon the RLM had a national development programme worked out and the main engine companies were all involved.

One of the most important sections within the RLM was the Technisches Amt, which supported the research, development, production and testing of all sorts of jet and rocket engines and covered, of course, turbojets. The affable Ernst Udet was in charge of the Technisches Amt from 1936 until his suicide in November 1941, when Erhard Milch was appointed in his place. As the war progressed and Germany's situation deteriorated, the SS increasingly took control and particularly watched all politically unreliable physicists and engineers. Increasingly, the way that a development could obtain higher priority and thereby make progress was by having the interest and support of Hitler. Finally, the sinister SS General Hans Kammler became responsible for ruthlessly pushing through all technological developments that might turn the tide of war against the Allies.

In among all these machinations and turmoil, the figure of Helmut Schelp shines through as the Technisches Amt engineer who intelligently devised and steered the country's comprehensive turbojet development programme from its earliest days through to the war's end. In particular this programme had to take into account, in the design of its engines, Germany's chronic shortage of metals suitable for alloying heat-resisting steels.

With the German collapse at the end of the war, the country's technicians and engineers were not idle for long, being put to work, chiefly by the Soviet Union, the USA and France, in the continuation of turbojet and turboprop development. A valiant attempt was later made to revive the aircraft and engine industry of East Germany, under Communist domination. Many of the hopes for the country's economy depended on this revival, but ultimately it failed. German engineers also came together in teams in West Germany and pursued turbojet projects both at home and abroad, although most of these programmes failed in the end, largely due to the faster pace of development in Great Britain, the USA and the Soviet Union. This led to further dispersal of German engineers, whose talents were always in demand.

CHAPTER EIGHTEEN

Ernst Heinkel A.G.

Credit must go to Ernst Heinkel and his aircraft company, inspired by the physicist Hans-Joachim Pabst von Ohain, for launching turbojet development in Germany. However, unlike the de Havilland aircraft company in England, which at least had experience of building piston engines before embarking on turbojet development, before von Ohain arrived the Heinkel company only had experience of airframe construction.

The diminutive and entrepreneurial Ernst Heinkel was extremely keen to pursue innovations in aeronautics and was particularly keen to increase the speeds of his aircraft. Heinkel's experience with aircraft design went back to the First World War and he worked for various companies until he formed his own on 1 December 1922 at Warnemünde. New factories followed at Rostock (1932) and Oranienburg near Berlin (1937). In 1939 Heinkel took over the P.Z.L. factory at Mielec, Poland, and in 1942 his design staff were transferred to Vienna-Schwechat.

Heinkel's first jet aircraft were, in fact, rocket powered. Although not the first to experiment with rocket aircraft, Heinkel nevertheless put considerable effort into them and made the first flights with liquid-fuelled rockets (as opposed to powder rockets). As a first step, autumn 1936 saw an adapted Heinkel He 72 Kadet biplane trainer being flown with a single Walter rocket motor.

Hans von Ohain

From 1935 the 24-year-old von Ohain was, meanwhile, beginning to formulate his ideas on jet propulsion. He was studying physics at the University of Göttingen, where eminent professors taught mathematics, physics and chemistry but nothing in the way of applied technology. Von Ohain's professors included Robert W. Pohl and Frederick Courant. Also in Göttingen was the Kaiser Wilhelm Institute for Fluid Mechanics (KWI), which was staffed by such famous men as Prandtl, Betz, Encke and von Kármán. Although von Ohain attended some of Prandtl's and Betz's lectures on fluid mechanics and aerodynamics, he learned nothing about compressors or turbines at that time, or any other practical matters. Only later did von Ohain learn of Encke's work on the axial compressor and, probably, of von Kármán's vortex theories.

While studying for his doctorate, von Ohain began thinking about jet propulsion. He also became familiar with aircraft and was interested in aerodynamics by summer 1931. He joined the Academic Gliding Group and learned to glide. Later, he compared the smoothness of such flight with the vibration and noise of airscrew-driven aircraft:

> I thought the reciprocating engine was a horrible monster and that the propeller lacked elegance. I thought that it should be possible to drive an airplane fast and smoothly through the air like a very fantastically fast glider . . . That, I think, stimulated my interest in the turbojet. From then on I started to think about what better thermodynamic process could be considered.

Following his usual practice, he did a certain amount of thinking on his own before conducting a survey of the available literature, so he knew nothing of the ideas of Lorin, Guillaume, Whittle and others. On being told about his ideas for a turbojet engine, Pohl, with whom von Ohain's relationship was unusually good (for a German university, where professors were normally aloof), advised that: 'When you have something which really revolutionizes technology by using physics this is fine, but don't do mediocre stuff. That would not be good.'

The 'garage engine': the first demonstration turbojet

Von Ohain then went ahead and designed a demonstration model turbojet that, according to his recollection, was about 1m (3.28ft) in diameter and about 0.35m (1.15ft) wide. It was designed to be made largely of sheet metal, for ease of production and cheapness, and had a primitive single-sided centrifugal compressor, an annular combustion chamber with copper burners and an inward-flow radial turbine. A compression ratio of only 3:1 was thought adequate but, as we have seen, Whittle considered at least 4:1 was necessary. Von Ohain was no engineer and knew little about combustion and so he took his design to his local garage, Bartels und Becker, where the skilled and talented mechanic Max Hahn took on the task of building the model engine and putting much of his practical knowledge of engineering into it. The initial cost was DM500, which von Ohain borrowed from his father.

Taking the completed 'garage model' engine to the backyard of the Physical Institute of the University of Göttingen, von Ohain attempted to obtain self-sustained running from the engine, using petrol as the fuel. This was injected from a pressurized tank through spray nozzles and lit by a small igniter. Some 200kp (441lb) of thrust was expected but success eluded him. Most of the combustion took place on the exhaust side of the turbine, due to the inadequate length of the combustion flow path. Flames up to 3m (10ft) long emerged from the exhaust nozzle, with the engine run up to about 8,000rpm by an overheating electric motor. Apart from poor combustion, the efficiencies of the various components of the engine were far too low.

Professor Pohl was mildly amused by the desperate efforts of his young student and said: 'Why don't you first develop the complete theory of this thing?' Von Ohain said that he had already done that in a sketchy way, but he then worked out a full theoretical analysis. Pohl looked over this and said of it:

> Well, Hans, this is sound. It should work but you are not familiar enough with technology.

Hans von Ohain's first demonstration turbojet model (the 'garage engine'). These photographs and captions were supplied by von Ohain:

(A) This was the first turbojet engine built by Max Hahn in the automobile garage Bartels and Becker, Göttingen, in 1935. It was designed and financed by myself. Total cost somewhat below 1,000 Marks.

(B) Max Hahn with the model engine.

(C) Rotor: radial outflow compressor and radial inflow turbine (in order to show rotor blades, outer circular shroud removed).

You must have a bit more respect for the formidable knowledge in technology that you are certainly lacking. I will tell you what I will do for you. Tell me which company you would like to work with on your engine and I'll write a letter of recommendation for you to its director.

Von Ohain replied:

Well, you know, when I studied in Rostock, I became familiar with the Heinkel company. Ernst Heinkel is the owner of the company and he has the reputation of being a little bit crazy. I am considered to be crazy too so maybe we'll make a good team!

Professor Pohl would have suggested the Messerschmitt company, on account of his family connections with Willy Messerschmitt, but in February 1936 he wrote a short letter of recommendation to Ernst Heinkel, which the latter received on 3 March. On 17 March 1936 von Ohain took his notes and photographs of his garage engine to a meeting set up by Heinkel. As he enthusiastically presented his ideas to Heinkel and some of his engineers and designers, he frankly admitted that so far his engine would not work. This presentation received a mixed reception from Heinkel's staff but, following a report from Siegfried Günter, Heinkel decided to go ahead with a turbojet project. Accordingly, on 15 April 1936 he hired von Ohain and the mechanic Max Hahn to form a very small research team under the supervision of Dipl.-Ing. Wilhelm Gündermann. Experienced in construction and engineering drawing, Gündermann had studied practical aircraft construction at the Technisches Hochschule in Berlin and also turbo machinery construction under Prof. Hermann Fottinger.

This small team was set up in a shed in a fenced-off, secret section of the works and was grandiosely known as Sonderentwicklung II or Special Development II. (So.I was concerned with secret airframe development.) The team's starting point was to ascertain why the garage engine would not run in a sustained manner. At first, using an electric motor, they ran the engine cold up to about 10,000rpm and took pressure readings. Then, a little petrol was used in hot tests. This time, although the combustion was not right, no flames emerged from the exhaust nozzle. It was discovered that the air flow from the compressor to the combustion chamber was very turbulent and there was even some backflow, which was causing a pressure drop so that not even a compression ratio of 2:1 was being realized. Following these early tests at Heinkels, which took about two months, it was decided that the cross-sectional area of the turbine was too small and that it needed inlet guide vanes.

A review of patents was then made and it was noticed that Whittle's 1930 patent indicated an axial turbine. However, due to the complexities of developing axial turbines and compressors, these could not be contemplated at that time. This was because Ernst Heinkel was impatient and wanted results as quickly as possible: he did not care whether the turbojet was going to be any good or not, so long as his company flew the first turbojet-powered aircraft. On the other hand, von Ohain's aim was to spend about a year at Heinkel's plant, see his ideas proved and a turbojet engine flown, and then return to an academic life of pure research physics as an assistant to Prof. Pohl.

At about this time, because Heinkel had no special interest in engine patents, von Ohain went off to the patent office in Berlin and made a patent search. The German patent examiner did not find Guillaume's French patent but did find Whittle's. Von Ohain, however, was granted two or three patents because his ideas differed from Whittle's.

HeS 2 demonstration turbojet

Armed with information from tests on the garage engine, it was decided in June 1936 to design another demonstration engine, which became designated as Projekt 2 or the HeS 2. Heinkel was pushing for a flight engine as early as possible, but von Ohain told him that a sheet metal engine could not develop enough thrust and, besides, the great problem of combustion was still to be solved. In addition, there were no engine accessories or controls available. Gündermann drew up a new design. Sheet metal construction was again chosen as suitable for Heinkel's facilities, but with more sophisticated curves than the first garage engine. A local shipyard was brought in to help. Here sheet metal was pressed over hardwood formers and then riveted together after the flanges had been machined true. Compared with the garage engine, a greater spacing between the compressor and turbine was used in order to give less abrupt passages between them and better flow characteristics through the combustion chamber. Therefore, the inside faces of the compressor and turbine were bolted to a short, hollow rotor.

The compressor was a simple centrifugal type, preceded by guide vanes. The simple annular combustion chamber burned hydrogen, fed from a battery of hydrogen bottles, sufficient for two to five minutes of running time. Inlet nozzles directed the hot gases into the radial-inflow turbine and there were anti-friction bearings in front of the compressor and behind the turbine.

In this form the HeS 2 first ran in spring 1937, but the hydrogen combustion caused considerable burning of the metal. In order to solve the problems of running on petrol, von Ohain and Hahn worked night and

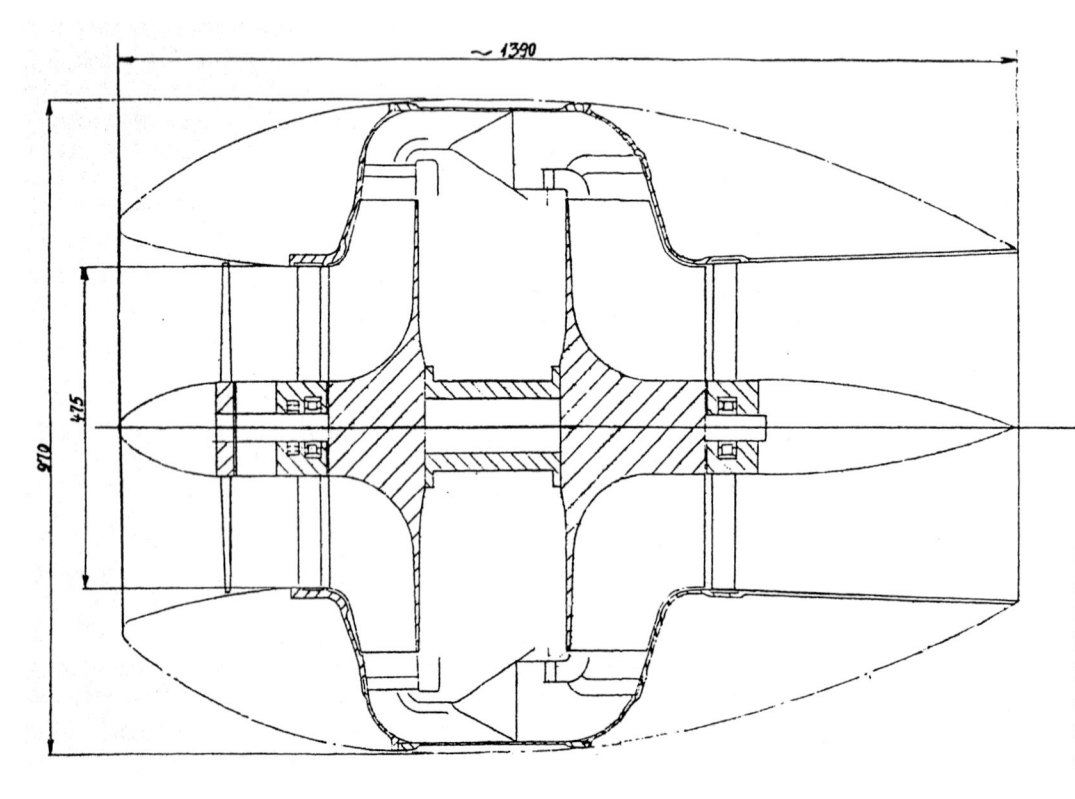

Dipl.-Ing. Wilhelm Gündermann's design sketch for Projekt 2 (or the HeS 2) turbojet. It was an improvement aerodynamically on the first 'garage model'; the compressor and turbine are spaced wider apart to provide a longer combustion space.

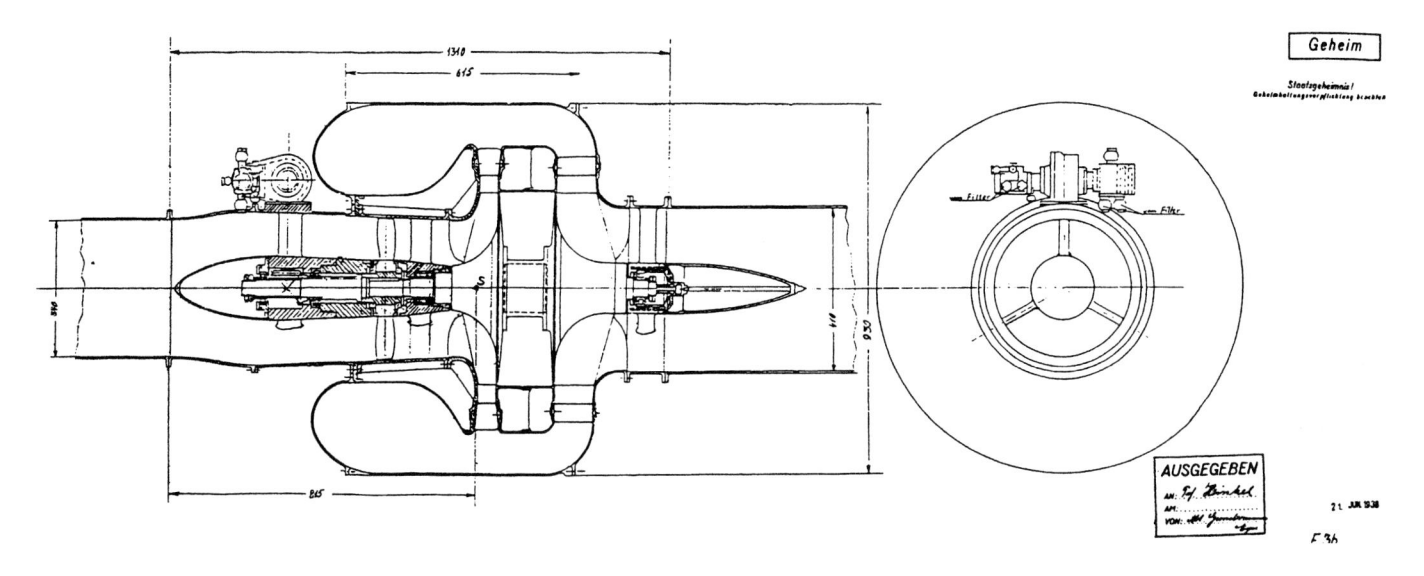

Gündermann's first layout for the F3B (HeS 3B) turbojet, designed to power the Heinkel He 178. The combustion chamber has been further lengthened by using the space in front of the centrifugal compressor. Also, secondary and primary air are now used in the combustion system.

day. Eventually Hahn came up with a scheme inspired by his gas soldering torch. In this, a boiler-type petrol vaporizer produced vapour that could be burned in the combustors, even though all the engine's airflow passed through the combustion chamber.

The first run of the HeS 2 using petrol (or gasoline) was made in September 1937, but, as von Ohain later said, 'That gasoline combustor was of a horrible design and clogged up quickly'. Nevertheless, owing to the lightweight rotor's small moment of inertia, the engine's throttle response was impressive and the feasibility of the turbojet was proven. Heinkel therefore decided that the company should press on with the development of a flight engine, designated HeS 3, and a demonstration turbojet aircraft, the He 178. Approximate data for the HeS 2 demonstration turbojet included a thrust of 136kp (300lb) at a speed of 10,000rpm. Its length was 1.39m (4ft 6¾in) and its overall diameter was 0.97m (3ft 2¼in); this large diameter was caused by the combustion chamber being outside of the compressor and turbine.

HeS 3 flight engine

The optimistic requirements for the HeS 3 flight engine included a reduced diameter, a petrol combustion system and a static thrust of 800kp (1,764lb). In truth, these aims exceeded the resources available. Instead of systematically developing the components and accessories with specialist companies and RLM support, Heinkel decided to keep them all out at this stage for the sake of secrecy. Consequently components had to be developed or selected by relatively inexperienced engineers. Heinkel's aim, however, was not to produce a production engine but simply one that could make the first turbojet flight. From this he expected to obtain enough kudos to enable him to purchase an engine factory and full RLM support.

Von Ohain spent much time studying combustion and in visiting industrial and scientific establishments, much as Frank Whittle did in England. Problem-solving was not helped by the fact that there were no facilities at Heinkels to test the compressor, combustion chamber and turbine elements separately. The first example of the HeS 3A was being bench-run around March 1938 but with poor results. Flight tests followed in May or July 1939 with an HeS 3A suspended beneath a Heinkel He 118 (D-OVIE), which had a conveniently long undercarriage giving good ground clearance for the test engine. The first flight tests, in the interests of secrecy, were begun just before sunrise, with landings made before the factories opened. The pilot was Erich Warsitz and the turbojet was switched on and monitored by engineer Kunzel following a take-off with the piston engine alone. Burning petrol resulted in a bluish exhaust jet from the turbojet, although the engine still had to be started with hydrogen. After several flight tests, one of the HeS 3As being tested caught fire and was destroyed after landing, the fire being caused by a leaking fuel line.

The attempt to minimize engine diameter resulted in an undersized compressor and poor combustion so that insufficient thrust was developed. The turbine soon burned out. While construction of the He 178 test aircraft was proceeding, the turbojet was redesigned as the HeS 3B. The extra engineers available to von Ohain at this time were, in the main, 'those others were glad to get rid of'. Max Hahn concentrated on the combustion problem. At one point he suggested using the free space in front of the compressor to create a longer route for the air and fuel mixture and also, very importantly, to bypass some of the air past the combustion zone. The essential idea of using primary and secondary air in the combustion process had thereby been arrived at and a patent was taken out in Hahn's name by the end of 1937. Combustion trials were carried out in secrecy at night using Heinkel's wooden wind tunnel, using hydrogen as fuel.

At some time during 1938 Heinkel decided to let Helmut Schelp and the higher officials in the RLM see the work going on with the turbojet. Schelp was very enthusiastic, but enmity within the RLM against Heinkel was aroused since minor officials were excluded. Heinkel, who had

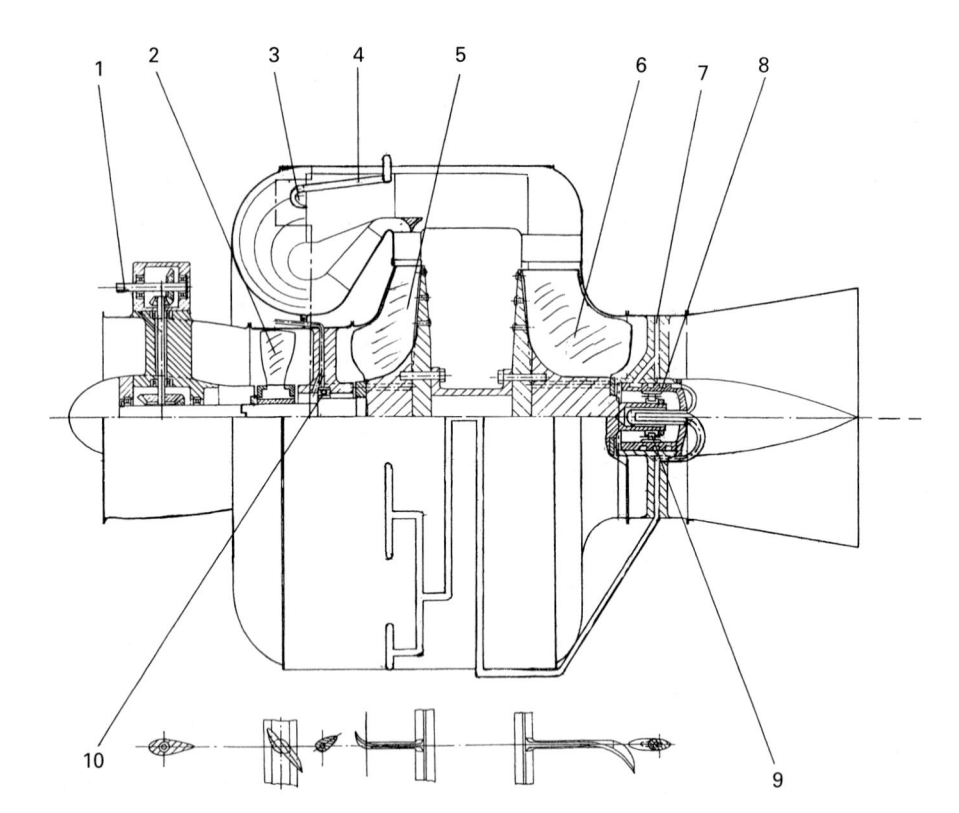

ERNST HEINKEL A.G.

directed inwards past the fuel burners to cool the inner wall of the combustion and mixing chamber.

A fuel vaporizing system was used, the fuel tubes being heated by combustion gases. The fuel was first pumped around spiral grooves to cool the roller bearings at the rear of the turbine. Heat-resisting steel with 38 per cent nickel content was used for the combustion chamber plates and vaporising tubes. Grease was used to lubricate the turbine roller bearing and, for the ball bearing in front of the centrifugal compressor, oil under pressure in a container was used, there being no recirculating oil system.

The radial-inflow turbine had de Laval bulb roots for its fourteen blades and curved inlet nozzles directed the hot combustion gases at right angles to the blade tips. Krupp's P.193 heat-resisting steel was used for the turbine and heat-resisting cast steel for the turbine housing and hub. There was an electric speed indicator behind the hub at the front of the rotor and a thermal

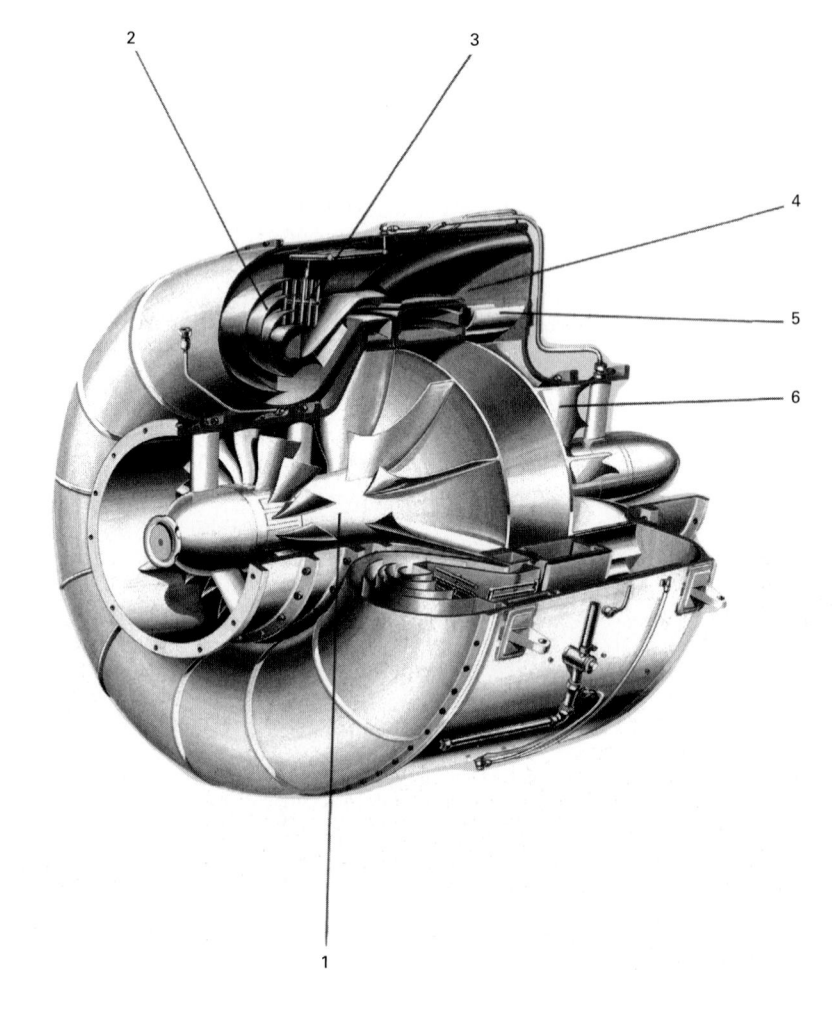

The Heinkel HeS 3B turbojet (based on Gündermann's original schematic drawing): (1) starter drive; (2) axial inducer; (3) fuel jet; (4) pipe grid (spray bar); (5) centrifugal compressor; (6) radial-inflow turbine; (7) fuel inlet; (8) spiral groove to cool bearing housing with fuel; (9) rear roller bearing; (10) front ball bearing.

considerable power and a wealthy company, turned down the money and engineers that the RLM could provide in order to keep full control of his projects.

By summer 1939 two HeS 3B turbojets were showing promise. The engine had a centrifugal compressor and diffuser made of heat-resisting 'Duralumin-W', its sixteen blades being held by de Laval bulb roots. This was preceded by a Duralumin eight-bladed axial inducer. From the compressor, most of the airflow was directed forward and then reversed with the aid of curved guide vanes or cascades. The airflow followed two paths, the inside path leading into the annular combustion chamber and the outer path being used to cool the outer wall of the combustion chamber. A third airflow path left the compressor and was not reversed but

The HeS 3B, Heinkel's first flight engine: (1) axial inducer; (2) ducts; (3) pipe grid and spray bar; (4) combustion chamber; (5) inlet nozzles; (6) radial inflow turbine.

sensor behind the turbine to read the exhaust gas temperature. There seems to have been no attempt to economize on heat-resisting steels in the design.

For starting, air was blown onto the turbine blades and the burner vaporizer pipes were heated with hydrogen until hot enough to vaporize the petrol, at which point the petrol was turned on. Two fuel pumps were used, driven by an electric motor and onboard battery, the pumps running at maximum speed all the time. To control the fuel flow and hence engine speed, the throttle was used to operate valves differentially on the fuel lines to the engine and the backfeed to the fuel tank.

He 178, the world's first turbojet-powered aircraft

From mid-1938 construction of the He 178 aircraft was under way at Sonderentwicklung I, where the He 176 rocket plane was also under construction. (The He 176, designed by Hans Regner, was first flown on 20 June 1939 with Erich Warsitz at the controls.) The He 178 was designed largely by Hans Regner and built under the direction of Heinrich Hertel, Walter Günter and his brother Siegfried Günter. When Walter Günter was killed in a car accident in September 1937, his place was taken by Heinrich Helmbold.

The He 178 was a small, shoulder-wing monoplane with the pilot's cockpit well forward of the wing leading edge. The wings were largely of wooden construction with very slight dihedral and an equi-taper planform, and were mounted about halfway along the fuselage. A symmetrical, Günter profile was used for the wings, which had a thickness/chord ratio of 12 per cent at the root with a linear taper down to 7 per cent at the tips. Controls were conventional and included large inboard trailing edge flaps. The fuselage was of Duralumin, stressed-skin, monocoque construction and enclosed a single fuel tank behind the cockpit, followed by the turbojet just behind the wing's main spar. Mounting of the turbojet was by means of four brackets attached to the main airframe longerons. Various air intakes were considered, including side intakes, but a nose intake was chosen. From this led a duct that curved under the cockpit and fuel tank to the engine.

At one stage a variable-area exhaust nozzle was considered but not proceeded with. Since a design speed of 800km/h (497mph) was aimed at, it was intended that the tailwheel undercarriage should be hydraulically retracted into the fuselage. However, there was no time to develop this system: as photographs indicate and von Ohain confirmed, the undercarriage was left down and its fuselage openings were faired over. As a back-up in case of an accident, a second prototype of the He 178 with a larger wing, V2, was begun but probably not completed. Perhaps inevitably, the He 178 was similar in layout to the Gloster E.28/39, the latter differing chiefly in its bifurcated air ducts from the nose, its low-mounted wing and its tricycle undercarriage. The Gloster aircraft was slightly larger than the Heinkel machine, but was lighter when empty and a great deal heavier when loaded, reflecting the fact that the Gloster aircraft was designed for flights of some endurance rather than just a few circuits.

On 28 August 1938 a mock-up of the He 178 was inspected by Heinkel and those chiefly responsible for the project. Discussions took place regarding desirable modifications, including cockpit and canopy alterations, control adjustments, the linking of the flap activation valve with the throttle lever and provision of an escape hatch in the right side of the cockpit. A planned hydraulic pump for flap and landing gear actuation was deleted since there was no power to spare from the engine to drive such an auxiliary. Instead, a compressed air cylinder was deemed sufficient for flap and undercarriage operation, although, as mentioned, the undercarriage was left fixed down.

By May 1939 the He 178 V1 was complete and the improved HeS 3B engine was soon ready for installation. The static thrust of the HeS 3B had been coaxed up to 450kp (992lb), but once installed in the aircraft this was reduced by about 15 per cent due to losses in the long exhaust duct. Improvements obtained with the HeS 3B were largely brought about by modification of the curved guide vanes between the compressor and combustion chamber and the turbine inlet nozzles, these being readily replaceable.

The weight of the HeS 3B was 360kg (795lb), its diameter was 0.93m (3ft 0½in) and length 1.31m (4ft 3½in), excluding inlet and exhaust ducting. At 13,000rpm its compressor's pressure ratio was 2.8:1, its air mass flow was 22.5kg/sec (49.6lb/sec) and its specific fuel consumption was 2.16.

Erich Warsitz taxis out and takes off on 27 August 1939 in the Heinkel He 178, powered by an HeS 3B turbojet, to make the world's first flight of a turbojet-powered aircraft.

Before the He 178's first flight, it was ordered to E-Stelle (Erprobungsstelle; 'testing or trial station/proving ground') Rechlin in June 1939 to be shown to Hitler, Goering and others. On a fine morning Erich Warsitz demonstrated the He 176 rocket plane in an extremely impressive flight. He then showed the He 178 to Hitler and the others and started up the engine, while von Ohain explained what it was all about. Hitler asked why a new aero-engine was necessary and appeared very unfriendly and cold, but it was later explained that he did not like to get up before 11.00 am!

In spite of its improvements, the HeS 3B was very far from being a reliable flight engine and would not have passed even a special category type test. Nevertheless, Heinkel insisted on getting the He 178 into the air and so Warsitz began taxiing trials, which culminated in a short hop along the runway at Marienehe on 24 August 1939. Then, early on the morning of 27 August, with the sun still quite low, Warsitz took off in the He 178 V1. He tells us that he tried to retract the landing gear but that 'something was wrong with it': perhaps, when pulling the lever on the instrument panel, he forgot that the retraction system was never completed. It has been said that the He 178 flew at 600km/h (373mph), but it is unlikely to have exceeded about 300km/h (186mph) due to the undercarriage being unretracted. In any case, Warsitz had been advised to hold the speed down and the engine was only allowed to run for six minutes. We are told that two large circuits of the airfield were made. By the time Warsitz was making his landing approach, one of the fuel pumps had broken but the engine was still controllable. To lose height quickly and get into the small Marienehe airfield,

Ernst Heinkel (left) and Hans von Ohain at a small celebration following the flight of the Heinkel He 178.

Warsitz had to sideslip the He 178 and he then made a smooth landing. To the joy of Heinkel and those watching, the world's first flight of a turbojet-powered aircraft had been made. Despite the technical shortcomings of the engine, all credit is due to Heinkel and von Ohain and his team for this significant event.

HeS 6 turbojet (109–001)

Heinkel now asked for a more powerful version of the HeS 3B to be developed. This was designated HeS 6. Thinking that the time was right to work with the RLM, he invited State Secretary (and head of the RLM) Erhard Milch, together with Ernst Udet, Wilhelm Lucht and others, to witness the He 178 in flight. This took place on 1 November 1939, after a delay caused by another fuel pump fault. Warsitz gave a spirited demonstration that included 'buzzing' the prominent visitors. Strangely, and much to the disappointment of Heinkel and his staff, these visitors did not appear impressed. Privately, however, they were and the RLM redoubled its efforts to get a reluctant German aero-engine industry interested in the turbojet. It is said that the He 178 V1 was flown about fifteen times altogether before being sent for display in the Berlin Air Museum. Unfortunately, this historic aircraft and the He 176 were both destroyed in a bombing raid in 1943.

Modification of the HeS 3B engine produced the HeS 6, which gave a static thrust of 550kp (1,213lb) at 13,300rpm, although the weight went up to 420kg (926lb). However, the fuel consumption, at 1.6, was much improved over the HeS 3B, so progress was being made. The HeS 6 was test flown beneath an He 111 bomber before the end of 1939, but this engine was not worked on for long since attention turned to developing a production turbojet, designated the HeS 8. This was the first turbojet from Heinkel, or any other company, to receive official backing and it received the RLM designation 109-001.

He 178 V1		
Span	7.20m	(23ft 3½in)
Length	7.48m	(24ft 6½in)
Wing area	7.90sq m	(85.0sq ft)
Height	2.10m	(6ft 10½in)
Empty weight	1,570kg	(3,462lb)
Loaded weight	1,950kg	(4,300lb)
Take-off distance	500m	(1,640ft)
Estimated maximum speed but probably about 300km/h (186mph) with extended undercarriage	600km/h	(373mph)
Landing speed	220km/h	(137mph)

CHAPTER NINETEEN

Towards an Official Programme

The Verkehrsministerium issued Germany's first official jet engine research contract from 1931, this being to Paul Schmidt of Munich who was working on pulsejets. This organization was replaced on 29 April 1933 when the RLM was established, headed by the Reichskommissär for aviation, Hermann Goering. From 1935 the Technisches Amt was set up within the RLM; this supported not only Schmidt's work but also, from 1937, the rocket and ramjet work of Helmut Walter at Kiel. In 1936 Ernst Udet was ordered to take charge of the Technisches Amt as its Chief Engineer, replacing Günther Bock. However, Udet, an excellent pilot, was not an engineer.

Helmut Schelp

In August 1937 Helmut Schelp was sent to work at the research section (LC1) of the Technisches Amt, from where he oversaw the work of Schmidt and Walter. Schelp was to become one of the key figures in Germany's turbojet programme. Following studies in Dresden, he received a master's degree in engineering at Stevens University at Hoboken, New Jersey, in June 1936 when he was twenty-four years old. The reason why Schelp studied in the USA was that it was a condition that a student had to spend at least one year studying abroad before enrolling in a new, advanced course in Germany. The subject of his Master's thesis at Stevens, for which he won first prize, was an investigation into the anti-knock effects of alcohol additions to petrol.

Returning to Germany in 1936, Schelp took the advanced course run by the Deutsche Versuchsanstalt für Luftfahrt (DVL; German Experimental Institute for Aviation) in Berlin and was trained in all aspects of aeronautical engineering. He also trained as a pilot. The aim of this course was to provide a much-needed nucleus of skilled personnel for positions of authority. Schelp emerged from the course with the technical degree of Diploma Engineer and, for entry into government service, the rank of Flugbaumeister. One of the requirements for a Flugbaumeister was that one had to be a pilot with a wide experience of aircraft types. Schelp's interest in aviation went back to his earliest years: when he was 17 years old, he and a few friends scraped together enough money to build their own glider.

For the advanced training part of the DVL course, Schelp chose to concentrate on aircraft powerplants. His first general assignment in this field was to investigate how contemporary aircraft speeds could be doubled. He quickly realized that the reciprocating engine and airscrew combination would soon approach its limit and that to try to push it further would give only diminishing returns due to airscrew inefficiencies and unacceptable engine weights. The limit for this system was thought to be Mach 0.82 or 1,006km/h (625mph) at sea level. Schelp therefore made a systematic study of jet propulsion engines, including rockets, ramjets and turbojets, and of turboprops, ducted fans and other hybrid types. Of course, most of these types of engine did not exist at that time, although the Walter company at Kiel was developing the rocket for assisted take-off (ATO) use and had begun experimenting with small ramjets.

As early as November 1930 Ferdinand Porsche was sketching ideas in his notebook for an exhaust gas turbine with water cooling and for a turbojet with a radial-flow compressor. Porsche's attentions, however, were directed exclusively to vehicle construction for the military and only in the later stages of the war could he return to the turbojet (see Chapter 30).

On the other hand Helmut Weinrich, a freelance engineer from Chemnitz, submitted plans to the RLM in 1936 for a very advanced scheme for a turboprop engine using a contra-rotating axial compressor and turbine (see Chapter 22). Weinrich's work, however, does not appear to have come to the attention of the Technisches Amt or, at any rate, to have received any official acknowledgement or sponsorship. It was only several years later that Hans A. Mauch learned of Weinrich's proposal by chance and, with official backing, introduced him to the Bramo engine company.

Because of his powerplant studies at the DVL, Schelp was asked to spend more and more time at the LC1 Forschung (research) department of the Technisches Amt. In September 1938 he was transferred to the LC8 engine development department at the request of Mauch, who had just taken over the department but, though a very good engineer and very methodical, had no knowledge of jet propulsion theories.

Schelp and Mauch then set about trying to get industry interested in developing the new powerplants, particularly the turbojet, but without success. German industry had its hands full making up ground lost to the USA and Great Britain in developing and producing powerful piston engines, following the restrictions on such developments after the First World War. Neither the Technisches Amt's LC1 nor LC8 knew of Heinkel's early turbojet experiments until Schelp and other higher officials were invited to see von Ohain's work in 1938. Once the He 178 had made its historic first flights in 1939, the German aero-engine industry became interested in the turbojet and the RLM was able to use Heinkel as a lever to get development moving. At the same time Flugbaumeister Hans Antz of the LC7 airframes section tried to get the aircraft companies interested in developing jet aircraft. Messerschmitt was the first to take an interest. Like Schelp, Antz had also been an exchange student to the USA, obtaining his Master's degree at the California Institute of Technology.

Because of the then radical nature of jet engines, Mauch considered that their development should be entrusted to established engine companies and not to airframe companies such as Heinkel and Junkers. Together with Schelp, he visited and consulted with the companies of BMW, Bramo, Daimler-Benz and the Junkers engine division. When, thanks to Heinkel's research demonstrations, these companies began to take an interest, there was much objection to the technical difficulties they faced in developing the axial compressor. Schelp insisted upon the axial compressor, since he foresaw minimal frontal area being of paramount importance. This was all the more so because a twin-engined aircraft with engines below the wings was considered the logical way to go in order to carry a useful payload with the engine sizes then in prospect. This view was influenced by the fact that the fuselage in the He 178 research aircraft had been taken up by the engine and its ducting, leaving no room for any payload.

Schelp was assured that axial compressor difficulties could be overcome by the promising research being conducted by Ludwig Prandtl, W. Encke and Albert Betz at the Aerodynamische Versuchanstalt (AVA; Aerodynamic Research Institute), Göttingen, and these gentlemen were most cooperative. Official favour was also given to the annular combustion chamber but not insisted upon. On the other hand, the great shortage of chromium, silicon, molybdenum, nickel and manganese with which to make heat-resisting steels had a highly significant effect on German turbojet design and led to a great reliance on air cooling of the turbine and other components. Such supplies of these alloying metals that were available, particularly chromium, went largely to the armaments industry.

By the end of 1939, encouraged by Heinkel's early successes, Mauch had launched a programme of jet engine study and development. This programme was opposed by Wolfram Eisenlohr, head of LC8, but he was overruled by Ernst Udet, head of the Technisches Amt. However, no sooner was the programme under way than Mauch, who could not work under Eisenlohr, left the Technisches Amt to work in industry. He was replaced by Ernst Beck, since Schelp was considered too young for the position.

The first companies to have turbojet engines sponsored by the official programme were Heinkel, BMW and Junkers. The sponsored engines were Heinkel's 109-001 (HeS 8), BMW's 109-002 and 109-003, and Junker's 109-004; only the last two of these achieved production status. Schelp's carefully worked out development programme, which was supposed to span some sixteen years, envisaged the progression shown in the following table (p. 177).

It should be noted that, after the Class I turbojets had been developed, turbojets were also to be developed into turboprop versions by the addition of an extra turbine stage and suitable gearing for an airscrew. The same basic engines were also to be developed into ducted fan engines to cater for ranges in between those economical for the turbojet and turboprop. Walter Brisken and Emil Waldmann, key members of Schelp's department, also carried out many studies of turboprops with heat exchangers in pursuit of maximum economy over long ranges.

Comparison of sizes of German Class I turbojet engines: (a) 1941 BMW P.3304; (b) 1941 BMW P.3302; (c) 1944 BMW 109-003A; (d) 1941 Junkers 109-004 V-series; (e) 1942 Junkers 109-004 A-0; (f) 1943 Junkers 109-004 B-0.

| Official development programme ||||||
| Class | Approx. thrust (kp) for turbojet | Pressure ratio | Turbine stages ||
			Turbojet (TL)	Turboprop (PTL)
I	Up to 1,000	3.5:1	1	–
II	1,300–1,700	5.0:1	2	3
III	2,500–3,000	6.0:1	2	3
IV	3,500–4,000	7.0:1	3	5

The official jet propulsion programme gained much support from Ernst Udet, who was always enthusiastic about radical aeronautical developments. His position as head of the Technisches Amt enabled him to overcome any opposition to turbojets and other developments but, following his suicide under great pressures on 17 November 1941, decisions on development and production were taken by Erhard Milch. Milch was another who was not initially in favour of jet aircraft and preferred to see the large production figures attainable for conventional aircraft, since this was the way to impress Hitler at that time. The impetus that might have led to the early introduction of jet aircraft slowed down, and it was not until November 1943 that another proponent of jet aircraft, the competent Oberst Ing. Siegfried Knemeyer, was made responsible for airframe and engine development (known from 1941 as, repectively, GL/C-E2 and E3) at the Technisches Amt.

Technically qualified and a pilot with operational experience, Knemeyer insisted that a speed advantage of at least 150km/h (93mph) over enemy aircraft was essential in view of Germany's critical air defence position. He therefore proposed the abandonment of most conventional aircraft in order to concentrate on jet fighters and bombers. Nevertheless, Hitler, who was badly advised by Goebbels and Bormann, insisted on continuing largely with piston-engined aircraft, the production of which soared to record heights. After September 1944, when Messerschmitt Me 262 jet fighters became operational, Hitler realized their importance, but by then the shortage of pilots and fuel, together with the overwhelming numbers of Allied fighters and bombers infesting German skies, meant that all advantage was lost.

CHAPTER TWENTY

Heinkel and the Struggle for a Production Turbojet

The HeS 8 was a centrifugal turbojet aimed at giving a static thrust of about 700kp (1,544lb) for a weight and diameter of less than that for the HeS 3B/HeS 6 series. As the first turbojet to receive an official contract, it was given the RLM designation of 109-001. By abandoning the previous reverse flow of air to the combustion chamber in favour of a straight-through flow, it was possible to have the combustion chamber behind instead of outside the compressor and thereby give a smaller diameter. Von Ohain had planned to enter into axial turbojet development, as favoured by Schelp, but Heinkel was ordered to keep von Ohain on advanced development work on centrifugal engines.

The design of a twin-jet fighter, the He 280, to be powered by HeS 8 engines, was also put in hand at this time and all this was now being financed by the RLM. There was considerable lobbying, however, mainly by Schelp, for Heinkel to acquire an engine company for turbojet development and this accorded very well with Heinkel's hopes. The only company possibly open to such an acquisition was Hirth-Motorenwerke at Zuffenhausen, Stuttgart. Mauch did not like the primitive conditions for engines at Heinkel's factory and tried to persuade him to hand over his turbojet team to the Daimler-Benz company, where suitable test and manufacturing facilities, plus engineers, could be made available. Mauch intimated that support for Heinkel's work would be withdrawn unless his wishes were complied with.

Not surprisingly, Heinkel stood his ground and complained about the situation at a meeting at the RLM. This resulted in full support from Eisenlohr and Udet, who was friendly with Heinkel. On 18 October 1939 Udet sent Heinkel a telegram setting out this support and the financing of the company's work on jet flight. Heinkel was to have access to scarce materials and skilled labour, was not to suffer any more interference from RLM officials, and was to be answerable only to Schelp and Lucht. By autumn 1939 Mauch had left the RLM and so was no longer a factor anyway. The way was now clear for Heinkel's development of various turbojet engines and the He 280 jet fighter. Also, the RLM cleared Heinkel's purchase of Hirth-Motorenwerke – providing that the first flight of the He 280 was made by April 1941.

On 20 June 1939 work began on the mock-up of the He 280 fighter. Within two months a start was made on detail designing under the direction of the company's new Technical Director, Robert Lusser. Regarding engine development, matters were not improved by the diversity of engine types begun without anything like sufficient engineers or facilities. This made the acquisition of the Hirth company even more necessary. By the end of 1939 only about 120 engineers were working on jet engines and aircraft at Marienehe. October had seen a group of eighteen arrive from Junkers at Magdeburg, where they had been part of a team working on an axial turbojet under the talented Herbert Wagner. Their leader was Max Adolf Mueller, former assistant to Wagner, and they continued the work started at Junkers, including the HeS 30 (or 109-006) axial turbojet, the HeS 40 constant-volume turbojet and the HeS 50 and HeS 60 ducted fan, motor jet engines. Von Ohain's group continued working on centrifugal turbojet developments and he was in overall charge of all engine groups.

HeS 8 (109-001) turbojet

Von Ohain was doubtful that Mueller's HeS 30 axial engine would make any fast progress, owing to its complexity, but thought his own HeS 8 engine could be developed in about fourteen months. By April 1940 the first HeS 8 was ready for testing on the bench, but the first runs revealed poor combustion with unequal temperature distribution. It was hoped to achieve acceptance run-ups with the V2 prototype and flight tests with the V3 and V4 prototypes in the He 280, but the engines were unreliable and low on thrust.

The HeS 8 had a nineteen-blade centrifugal compressor, preceded by a fourteen-blade axial inducer in the intake. The inducer blades were of aerofoil section and were forged from aluminium alloy. The compressor blades, also of aluminium alloy, were retained in a steel hub and riveted to a rear disc. The fourteen-blade radial-inflow turbine was of similar construction to the compressor but the blades were in steel. A drum-type rotor with flanged ends connected the compressor and turbine components by bolting into the disc and hub at each end. Anti-friction bearings consisted of a roller bearing behind the turbine, two ball bearings in front of the compressor, and ball bearings in front of the inducer.

Air from the compressor passed through two sets of diffuser vanes before entering the annular combustion chamber. Fuel was injected into the combustion chamber through sixteen sets of eight nozzles (giving 128 nozzles in all), each nozzle being a small tube. Half the nozzle tubes were of a different length to the others so that fuel could be sprayed onto two annular rings at the combustion chamber entrance. These rings were of steel and had grooved surfaces from which the fuel vaporized, mixed with the air and burned. Not until the V16 prototype was there division of the air into primary and secondary flows.

The main casing of the engine consisted of castings with external ribbing for increased rigidity. The exhaust nozzle was of the fixed-area type and all auxiliaries were grouped around the smaller diameter of the air intake section. Drive shafts for the auxiliaries led out through streamlined

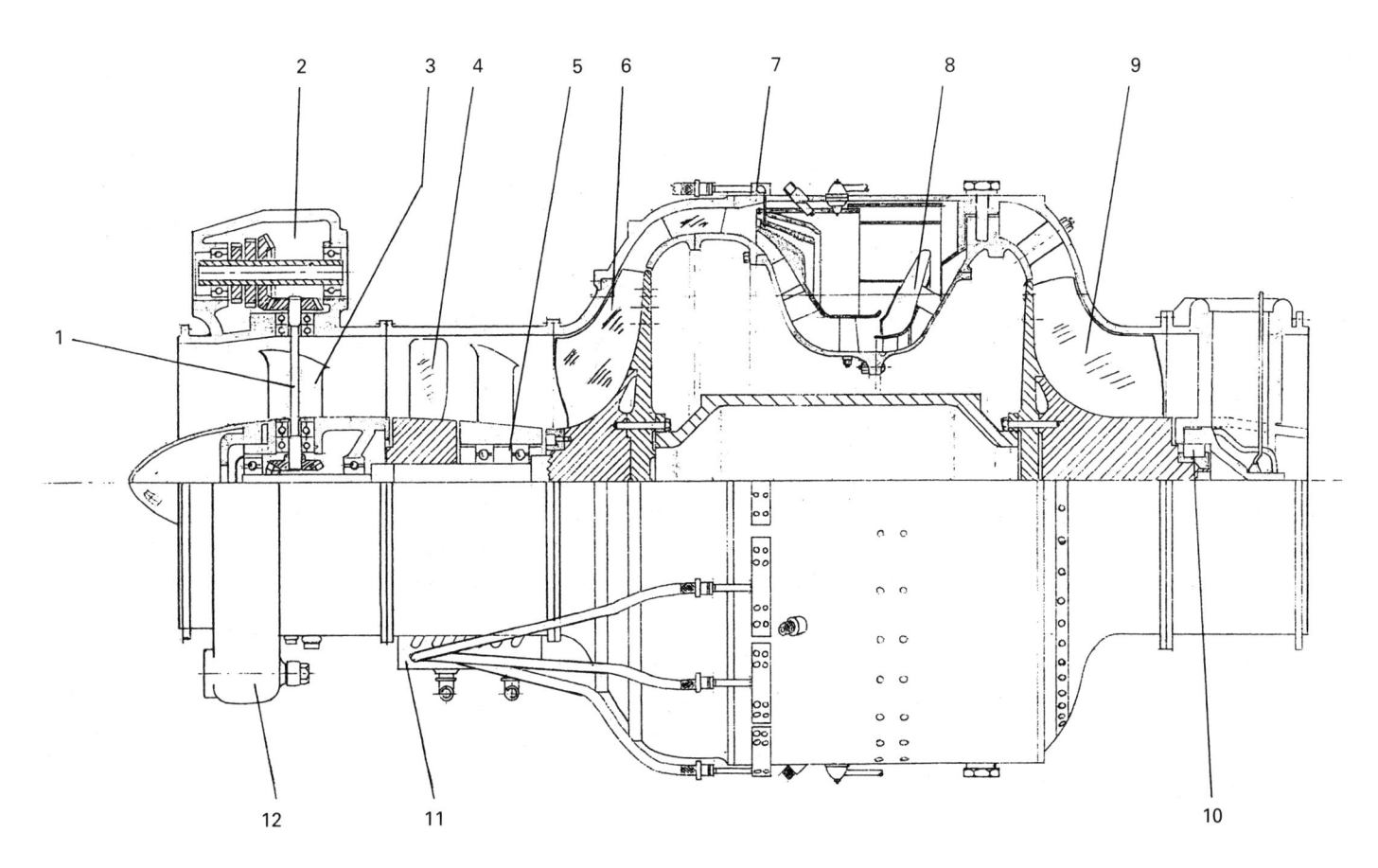

Heinkel HeS 8 V16 turbojet with primary air to combustion chambers and secondary air: (1) drive shaft; (2) auxiliaries gearbox; (3) front support vanes; (4) axial inducer; (5) front ball bearings; (6) centrifugal compressor; (7) fuel injector; (8) sandwich secondary air mixer; (9) radial inflow turbine; (10) rear roller bearing; (11) fuel pump; (12) auxiliaries oil tank with pump. Author

Heinkel HeS 8A centrifugal turbojet: (1) rear mounting; (2) forward mountings; (3) sparking plug cables; (4) control rods; (5) starter motor; (6) oil pump; (7) oil return compressor; (8) rpm indicator; (9) oil feed pipes; (10) air inlet duct; (11) inlet cone; (12) ignition magneto; (13) oil sump; (14) exhaust duct; (15) fuel rear support cooling and pre-heated fuel return pipes; (16) main fuel pump; (17) fuel injector pipes; (18) pre-heating fuel pump; (19) throttle control; (20) fuel lines; (21) four sparking plug cables; (22) gearbox; (23) axial inducer; (24) centrifugal compressor; (25) rotor drum; (26) radial inflow turbine.

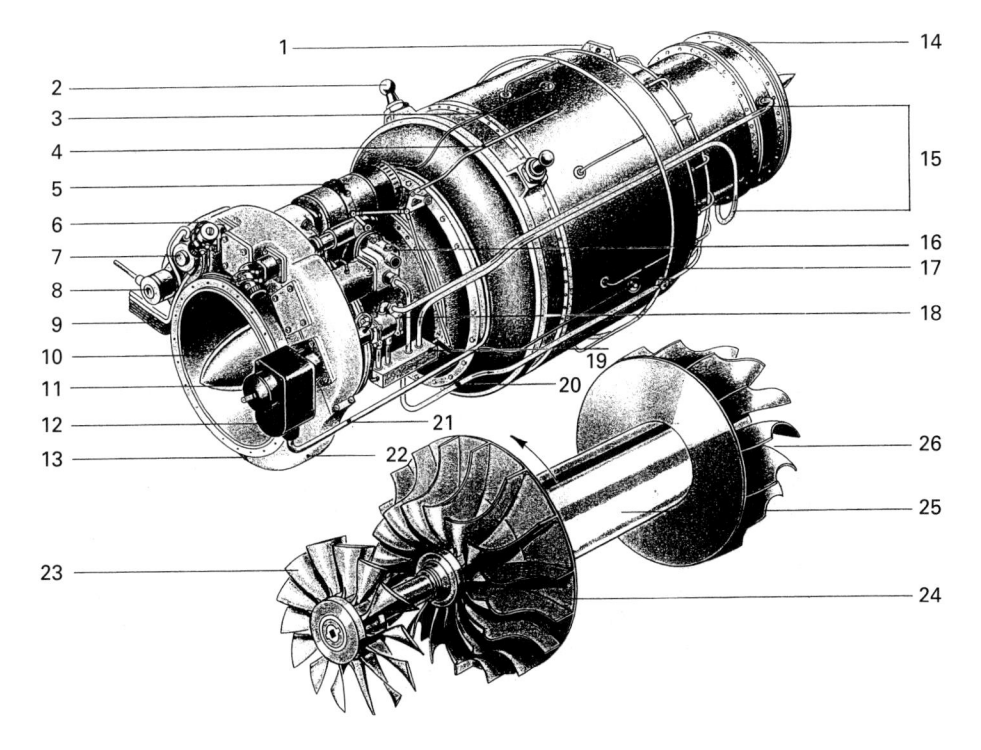

Taking off from Marienehe airfield is the Heinkel He 280 V2 on its first flight on 5 April 1941. Its HeS 8 turbojets are completely uncowled because these early engines leaked fuel and thereby posed a fire hazard. The He 280 prototypes were Heinkel's bid to forge ahead of his rivals in the fighter and jet aircraft field.

fairings in the air intake and made their connections through bevel gears. One shaft, however, was simply in the airstream.

By September 1940 the airframe of the V1 prototype of the He 280 (c/n: 280 000 001, DL+AS) was completed but HeS 8 engines were still not ready for powering it. Therefore, gliding trials were carried out with He 280 V1 at E-Stelle Rechlin, and in November at Marienehe, while work continued on the airframes of other prototypes. By March 1941, when the He 280 V1 had made about forty gliding flights, the HeS 8 was producing about 500kp (1,102lb) of static thrust and so two of these engines were installed beneath the wings of the He 280 V2 (c/n: 280 000 002, GJ+CA). This aircraft made its first short, powered flight before RLM officials on 30 March 1941, Fritz Schäfer being the pilot. At this stage the engines were completely uncowled, due to leaking fuel posing a fire hazard by pooling inside the cowlings.

The twin-jet He 280 fighter was designed as an all-metal mid-wing monoplane. The wing had a straight leading edge, a curved trailing edge and slight dihedral outboard of the engines, while the tail unit consisted of a high-mounted tailplane carrying a fin and rudder at each tip. Controls comprised ailerons, flaps, twin rudders and a one-piece elevator. A slender fuselage, which carried the fuel and nose-mounted armament, had a stepped cockpit with rearwards-sliding canopy. A radical feature was the ejector seat, which was operated by compressed air. For the tricycle undercarriage, the main members were attached to the wing just inboard of the engines to retract inwards (partly into the wing and partly into the fuselage), while the nose member retracted rearwards into the fuselage.

With the first full flight of the He 280 V2, before RLM officials on 5 April 1941, permission was given for Heinkel to buy the Hirth works, despite protests from other interested companies (especially Messerschmitt), but the market price was inflated by 50 per cent. The HeS 8 engine, however, was still a long way from being ready for production and was suffering from many faults, including bearing breakdown due to poor lubrication and the corrosion and carbonization of the fuel jets. Also, the engine had no automatic speed governor, which the RLM demanded. During 1942 von Ohain accordingly worked on a new design, the HeS 8B, which had a new compressor and turbine, and had few parts in common with the original HeS 8A. The RLM, however, was in favour of continued development of the HeS 8A, which resulted in an equalization of the temperatures around the combustion chamber, acceptable cooling and lubrication of the bearings, and the elimination of fatigue cracking of the turbine blades. Despite these improvements, the thrust was raised to only about 550kp (1,213lb), although, very importantly, reliability was improved. By spring 1942 only fourteen V-series engines had been built, although up to thirty examples had been planned.

In the HeS 8 V15 an attempt was made to improve the compressor system by fitting a single-stage axial compressor just after the centrifugal compressor. A design study attributed about 16 per cent of the 3.2:1 total pressure ratio of the compressor system to the new axial stage. This layout was to be continued as the HeS 9 turbojet with a two-stage axial compressor after the

Heinkel HeS 8A (109-001A) centrifugal turbojet.

centrifugal compressor, but this remained as a project.

In the HeS 8 V16 prototype, sandwich secondary air mixers were introduced, previous engines having had no secondary air system for the combustion chamber. Adjustable exhaust nozzles were test flown on the V16 and V17 prototypes. A converted He 111 bomber (c/n: 5649) was used as a flying test-bed, with the turbojet under test hung from the fuselage by two struts. These struts were streamlined and telescopic so that the engine could be lowered into less disturbed airflow once the aircraft was airborne. This aircraft was also useful for the testing of speed regulators.

It was not until summer 1942 that more HeS 8A engines were available to power He 280 prototypes. First to fly was the He 280 V3 (c/n: 280 000 003, CJ+CB) on 5 July 1942. On 8 February 1943 the starboard engine failed on take-off, precipitating an emergency landing. Next, the He 280 V5 (c/n: 280 000 005, CJ+CD) made its first flight in July 1943 with HeS 8As (the V14 and V15 prototypes), but was almost destroyed in an emergency landing following an engine fire.

Because Heinkel's jet fighter development was not proceeding very well, it was decided to use the He 280 to help develop the Junkers 109-004 turbojet. The He 280 V2 was retrofitted with two of these engines and first flew in this form on 16 March 1943. Later in the year the He 280 V6 (c/n: 280 000 008, NU+EA) and He 280 V8 (c/n: 280 000 006, NU+EC) were fitted with Junkers 109-004 turbojets from the outset. The latter aircraft was also to have tested a V-tail (as part of the He 162 programme), but this did not happen.

After using the problematical and underdeveloped HeS 8A engines, operating the He 280 with the Junkers turbojets was very agreeable. Already the Junkers engines had an operating automatic speed governor and thrust control. However, although the He 280 with these engines attained a level speed of 800km/h (497mph) at 4,000m (13,000ft) altitude, the airframe was not strong enough, especially the tail empennage, and vibration would force the pilot to reduce speed. In addition, the aircraft had insufficient ground clearance for these engines.

BMW 109-003A-1 turbojets were to have been fitted to the He 280 V9 (c/n: 280 000 009, NU+ED), but this aircraft was not assembled. The He 280 V1 and the He 280 V4 (c/n: 280 000 004, GJ+CC) were each tested with six Argus 109-014 pulsejets at E-Stelle Rechlin during 1943, these simple engines giving considerable and damaging acoustics and vibration. Their only successful employment was in the powering of the V1 flying bomb (also designed by Robert Lusser) and the He 280 pulsejet tests were actually part of the flying bomb development programme.

On 26 June 1943 the Argus test pilot Schenk made the first ejection seat bale-out in history when he was forced to abandon the He 280 V1; this was due to an iced-up towing cable, which could only be released from the two Me 110 fighters that had towed the He 280 to altitude. Pulsejet tests continued with the He 280 V4 from August.

This Heinkel He 111 bomber (c/n: 5649) was used as a flying test-bed for the HeS 8A turbojet. Once airborne, the test engine was lowered on streamlined telescopic struts into the less disturbed airflow.

He 280 V5 with two HeS 8A (109-001A) engines			
Span		12.2m	(40ft)
Length		10.4m	(34ft 1½in)
Wing area		21.5sq m	(231.5sq ft)
Height		3.06m	(10ft 0½in)
Empty weight		3,215kg	(7,073lb)
Fuel		810kg	(1,782lb)
Pilot		100kg	(220lb)
Ammunition		185kg	(407lb)
Loaded weight		4,310kg	(9,482lb)
Maximum speed (maintainable)	at 6,000m (19,680ft)	820km/h Mach 0.72	(510mph) or
Landing speed		140km/h	(87mph)
Initial climb rate		1,145m/min	(3,755ft/min)
Theoretical service ceiling		11,500m	(37,700ft)
Range		650km	(404 miles)

Despite the fact that Heinkel had the kudos of flying his He 280 jet fighter a year before Messerschmitt's Me 262, his fighter, and with it the languishing HeS 8A engine, was cancelled by the RLM on 27 March 1943 in favour of the Me 262. This decision came from the highest authority, Erhard Milch, and was doubtless influenced by the He 280 test flights made with Junkers 109-004 engines, which showed that the airframe was flawed. Also, its inbuilt tankage gave a poor range, a consequence of underestimating the fuel consumption of the engines. The order was made that the He 280 prototypes were to be employed as flying test-beds for various engines, as we have seen, and so the He 280B production fighter was not built.

HeS 10 project

During the early stages of the HeS 8 there were projected developments designated HeS 9 (see above) and HeS 10. In the latter, officially designated 109-010, a modified HeS 8 was externally faired into a suitable shape and was supported by struts inside a cowling forming a duct around the whole engine. A fan near the cowling intake was driven by an extension of the engine rotor through gearing. Air from the fan emerged together with the engine's exhaust gases from the cowling exit nozzle. A radial-inflow turbine was used, which in the normal HeS 8 engine was only sufficient to drive the axial inducer and centrifugal compressor. For the HeS 10 the previous axial inducer was omitted, but it was necessary to provide more turbine power to drive the ducted fan. Therefore a second turbine, of the axial type, was fitted inside a new exhaust nozzle behind the radial-inflow turbine. The second turbine was not connected independently to the ducted fan, as has become modern practice. The HeS 10 had a combustion chamber along the lines of the early HeS 8 design. With the HeS 10 it was hoped to provide an engine more economical at lower speeds for longer flights. However, the development was not pursued, although at least one test engine may have been built during the early development phase of the HeS 8. The planned thrust of the HeS 10 was 900kp (1,985lb) at 600km/h (373mph), the engine speed being 13,500 rpm; its weight was 500kg (1,102lb), diameter 1.01m (3ft 3¾in) and length 2.86m (9ft 4½in).

Heinkel-Hirth at Zuffenhausen

On 9 April 1941 Heinkel obtained control of the Hirth Motoren GmbH, which had specialized in the development and production of small piston aero-engines and auxiliaries such as oil pumps and turbo superchargers. The Hirth organization comprised the main factory on Marconi Strasse in Stuttgart/Zuffenhausen and a smaller factory known as Werk Waltersdorf in Berlin/Grünau. While the Hirth factory at Zuffenhausen was being refitted to suit Heinkel's work, von Ohain and Max Adolf Mueller continued work with their groups at Marienehe. The refitting was completed by the end of 1942 and complete absorption of the Hirth company was finally achieved by April 1943.

The Hirth company's experience of turbo superchargers and also of gas turbine auxiliary powerplants was certainly a help in the turbojet work that was to come. The company also had a very useful pool of experienced and leading engineers, as well as machine tools and test stands. Of the scientific personnel at Hirth, particular mention should be made of Werner Hartenstein and its Chief Engineer, Dr Max Bentele, who became famous throughout Germany for solving problems of vibration fractures in engines. During 1941 Mueller's group moved to Zuffenhausen and was then followed by von Ohain's group the following year. By May 1942, however, Mueller, who did not care to be organized, fell into dispute with Heinkel and resigned.

One of the conditions under which Heinkel was allowed to buy the Hirth company was that a new Class II turbojet of 1,300kp (2,866lb) static thrust would be developed. Development of this turbojet, which was designated the HeS 011 or 109-011 and incorporated Schelp's basic ideas, became the main job for von Ohain once the small HeS 8A was abandoned. Mueller's group, on the other hand, spread its efforts over the many engine types which will be described next.

Mueller's engines

Before Mueller arrived at Rostock-Marienehe, he and his group had been working since 1936 on various engine projects at the Junkers Fluzeugwerke in Magdeburg. One of these engines was an axial turbojet known as the RTO (see Chapter 21). Later, at the Heinkel works, Mueller and his group used this experience to design a very advanced axial turbojet known as the HeS 30 (or 109-006), with a planned thrust of 1,125kp (2,481lb) at 800km/h (497mph). Of particular interest was the reaction blading of the axial compressor, in which the work of air compression was shared roughly equally between the stator and rotor blades. This differed from German practice in most other axial compressors, which used impulse blading, whereby the stator blades performed almost all the work of compression; such compressors were chiefly to the designs of Encke and Betz of the AVA. The HeS 30 compressor, which achieved a pressure ratio of about 3:1 with only five stages, was designed by the aerodynamicist Rudolph Friedrich.

HeS 30 (109-006) axial turbojet

Mueller promised to have the HeS 30 running within one year of his arrival at Marienehe in October 1939. In fact, the first run was not made until more than two years had passed. It was planned to develop the HeS 30 alongside the HeS 8 as alternative powerplants for the He 280 fighter. For both engines, small frontal area was a prime aim in order to give acceptable wing nacelle mountings and thus avoid troublesome fuselage air ducting, such as that used in the He 178 aircraft. Of the two engines, the HeS 30 had the smaller diameter.

The five-stage axial compressor of the HeS 30 was driven by a single-stage axial turbine. Ten individual combustion chambers were provided, each varying from a circular cross-section at its entry to a heart-shaped cross-section at the turbine entry nozzles. Here a particularly novel feature was introduced in the form of variable-incidence turbine inlet guide vanes. Secondary air was admitted to each combustion chamber by allowing it to pass under five overlapping, progressively larger, tubular sections. The exhaust nozzle was of the variable-area type, a feature that was to become common on German axial turbojets to provide, as far as possible, optimum operating conditions. The area was varied by moving the sloping inner wall of the tail cone in or out. This inner wall was attached in sections to three streamlined struts attached to a central bullet. The bullet was held out by a spring

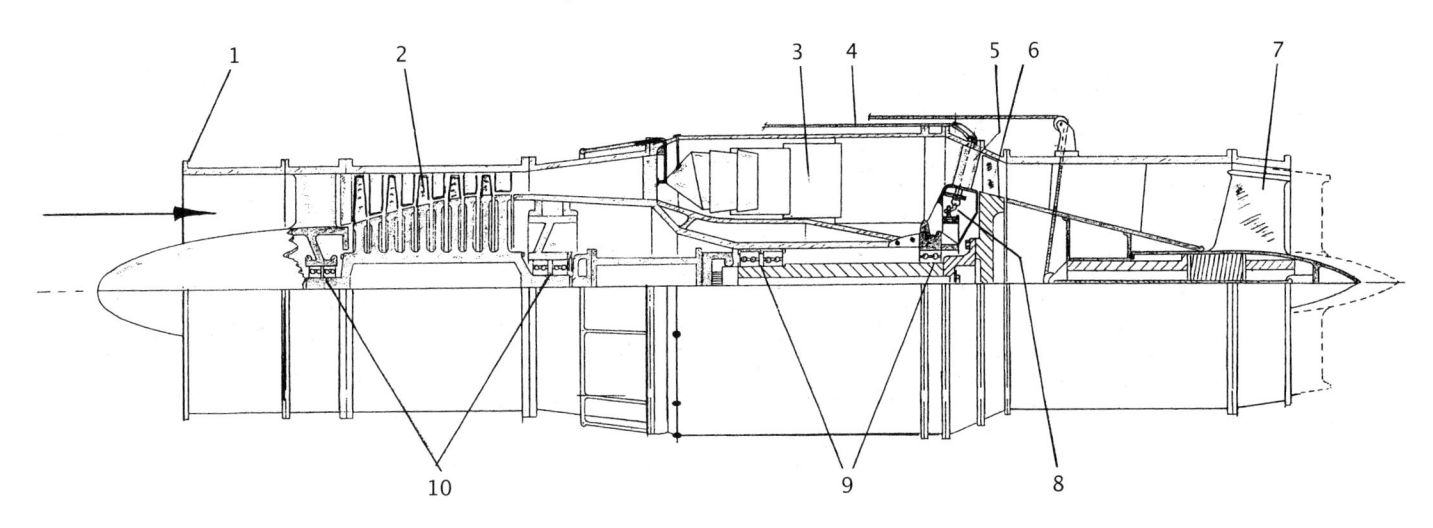

Heinkel-Hirth HeS 30 (109-006) axial turbojet: (1) intake cowling or duct attachment flange; (2) 5-stage axial compressor; (3) ten cannular combustion chambers; (4) cable and lever linkage to (5) variable turbine inlet guide vanes; (6) single-stage turbine; (7) cable-operated variable-area exhaust nozzle; (8) linkage mechanisms connecting turbine inlet guide vanes; (9) turbine rotor bearings; (10) compressor rotor bearings. (Auxiliaries gearbox is not shown.) Author

and was pulled in by a system of pulleys and cable.

Three experimental engines were ordered under the official designation of 109-006, but facilities at Marienehe were still not conducive to fast progress in building experimental engines and so the first HeS 30 progressed slowly. A major problem lay in trying to match the turbine to the compressor, which gave an air mass flow greater than designed for and an efficiency of 87 per cent. The pressure ratio was 3:1. The first engine was running under its own power by about April 1942, but the next month Mueller resigned after an arguement with Heinkel. By the time he left, however, the HeS 30 was running successfully and developing a static thrust of about 820kp

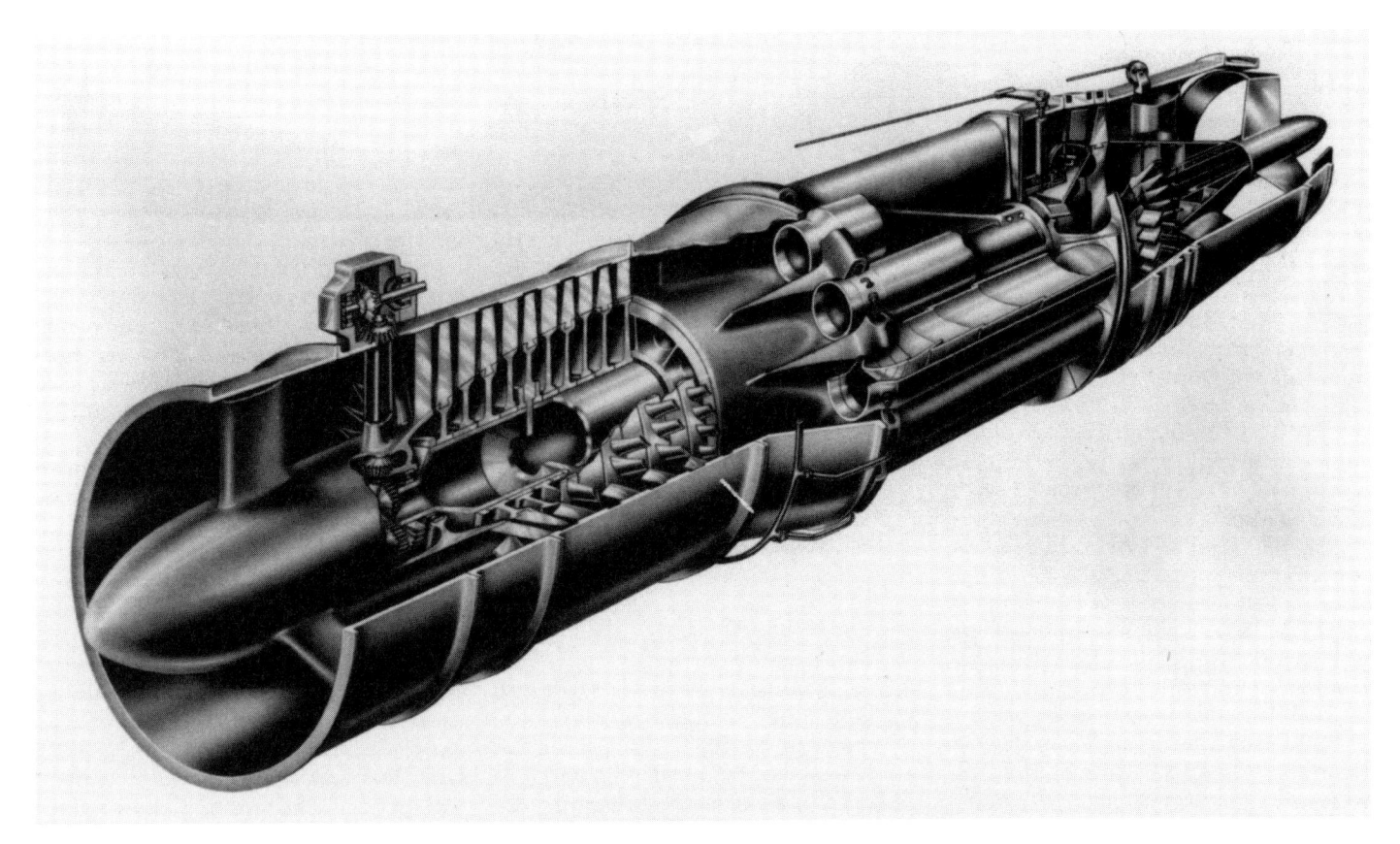

An artist's impression of the Heinkel-Hirth HeS 30 (109-006) axial turbojet, one of Germany's most promising engines.

The 5-stage axial compressor of the HeS 30 (109-006) turbojet, designed by Rudolph Friedrich to have 55 per cent reaction on the rotor blades.

TOP: Rotor blades and shaft with labyrinth seal – the diameter is given as 0.46m and the length 0.27m.

BOTTOM: Stator blades and rings with half of the casing removed – the length of the casing is given as 0.35m.

bearings, for example, being somewhat extravagant. Schelp also considered the HeS 30 too small for future requirements and thought its place would soon be adequately filled by the more conservative Junkers 109-004 and BMW 109-003 engines.

The official RLM line was that the HeS 30 was excellent but too late, since the Junkers and BMW engines were further along the development path, and they were not prepared to waste money just to help Heinkel catch up. According to von Ohain, Ernst Heinkel almost wept when he heard this. The RLM said it was his own fault for not listening in the first place when they were trying to run a national programme. It appeared that Heinkel was punished for his early, secretive, pioneering efforts.

Schelp, in fact, wanted work to start on the Class II HeS 011 engine and by the end of 1942 orders had been given for Heinkel to concentrate on this engine to the exclusion of all else (*see* Chapter 25). Although the HeS 30 lost RLM sponsorship after only a few test runs, it seems that tests were still made from time to time, probably in the hope that the engine would be ordered back into development: during a test in 1945 a static thrust was obtained as high as 910kp (2,006lb) at a speed of 10,500rpm, the weight of the engine at that time being 390kg (860lb).

HeS 40 constant-volume turbojet

At some time during 1940/41 Mueller's group worked on the interesting design of the HeS 40 turbojet engine. Its special interest lay in that it was designed to operate on the constant-volume principle, as opposed to the constant-pressure principle universally employed for turbojet and gas turbine engines today. Indeed, during the early history of the turbojet, active protagonists of the constant-volume

(1,808lb), thanks to a better matching of the turbine to the compressor. By October 1942 the static thrust was up to about 860kp (1,896lb). The engine's thrust to weight ratio was far better than that of, say, the Junkers 109-004 and, in terms of specific fuel consumption and frontal area per unit thrust, it is claimed that it was not equalled anywhere until 1947.

Unfortunately this remarkable engine was not favoured by Schelp, who had some objection to the reaction-type compressor blading. Also the engine was not going to be cheap to produce, the use of anti-friction

cycle were in the minority, and some explanation of the cycle may be worthwhile at this point.

In the conventional constant-pressure turbojet, the compressor delivers air at constant pressure to the combustion chamber, where it is accelerated by the application of heat, the reaction to this acceleration providing the thrust. In such a turbojet, the combustion chamber is open at both ends. If, however, air at little more than atmospheric pressure is mixed with the fuel in a combustion chamber with the ends closed, combustion takes place at constant volume and density. The pressure then rises in proportion to the temperature, and release of the combustion gases through a nozzle gives a propelling impulse. Complication then arises when attempts are made to give a continuous process by using a number of combustion chambers firing at different times to even out the impulses. Such a system, albeit not for a turbojet engine, was followed by Holzwarth. A simplified constant-volume jet engine is the pulsejet, whereby only one end of the combustion chamber (the intake) is closed during combustion.

Operating with a compression ratio of less than 2:1 and with considerably increased combustion speed, the constant-volume, intermittent-combustion turbojet was regarded by some as a competitor of the constant-pressure turbojet, having a higher compression ratio and slower combustion. There were, however, a number of problems to solve, not least of which was the method of control. Mueller therefore decided to build an experimental engine, the HeS 40, using the axial compressor (modified for a lower compression ratio), variable-area exhaust nozzle and other mechanical parts from the HeS 30 turbojet engine, which was then making slow progress. By having the HeS 40 and HeS 30 similar in as many ways as possible, useful comparisons of the two systems were made easier and more accurate.

Thus, in the HeS 40 compressed air was delivered to the combustion chambers by a five-stage axial compressor driven by a single-stage turbine. Six individual combustion chambers were provided, each being open at the exit to the turbine and closed at the opposite, inlet end during combustion by a poppet-type valve. Each inlet valve stem extended forward into a streamlined guide mounted in the diffuser section leading from the compressor. It is not known if the experimental HeS 40 was even built, but in any case the project was abandoned in 1942 when the HeS 30 was making excellent progress. The only data known for the HeS 40 is that the planned thrust was 940kp (2,072lb) at 13,400rpm. There would have been some weight increase over the HeS 30, owing to the valve gear and larger combustion chambers, but this could have been offset by lightening the less-loaded compressor.

HeS 50 and HeS 60 ducted-fan motor engines

A particular interest pursued by Mueller's group at Marienehe was the development of ML and MTL ducted-fan units powered by piston engines. These had been studied during Mueller's former days at Junkers, where the stage of a wooden mock-up was reached. It was thought that these units, with their airflow rate somewhere between that of the conventional airscrew and the turbojet, would prove more advantageous than the piston-driven airscrew unit at speeds around 720km/h (447mph) on long duration flights of up to ten hours. It was thought that they would have about half the fuel consumption rate of the turbojet. Such compound engines also avoided the problems of developing a compressor.

The HeS 50d, the first Motorenluftstrahl (piston-driven ducted fan) or ML unit, commonly known as Marie Louise, was designed to use a diesel engine because, despite its high specific weight, it was hoped the low specific fuel consumption would be an advantage during long flights. However, the prime requirement of an ML or later an MTL unit, the latter a combination of ML and TL (Turbinenluftstrahl, or turbojet), was seen to be an engine with a high power-to-weight ratio and very high rpm. The first requirement could be fulfilled by a two-stroke engine, but there were unsolved problems involved in increasing the rpm of such an engine. Thus it became necessary to develop a piston engine specially suited to the ducted fan units, but such specialized work was outside the scope of the Heinkel company (this was before the acquisition of the Hirth company). Research was therefore put in the hands of Prof. Wunibald I.M. Kamm of the Forschungsinstitut für Kraftfahrwesen und Fahrzeugmotoren of the Technisches Hochschule Stuttgart (FKFS). This research reached the stage of a single-cylinder unit being built and tested. This formed the basis for the design of multi-cylinder, two-stroke petrol engines. The piston engines subsequently designed for ducted fan units, such as the HeS 50z and HeS 60, had a stroke shorter than the cylinder bore. Concerning the ducted fan unit as a whole, it was calculated that the pressure ratio of the fan should be between 1.5 and 1.9:1, while the piston engine should have a compression ratio of between 3 and 6:1.

Piston-driven ducted fan units, however, did not progress beyond the research and mock-up stage at Marienehe. Between 1940 and 1942 three designs were worked on, these being the HeS 50d, 50z and 60. The two former designs at least had official backing. The HeS 50d was to have a static thrust of 950kp (2,095lb), but this was expected to fall off to about 325kp (717lb) at 800km/h (497mph). For this, a 24-cylinder, liquid-cooled diesel engine rated at 1,000hp was to be used. Following the FKFS research, the HeS 50d was superseded by the HeS 50z, which was simpler, lighter, smaller and had a far more even thrust curve, with more thrust at the high end of the speed range. The projected HeS 60 engine differed in having a turbine to extract a certain amount of power from the exhaust gases and air before discharge. In addition, a gear-driven supercharger was considered essential for the piston engine. The static thrust of the HeS 60 was to be 525kp (1,158lb), or 425kp (937lb) at 800km/h (497mph). Supercharging was estimated to increase thrust by about 675kp (1,488lb) at all speeds. The speed of the geared, three-stage fan was to be 6,000rpm and weight of the engine 800kg (1,764lb) for a diameter of 0.90m (2ft 11½in).

CHAPTER TWENTY-ONE

Junkers Flugzeug- und Motorenwerke A.G.

Born in 1859, Hugo Junkers was a scientist, engineer and entrepreneur. The founder of Germany's largest aircraft and engine company, he was responsible for great advances in aerodynamics and aircraft structures. The Junkers Motorenbau (Jumo) in Magdeburg, near Dessau, and the Junkers Flugzeugwerk (JFA) in Dessau, founded in 1913 and 1918, respectively, were merged on 15 July 1936 to form the Junkers Flugzeug-und Motorenwerke A.G., with separate airframe and engine divisions within the same company. Many other factories were later established, mostly plants dispersed to avoid Allied bombing. It would be true to say that Germany could not have prosecuted a modern war without Junkers aircraft and engine production.

Production of aircraft for the Luftwaffe was not Hugo Junkers's first choice (he was mainly interested in long-range airliner development), but by October 1933, as an opponent of the Nazis, he had been ousted on a trumped-up charge of treason. Junkers died on 3 February 1935. The RLM took over all the shares in his company and sold most of them to Nazi-approved companies in the armaments industry, while his many patents were distributed throughout industry. The great company had been stolen by the State. Meanwhile, after Klaus Junkers briefly held his father's former position, the RLM appointed Heinrich Koppenburg as Chairman of the Supervisory Council in Dessau.

As early as 1913 Hugo Junkers considered using a free-piston gas generator of his own design to provide hot gases to drive a turbine connected to an airscrew. This was the company's first proposal to utilize a gas turbine to power an aircraft. Other studies, soon rejected because of anticipated high fuel consumption, included turboprop engines with constant-pressure combustion and also an axial fan built with a closed channel and utilizing the waste heat of the propulsion engine.

Herbert Wagner

When Koppenburg became the General Director of Junkers in 1935, he brought Herbert Wagner to Dessau to 'breathe new life into the somewhat stagnant field of aircraft construction'. Wagner was born in 1900 in Graz. From 1924 to 1927 he gained experience in engineering at the Rohrbach-Metallflugzeugbau. Then, from 1927 to 1930, he earned his doctorate in aeronautical design at the Technisches Hochschule in Danzig. From 1930 Prof. Dr-Ing. Herbert Wagner taught at the Technisches Hochschule in Berlin, where his assistant was Max Adolf Mueller (see Chapter 30). From 1934 Wagner, while in Berlin as Professor of Aeronautics, projected an axial turboprop engine. When, in May 1935, Wagner was invited to work at Junkers, he took Mueller with him. There, as head of *Sonderentwicklung* (special development) at the company's Magdeburg factory, Wagner evaluated several types of engine.

After intensive study, Wagner concluded that the gas turbine was the right type of engine for high-altitude propulsion. Despite resistance from his colleagues, he convinced Dr Koppenburg that his novel concepts were the way forward. From 1 April 1936 Koppenburg set three young engineers to work in a secret department at Magdeburg, remote from Dessau. There, under Wagner's guidance, they started developing an aeronautical gas

Herbert Wagner's pioneering RTO axial turbojet on the test stand in 1939 at Junkers's Magdeburg plant. Half the compressor casing has been removed to reveal the 12-stage compressor's reaction blading. Also, without an exhaust nozzle at this stage, the 2-stage turbine is seen on the right. The shortness of the annular combustion chamber is noteworthy.

turbine. It seems likely that this work was unknown to the Jumo piston engine works or its Director, Otto Mader. Soon, about thirty designers and draughtsmen were involved in the work and, besides Mueller, Wagner's group included Wilhelm Papper (one source says Peppler) and Rudolf Friedrich. Later they were joined by Hans Stabernack and the team included 14 engineers. The whole team was known as a special Konstruktionbüro and were designated as Kobü II which was supervised by Prof. Wagner from Dessau on a weekly basis. (There were other Kobüs for Jumo piston engines at Dessau.)

Wagner's early turboprop and RTO/RT 1 turbojets

Wagner continued work on the axial turboprop that he had projected in Berlin in 1934 and which was to deliver its power approximately equally shared between the airscrew and residual thrust. Later he modified this to 67 per cent of the power to the airscrew and 33 per cent in residual thrust, a more realistic proportion. This turboprop concept was patented in Great Britain in 1937 (No. 495469) with German priority in 1936. In April that year, Wagner also began work on an axial turbojet known as the Rückstoss-Turbine 0 or RTO. This is believed to have been of similar layout to the turboprop, that is, with a 12-stage compressor, annular combustion chamber and a 2-stage turbine. Both types of engine were designed to have approximately 50 per cent reaction blading for their compressors, whereby the work of air compression was shared roughly equally between rotor and stator blades. Friedrich was responsible for this compressor design and preliminary investigations were made using one to three compressor stages with various blade profiles and blade twists. The basic layout of this axial turbojet was given in Wagner's German patent 724,091, submitted on 14 August 1938 and entitled 'Thrust installation for aircraft'.

Friedrich was the arch exponent of reaction compressors in Germany. He was born in 1909 in Waldenburg in Silesia and studied mechanical engineering at the Technisches Hochschulen in Breslau and Hannover. He joined Junkers at Dessau in the autumn of 1933. Friedrich was familiar with aerofoil theory as it was applied in 1934 by Curt Keller of Zürich in his dissertation on single-stage axial fans. It was also appreciated by Friedrich and others that flow losses, for example in airscrews, rise intolerably as the speed of sound is approached. Therefore, tip speeds of blades had to be carefully controlled. Also helpful to Friedrich was the data on aerofoil investigations made by Betz and Encke of the AVA. After much study and work, Friedrich's final design for the RTO turbojet had 55% reaction rotor blading.

Combustion tests were made using one tenth (36°) of the annular combustion chamber and both gaseous and liquid fuels were used. Air cooling of the turbine blades was also investigated and a patent was filed in Mueller's name and that of the Junkers Flugzeugwerk on 29 July 1939 for turbine rotor cooling. As we shall see, air cooling was to play a major role in the design of German turbojets, turboprops and gas turbines due to the increasing scarcity of metals to alloy heat-resisting steels as the war developed.

In the summer of 1938, Wagner approached the RLM to obtain official funding for the work but was turned down. As we have seen, Hans Mauch took the view that engine development was a job for engine manufacturers and so Junkers continued to finance Wagner's work.

The turbojet development in Magdeburg, which Wagner actually monitored from Dessau, continued unhindered from 1936 to 1938. Blades for the compressor were manufactured from duraluminium on a specially-built rotatable milling machine after which the trailing edges and surfaces of the blades were finished by hand. Although it was not possible to determine blades angles in advance, suitable measuring instruments being unavailable, great care was taken in setting the blades. For preliminary tests, the fitting of an exhaust nozzle was not required.

Early in 1939, the RTO turbojet was complete enough for preliminary testing and was mounted on a test stand or pedestal at Magdeburg. It was set up to burn propane gas. Testing went on until July 1939 by which time the RTO had been modified as the RT 1 unit. The combustion chamber was lengthened by 120mm (4¾in) and was still run on propane gas. Some 6,500 rpm was attained but, unfortunately, results were no better and Kobü II was closed down on 31 July 1939. Testing of the RT 1 was continued at Dessau but without success and the ultimate fate of this historic engine is unknown. However, according to Martin Graalmann, a Junkers aero-engine mechanic, several items which apparently included a compressor and a turbine and had come from Magdeburg, were seen in an isolated room at Dessau. The status of the project at the end of the Magdeburg work is best summed up by Anselm's Franz's report dated 5 August 1939, shown on p. 188.

It is quite clear from Franz's report that all was not well with the original RTO engine or its modified variant, the RT 1. A temperature of about 800°C could not be handled by the turbine. Any enthusiasm that Franz had for this project was greatly dampened by the large amount of research and development required to make it work. Mueller's commitment to Messerschmitt, at such an early stage, that turbojets could be delivered by April 1940 was incomprehensible. Paradoxically, the RLM decided to finance the RT 1 turbojet work but demanded that it be done with the full knowledge of the Jumo works at Dessau. Furthermore, the RLM ordered that the Magdeburg turbojet department be transferred immediately to Dessau.

None of this was to the liking of the team, who felt they were on the road to success. Already, at the end of 1938, Wagner had left and returned to Berlin, while Mueller soon departed to Heinkel's works. In August 1939 about fifteen personnel, comprising the RT 1 turbojet group, also left and joined Mueller at Rostock-Marienehe to work on the HeS 30 axial turbojet, based on the RTO/RT 1 engine studies.

It is likely that Herbert Wagner was not too dismayed at the demise of his turbojet project since he was a physicist and engineer of many interests, including structures, hydrodynamics, aerodynamics, missiles, aircraft, guidance and control systems. He achieved advances in all these spheres and is particularly remembered for his work on the first guided missiles as produced by the Henschel company, which Wagner joined in January 1940. The main effort under Wagner was directed towards developing air-to-surface guided missiles, of which the Hs 293A-1 was developed in time for operational use. This began on 25 August 1943 when the missiles were used to attack destroyers in the Bay of Biscay.

Anselm Franz

Late in 1938 Anselm Franz began making his own turbojet studies at the Otto Mader

Report on the RT0/RT1 turbojets by Anselm Franz, 5 August 1939

Work was still being conducted in Magdeburg in July. We had a series of tests demonstrated to us by Friedrich from which a picture on the status of development and important aspects for further development have resulted.

I. Available are:

1 – complete test stand unit (RT0/RT I)
1 – single- as well as a 4-stage compressor for preliminary tests
Various combustion chambers for preliminary tests

II. Test results – Status of development:

1) The *first* test-stand turbojet (RT0) did not run on its own, but only with external power supply.

2) The consequently *modified* turbojet (modified combustion chamber and turbine) now demonstrated to us by Friedrich *likewise* did not run unassisted. Even with the introduction of external power (with pre-compressed air through an AG blower), as is recorded in the Friedrich test report, maximum revolutions only attained a figure of 6,500rpm (required is 12,900rpm). [This is obviously the RT I engine]

3) In the attempt to raise the rpm, unacceptably high rotor-blade temperatures occurred (for a short period, even above 800°C) and combustion took place partially outside.

4) A thrust could not, up to the present, be measured.

Discussion of the results:

1) The lack of success with the modified turbojet [RT I], in our view, is principally due to the incorrect working of the combustion chamber (combustion partially outside the chamber, in and aft of the turbine). These problems may be able to be overcome through the alterations planned by Mueller to lengthen and sub-divide the chamber (already intended for the Messerschmitt turbojet). We shall give priority to combustion chamber testing with such chambers and especially with an effective pressure. The tests conducted up to now on combustion chamber test stands have been inadequate (only a small excess pressure, no degree of efficiency/effectiveness determinable).

2) The 12-stage axial compressor of the complete turbojet presents enormous difficulties. It does not work correctly at full rpm – at least, at the revolutions run so far (pressure drop in the last stages). For the tests therefore, the first four stages were removed; even so, a pressure drop was still shown in the last stages. At 6,000rpm, for example, the overall pressure rise in the compressor was only 0.087 atmospheres.

A clarification of such matters through correct testing of the compressor at the present time is not possible for the following reasons:

a) Individual tests on the compressor cannot be carried out, since the required power to drive it [4,000 hp] is not available.

b) Testing of the complete unit is not possible, since the [required] rpm cannot be attained.

c) Transference of the results of the preliminary tests on the 1-stage and 4-stage compressors is insufficient to provide clarification.

With regard to the correct layout and design of the compressor, there still exists a considerable amount of uncertainty.

In order to eliminate these problems, we shall build a smaller version of the compressor for which adequate power is available to drive it, so that the correct examination of the open questions can be carried out.

3) Completely unclarified are questions relating to the:
Regulation
Starting
Increasing the starting peak [*Startspitze*] and
The axial bearing problem (3.5 tonnes load; 80–90m/sec glide speed).

III: Immediate measures [necessary] for further development

1) Construction of a new test combustion chamber and testing it on the combustion chamber test stand.

2) Construction of a new combustion chamber for the RT0 and testing the complete unit.

3) Manufacture of a new 4-stage compressor on a reduced scale and testing it on the supercharger test stand.

4) In the design of the turbojet for Messerschmitt, in order to maintain the smallest dimensions, the allowable limits in accordance with present-day knowledge have been exceeded in various respects. A justification for this will not be produced by such a test. In the interest of being on the safe side we shall undertaken a redesign, whereby the rather larger dimensions shall be regarded as acceptable, against which Messerscmitt has no objections.

IV. Deadline matters

For the delivery of both the Messerschmitt turbojets, the deadline of April 1940 was quoted [to them]. We shall attempt with all our means to maintain this deadline, but a binding commitment is not possible at the present time for the following reasons:

From the details outlined under II, it transpires that up to now no test results were able to be achieved, upon which the expectations and that of the current design, on whose theoretical considerations it was based, have been confirmed. The design of the Messerschmitt turbojet must therefore be carried out without being supported by results.

In view of these facts, the abovementioned deadline commitment of Mueller's [April 1940] is incomprehensible, since a basis for the responsible commitment of such a deadline, according to the status and the work performed to date, has in any case not been achieved.

V. Remarks on the fuel consumption

It should be mentioned once again at this point that the fuel consumption that was advised to the Messerschmitt firm of 630gm/hp.hr in the calculations that were made under Friedrich cannot be attained, and that a figure of around 800gm/hp.hr must be reckoned with.

MSD Kobü – Entwicklung
Franz

Werk at JFA, Dessau. At the end of 1938, Mauch and Schelp from the RLM visited Otto Mader and persuaded him to at least accept a study contract for gas turbines and jet propulsion. Mader, however, was far from enthusiastic about developing new propulsion systems, despite the fact that he was an outstanding engineer. He saw his job as fully developing the piston engine and thereby catching up with the British and Americans in the aero-engine field. Much ground had been lost in Germany after the First World War, due to the Allied ban on the development of large engines.

Franz made a start on the official study contract in August 1938. Since he directed work on piston engine aerodynamic and thermodynamic problems, as well as the development of gear-driven superchargers and piston engine jet exhaust systems, he naturally first looked at free-piston jet engines. He soon concluded that these would have a power-to-weight ratio too low for aircraft use.

About May 1939 Franz looked at the work undertaken by Wagner at Magdeburg, but decided that none of the engines were worth pursuing. Earlier, Otto Mader had also visited Magdeburg and said he was only willing to allow development of one engine to continue. This led to Mueller's departure to the Heinkel works in October 1939. However, the development work could not continue, since at least seventeen of the Magdeburg team had left.

Only a few of the previous Magdeburg group stayed on to work under Franz, one such notable being Dr Heinrich Adenstedt, who headed metallurgical research. Also included in Franz's group were Dr Siegfried Decher as Chief Technician, Dr Wolfgang Stein as Chief Designer, Dr Beck as Design Engineer for turbines and Dr Hermann Hagen and Dr Bates for the aerodynamic research on compressors. It is noteworthy that, after Hans Rudolph Friedrich departed to the Heinkel company, reaction blading for compressors was dropped in favour of impulse blading.

Origin of the 109-004 axial turbojet

The outcome of all these changes was that Franz favoured making a fresh start and proposed a turbojet of 600kp (1,323lb) static thrust, designated the T.1. By summer 1939 this proposal had led to Junkers receiving an RLM contract for a turbojet, officially designated the 109-004, to be developed and put into production as soon as possible. In the design of the new turbojet, Junkers was given an almost free hand, the only official requirement being that the engine should develop 600kp (1,323lb) thrust at 900km/h (559mph) at sea level, which is equivalent to about 680kp (1,500lb) static thrust at sea level. Having fixed the thrust, the next important requirement was that the 109-004 should be rapidly developed up to production status, even, if necessary, at the expense of the engine's performance, which could be improved once production was under way.

In choosing the essential features of the engine, it was decided to employ an axial-flow compressor (instead of the centrifugal type familiar to Junkers from its supercharger work) to gain the benefits of a less devious flow path through the engine and also a smaller frontal area. The compressor's impulse blading was designed at the AVA, Göttingen, while the single-stage turbine was designed largely by Prof. Kraft of the Allgemeine Elektrizitäts Gesellschaft (AEG), Berlin. For the combustion system, individual can-type combustion chambers were chosen, instead of the preferred annular combustion chamber, chiefly on the grounds of the more rapid development that these promised, since one can-type chamber could be more easily tested and modified on its own. In any case, equipment to provide sufficient airflow to test an annular combustion chamber was not available at the time. (It does not seem to have been realized that a segment of an annular combustion chamber could have been tested, as Kobü II had done.)

To obtain data for the first engine design, it was decided to construct and test a scaled-down model, as proposed in Franz's report on the RTO turbojet (*see above*). The scale chosen was such that the compressor was to absorb only one-tenth, or about 400hp, at a speed something over 30,000rpm, and thus to allow the use of testing facilities then extant. The model was completed late in 1939 but the small scale gave inadequate combustion and the engine suffered from vibration. Independent tests were then made on the compressor alone but this burst asunder during one test at high rpm. With the failure of this model and the urgency brought about by the outbreak of war, it was decided that studies would proceed better if performed on a full-sized engine and so work on this, the 109-004 A, was begun in December 1939.

The 109-004 A had an eight-stage axial compressor, six separate combustion chambers and a single-stage turbine. This engine was intended to be rapidly developed into a pre-production engine with little regard to weight saving. Also, it used solid turbine blades in heat-resisting steel, but, since great economies in such steels were already foreseen as necessary, the development of hollow, air-cooled turbine blades was simultaneously begun for future, production engines. At that stage, weight-saving measures were also to be taken.

The prototype of the 109-004 A was first test run on 11 October 1940, without an exhaust nozzle, and was giving 430kp (948lb) of thrust at 9,000rpm by the end of January 1941. All was not well, however, and this engine was almost destroyed following a vibration failure of some of the compressor stator blades. The blade vibration specialist Dr Max Bentele was brought in to solve the problem. He changed the design of the stator blades, which had previously been cantilevered from their outer ends. The material of the stator blades was also changed from aluminium alloy to steel. A good six months of work and the testing of several engines produced a partial remedy of the vibration problem. Although these changes gave only a partial increase in life, the design thrust of 600kp (1,323lb) was reached during a test on 6 August 1941. At one time, consideration was given to using an axial compressor with contra-rotating stator blades, but the resulting overall diameter was considered too large for aircraft use and there were so many problems to be solved in the standard design that the contra-rotating compressor scheme was not pursued very far.

First 109-004 A flight tests

On 15 March 1942 a 109-004 A was first flight-tested, using a piston-engined Messerschmitt Bf 110. Soon after, two of these turbojets were sent for fitting to a prototype of Messerschmitt's Me 262 fighter. The previous year attempts had been made to fly the Me 262 V1 (PC+UA) with two Heinkel HeS 8 (109-001) turbojets and then with two BMW 109-003 turbojets, but without success. The first all-jet flight of this fighter type was therefore made on

A mechanic adjusts an early Junkers 109-004 A-0 (T.1) fitted to the Messerschmitt Me 262 V3 prototype, prior to this jet fighter's first flight on 18 July 1942. Note the extra large, rectangular holes giving access to the combustion chambers; much smaller circular holes were later used. Earlier attempts to fly the Me 262 V1 prototype, first with two HeS 8A (109-006) and then two BMW 109-003 turbojets, had failed.

18 July 1942 by the Me 262 V3 (PC+UC) fitted with two 109-004 A-0s rated at 840kp (1,848lb) static thrust each. By this time the engine was in pilot production and about thirty engines were finally built; a pair of these was used to power the Me 262 V2 (PC+UB) on 2 October 1942.

Arado's Ar 234 V1 (TG+KB), the initial prototype of the world's first jet bomber, made its maiden flight on 15 June 1943. The first four prototypes of this bomber were each flown on the power of two 109-004 A-0 engines. The Ar 234 V3 also used rocket-assisted take-off. The Me 262 V5 (PC+UE), first flown on 26 June 1943, also used two solid-fuelled rockets to boost the take-off power of its 109-004 A-0 engines. Previously flown on HeS 8 engines, Heinkel's He 280 V2 prototype (GJ+CA) was flown in April 1943 on 109-004 A-0 engines (some sources say B-0 engines). Junkers used a few of the A-0 engines for intensive experimental work.

On the test stand the V5 prototypes of the 109-004 A-0 developed as much as 1,000kp (2,205lb) of thrust by operating at an unacceptably high temperature for a short time. However, average data for the type included a static thrust of 840kp (1,848lb) at a speed of 9,000rpm and for a weight of 850kg (1,874lb), the specific fuel consumption being 1.4. The engine's diameter was 0.960m (3ft 1¾in) and its length 3.80m (12ft 5⅝in).

109-004 B turbojet

While experience was being gained with the 109-004 A, the design of a production version, the 109-004 B, went ahead. Some of the larger parts were also being made at an early stage, since these were expected to differ little from the 109-004 A-0 pre-production engines. The layout of the 109-004 B was chosen in December 1941 and its detail design was completed by October 1942. The first runs on the test

The first all-jet flight of the Messerschmitt Me 262 fighter took place with this V3 prototype (PC+UC) on 18 July 1942. It was powered by two Junkers 109-004 A-0 turbojets, each rated at 840kp (1,848lb) static thrust. Owing to the tailwheel undercarriage, the exhaust from these engines tore up the runway surface. The great improvement of a tricycle undercarriage was introduced with the Me 262 V5 prototype (PC+UE). Note the protective baskets over the air intakes, similar to the 'Daunt stoppers' used by early British Meteors.

stand were made in December 1942 and pre-production 109-004 B-0s were ready for delivery from January 1943. The chief alterations made for the B-0, mainly intended to aid mass-production, included using a compressor construction with separate discs and replacement of castings with sheet metal as far as possible. More than half the weight of strategic material used in the 004 A model was substituted with more readily available material, although solid turbine blades were still used in the 109-004 B. The aerodynamics of its air intake was improved by modifying the entry.

The 109-004 B was not entirely satisfactory but the small numbers built were released for use with prototype aircraft, some A-0 powered aircraft being retrofitted with B-0 engines. The Me 262 V1 (PC+UA), which had first flown on 18 April 1941 with a piston engine and airscrew in the nose only, was refitted with two 109-004 A-0 engines and flew on 2 March 1943. The similarly powered Me 262 V4 (PC+UD) flew the next month. On 22 April this aircraft was flown by Gen Lt Adolf Galland as part of the effort to convince the non-technical in the Luftwaffe and official personnel of the desirability of jet aircraft. Galland was immediately won over by the jet fighter and, following a meeting with the RLM, he was influential in getting much of Messerschmitt's production switched from piston to jet aircraft. These first prototypes of the Me 262 fighter, unlike subsequent examples, had tailwheel undercarriages that directed the line of the jet exhausts towards the runway: the result of this was that chunks of tarmac and even concrete were torn up and blown away during take-off. Beginning with the V5 prototype, the Arado Ar 234 began flying on 109-004 B-0 engines by the end of 1943.

Salient data for the 109-004 B-0 included a static thrust of 840kp (1,848lb) at a speed of 8,700rpm and for a weight of between 730 and 750kg (1,610–54lb). Its fuel consumption and external dimensions were the same as for the A-0 model. Compared with the A-series, the chief gain made with the B-0 engine was the considerable reduction in weight and strategic materials, while still maintaining the engine size and performance. The large variation in weight between different engines was due to the sand casting of parts, which were then only machined where necessary.

The major production model was the 109-004 B-1. Modifications to the compressor and turbine entry nozzles decreased vibration and increased thrust to 900kp (1,984lb). Although steel compressor blades had been tried previously, failures still occurred owing to resonance in the third row of stator blades. Thus, while Prof. Encke preferred to use compressor blades with thin aerofoil sections (such as

A pair of Junkers 109-004 A-0 turbojets was delivered to Arado's Warnemünde factory in February 1943 and fitted to the V2 prototype of the Arado Ar 234 jet bomber. The engine shown is c/n: 040. E.J. Creek

Arado Ar 234 V5 (GK+IV) powered by two Junkers 109-004 B-0s. The jettisonable take-off trolley was used pending the development of a retractable, tricycle undercarriage.

Göttingen 684) throughout, it was found better in the existing design to use thin-sectioned, wide-chord blades for the first two stages and comparatively thicker-sectioned, narrow-chord blades for the last six stages. Although the increased rigidity of the blades ensured a longer life, it did not cure the resonance problem and so other changes were made in later engines.

The first 109-004 B-1 production engines were delivered in May or June 1943 and the first aircraft to fly on the power of these engines was the Me 262 V6 (VI+AA, c/n: 130 001). This aircraft was the first in the series to have a fully retractable nosewheel undercarriage and it made its maiden flight in October 1943. The same aircraft was demonstrated on 26 November 1943 before Hitler, who set the jet fighter programme back by about six months by his insistence that the aircraft should be developed to carry a bomb load. The 109-004 B-1, although in slow production by then, was also suffering from vibration of the turbine blades. To obtain data, Anselm Franz decided to discover the ringing natural frequency of the turbine blades and asked a musician to stroke the blades with his violin bow and use his trained musical ear to determine the frequencies. Dr Max Bentele was again consulted and he discovered that the turbine rotor blade vibration problems originated with the six combustion chambers and also the three struts of the jet nozzle housing of the turbine.

By early 1944 the problem was cured by increasing the turbine blade taper, shortening the blades by 1mm, modification of the turbine entry nozzles and by reducing the maximum speed from 9,000 to 8,750rpm. By April that year, thirteen pre-production Me 262 A-0 fighters had been manufactured, in addition to some twelve developmental prototypes, the first of which was delivered in the summer to E-Stelle Rechlin for official tests. At the same time, the first pre-production Arado Ar 234 B-0 bombers were leaving the assembly line, some also going to E-Stelle Rechlin. Although jet fighters and bombers, using the 109-004 B-1 turbojet, were soon in production and Luftwaffe service, it was not until Germany's war situation deteriorated further that full prority was given to jet aircraft over conventional piston aircraft. A short résumé of jet operations, largely with the 109-004 B-1, is given below.

This Junkers Ju 88 A-5 bomber (GH+AQ) was used as a flying test-bed for the Junkers 109-004 turbojet. The test engine is mounted below the bomb shackles.

The 109-004 B-1 was the principal operational German turbojet during the Second World War. It developed 900kp (1,945lb) static thrust, or 730kp (1,610lb) at 900km/h (559mph) at sea level, at a speed of about 8,700rpm. Its net dry weight was 720kg (1,588lb), diameter 0.80m (2ft 7½in) and length 3.86m (12ft 8⅜in). Its eight-stage axial compressor had a compression ratio of up to 3.5:1 and an air mass flow of 21kg/sec (46.3lb/sec). Since the compressor's exit velocity was a relatively low 80m/sec (262ft/sec), no diffuser passages were needed before the combustion chambers.

The compressor used impulse-type blading whereby almost the whole of the stage pressure rise occurred across the stator blades. This type of blading was chosen in preference to reaction blading (advocated previously by Friedrich) to allow simple, although heavy, construction for the stator blades, and did not require fine axial clearances. A compressor efficiency of up to 80 per cent was achieved but stalling occurred on the low-pressure stages along the full length of the blades.

Air from the compressor was delivered to six separate combustion chambers, which were largely made from mild steel with air-cooled liners. For each, primary air passed through a fixed swirler and then on to a fixed upstream fuel nozzle. The combustion gases then passed through a conical grid assembly to join the secondary air flow.

The single-stage turbine followed contemporary steam turbine practice, owing to AEG's collaboration in its design. About 20 per cent of reaction was chosen for the blades in order to obtain sufficient work from them at the quite low rotational speed. Air cooling was used for the blade roots. Although solid blades were initially used for quick results on the 109-004 B-1 and B-2, the development of hollow, air-cooled blades was put in hand from the outset so that higher operating temperatures and lower fuel consumption could be realized in later engines. Also, the hollow blades would save on heat-resisting steels.

For the B-1 engine, the specific fuel con-

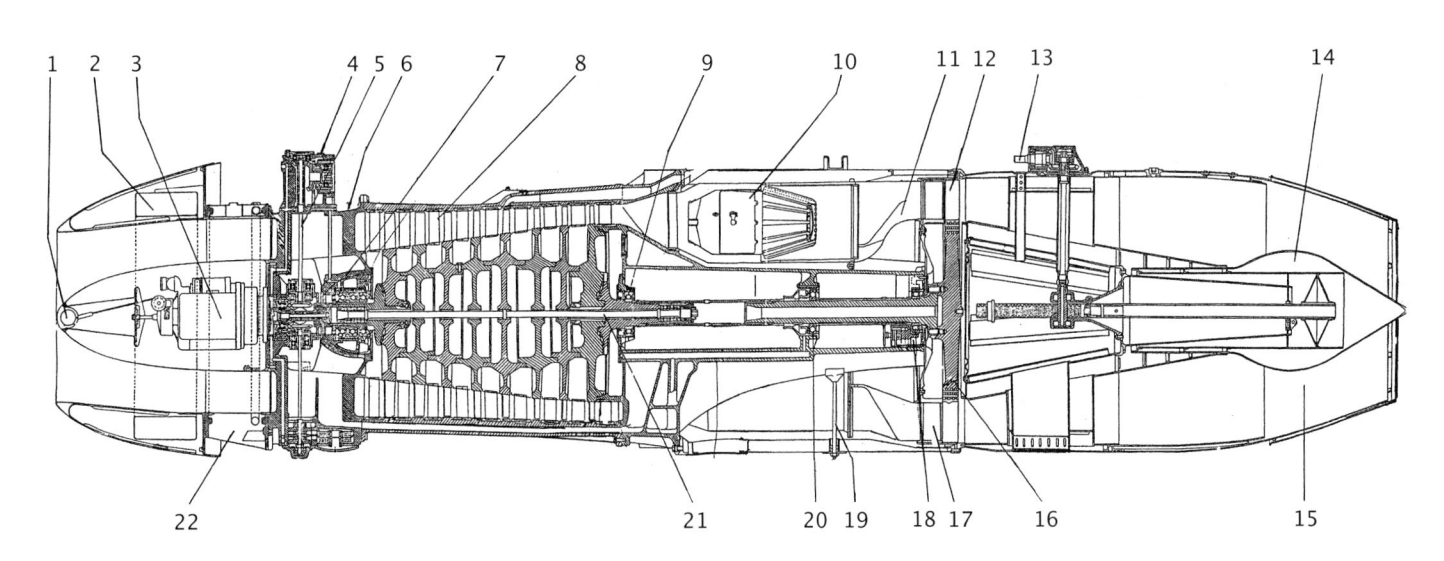

Junkers 109-004 B-1 axial turbojet: (1) pull cable for Riedel starter motor; (2) starter fuel tank; (3) Riedel AK-11 starter motor; (4) accessory gearbox; (5) accessory gearbox drive shaft; (6) intake casing; (7) front ball bearing assembly; (8) 8-stage axial compressor; (9) rear compressor roller bearing; (10) six combustion chambers; (11) turbine inlet ducting; (12) single-stage turbine; (13) servo motor shaft; (14) movable cone or bullet; (15) variable-area exhaust nozzle; (16) turbine wheel; (17) turbine inlet nozzle; (18) rear roller bearings; (19) six supports for outer casing; (20) front turbine ball bearing; (21) compressor tie rod; (22) annular oil tank with internal cooler.

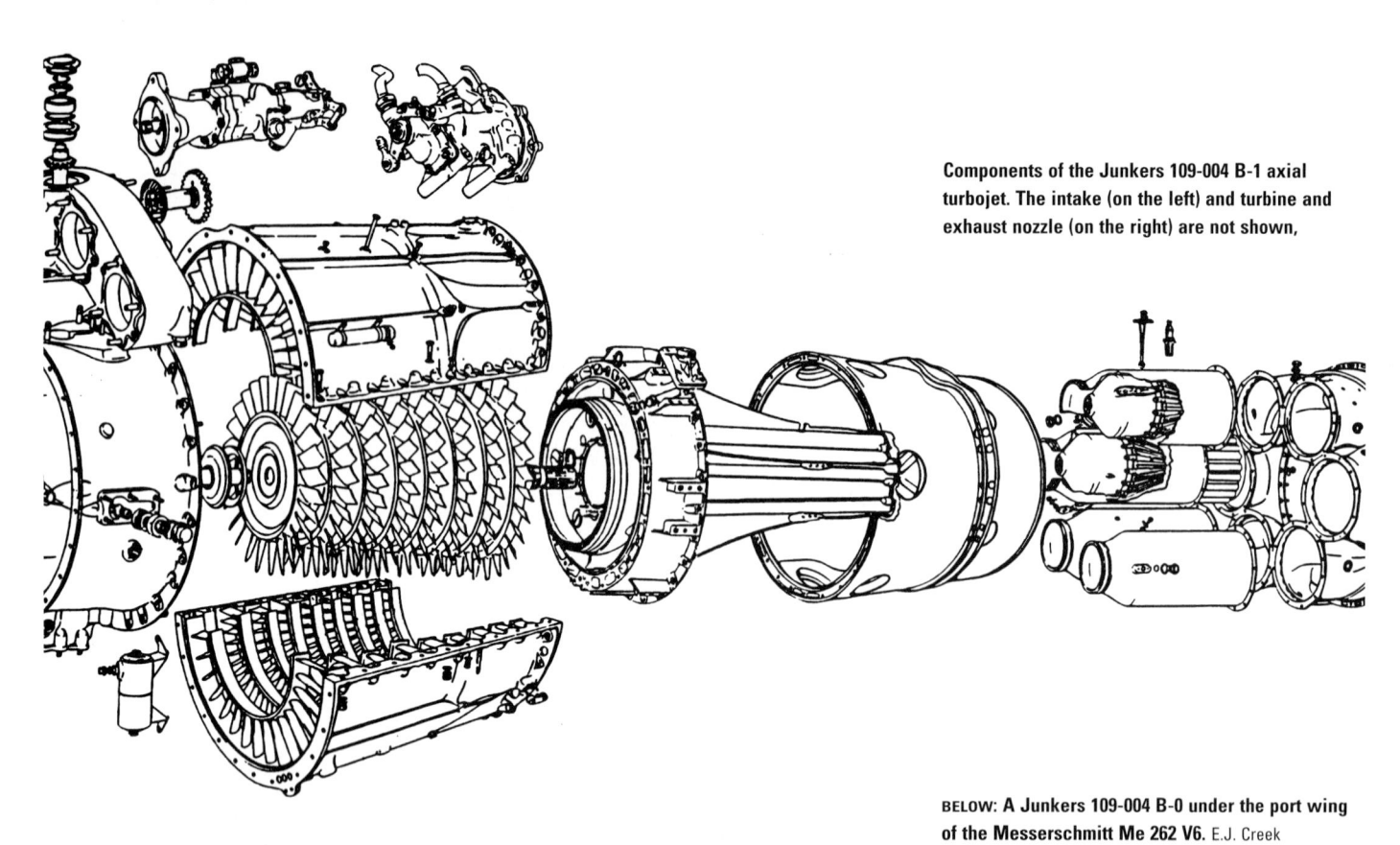

Components of the Junkers 109-004 B-1 axial turbojet. The intake (on the left) and turbine and exhaust nozzle (on the right) are not shown.

BELOW: **A Junkers 109-004 B-0 under the port wing of the Messerschmitt Me 262 V6.** E.J. Creek

sumption was, at its best, 1.38 but this could increase to 1.45 within a few hours due to the build-up of dirt on the compressor. All versions of the engine used air-cooled, hollow turbine inlet nozzles, the air for this being tapped off the fourth stage of the compressor. Most B-1 engines were fitted with sixty-one turbine rotor blades, each having a forked root secured with two rivets in the turbine disc. The turbine blades were made from Krupp's Tinidur, which was similar to the British Nimonic 80 except that, due to a shortage of nickel, its creep strength fell off sharply at about 580°C. Therefore, centrifugal and gas bending stresses on the blades had to be kept as low as practical.

Construction of the 109-004 B-1 was built around a central casting of silicon-aluminium alloy, which carried the three main points for attaching the engine to the aircraft. The central casting carried the rear compressor bearing, the two bearings in front of the turbine, the six combustion chambers and the turbine inlet nozzles. The forward end of the central casting changed the air passage from an annular shape at the end of the compressor into six circular entries to the combustion chambers. The rotor shaft connecting the turbine to the compressor ran inside the central casting, which also had, at the bases of its six ribs, five cooling air holes and one lubrication hole.

Silicon-aluminium castings formed the intake and compressor stator casing of the engine, the latter being in two halves longitudinally, and all were bolted together and then to the central casting. A mild steel shell of substantial gauge surrounded the combustion chambers and was provided with large holes to give access to the fuel injectors and igniters. This shell was bolted to the forward end of the central casting, while at the rear it was bolted to flanges on the turbine inlet ducting, nozzle ring assembly and the exhaust casing. The exhaust casing supported the entire exhaust system, which included a movable bullet (or onion) to vary the outlet area. Much of this end of the engine was made of aluminized mild steel sheet. For starting, a small two-stroke petrol engine with integral gearbox, made by Riedel, was located inside the intake central fairing. Auxiliaries included a gear-type main fuel pump, an electric petrol pump for starting, a servo motor for the variable-area exhaust bullet and an all-speed governor. These last two items, particularly the all-speed governor, were masterpieces of complexity and were later adopted by Heinkel and BMW for their turbojets. Today electronics and computers are used in engine management systems, but in those early days this job had to be largely performed by mechanical means.

A captured Junkers 109-004 B-1 turbojet at the RAE Farnborough, 23 April 1945.
TOP: Engine mounted on test frame, with nose cowling and exhaust system and nozzle removed.
BOTTOM: Engine with nose cowling, petrol and oil tanks, upper half of compressor casing and exhaust system all removed. On the left the Riedel starter motor is exposed. The casing on the right, with access holes, encloses the six combustion chambers.

For starting, the exhaust nozzle was set fully open, the engine was run up by the Riedel starter and petrol was fed to the burners. Once at 2,000rpm, the throttle was moved to the idle position and the main fuel system took over from the petrol system. Because of the low air speed through

One of the prototypes of the world's first jet bomber, the Arado Ar 234 V3, taking off with the assistance of jettisonable rocket units. This prototype, powered by two Junkers 109-004 B-0 turbojets, made its first flight on 25 August 1943 and had an ejector seat and a pressurized cockpit. It was destroyed on an early flight.
Philip Jarrett

the engine, over-rapid throttle movement between 3,000 and 6,700rpm could burn up the combustion chambers. When the throttle was moved to fully open, the exhaust bullet was automatically moved rearwards to decrease the exhaust nozzle area for take-off.

The design of the 109-004 B-2 was completed by about mid-1943, incorporating a new compressor resulting from cooperation between Junkers and the AVA. By this time Junkers was conducting its own aerodynamic studies and Dr Bates was working closely with Prof. Encke of the AVA. The new compressor design aimed at finally eradicating vibration failures while at the same time giving an improved altitude performance. Firstly, the number of rotor blades in each stage was reduced to put the critical rotational speed (when trouble occurred) outside the engine's speed range. For better altitude performance, the rotor blades were redesigned with increased blade chords to give six stages of small thickness-chord ratio plus two stages with comparatively thicker sections (instead of the previous two stages of thin blades plus six stages of thicker blades). The profiles of the stator blades were unaltered, but the blades were reset at slightly different angles, while the axial clearances between rotor and stator blades remained unaltered. Despite all the effort, the new compressor suffered its own form of blade vibration and so the 109-004 B-2 did not go into production. Evidence indicates that it was about 20kg (44lb) heavier than the B-1 model.

Research continued in order to find improvements for the compressor within the space limitations of the engines in production, and various modifications were introduced from time to time. One late scheme aimed at altering the frequencies of the rotor blades by introducing a slot cut down the centres of the blades of the first, third, fifth and seventh stages, but it is not known if this was worked out in time for production engines. The B-3 model could have been a developmental model prior to putting into production the B-4 with air-cooled, hollow turbine blades.

The need to develop such turbine blades arose, of course, from the German shortage of metals used in heat-resisting alloys, requiring the best results to be obtained with the inferior alloys available. Junkers' experiments with high-temperature ceramic blades failed to produce sufficiently robust blades and so hollow metal blades had to be the answer. Looking more to the future, it was realized that, for a given limiting stress in a turbine wheel, a higher rotational speed and thrust was possible with the lighter, hollow blades.

Using Krupp's Tinidur sheet metal, Junkers first tried making hollow blades by folding and welding, but Tinidur proved unsuitable for welding. Junkers therefore sought the assistance of outside specialists and, in February 1943, asked the firm of William Prym at Stolberg/Rhld (near Aachen) to develop a process for the manufacture of hollow turbine blades for the 109-004 B-4. Earlier, Prym had made hollow blades for Junkers piston engine exhaust turbochargers and had established a separate deep-drawing department. In this work Prym was highly successful.

After a couple of months' work, Prym delivered the first 70 hollow blades for the 109-004 on 24 April 1943, but these had an error in root dimensions. Nevertheless, it was plain that a satisfactory job could be done. On 11 May 1943 an important meeting was held at the Stolberg works to discuss blade problems. Not only was Junkers represented but also personnel from the other engine companies and the RLM. Soon, Prym was making hollow turbine blades for all and sundry, and it was necessary to open another, much bigger factory at Zweifall. While Prym was producing hollow blades in Tinidur, the Sächsicher Metallwarenfabrik Wellner Soehner A.G. of Aue was working out how to make hollow blades in another heat-resisting alloy, known as Cromadur. This was another Krupp alloy, intended to replace Tinidur since it used manganese instead of the scarce nickel. Since Cromadur was easy to weld (unlike Tinidur), Wellner produced its hollow blades by folding and welding down the trailing edge. Although the creep strength of Cromadur was inferior to Tinidur, the Wellner blades actually proved superior to the Prym drawn blades in service. However, neither company could achieve the requisite production rate and so both Tinidur and Cromadur hollow blades were kept in production for the 109-004 B-4.

By the end of 1944 the 109-004 B-1 was replaced in production by the B-4 version, which differed mainly in having hollow, air-cooled turbine blades. Other minor differences included changing the cooling air bleed after the last compressor stage from an incline to a right angle with the engine axis. Although, by a minor modification of the combustion system, the B-4 was capable of a higher working temperature and thrust, the previous static thrust of 900kp (1,984lb) was maintained in order to gain longer service life and greater reliability at the same temperature as the B-1. Apart from their economy, a possibly unforeseen benefit of the hollow turbine blades was their faster and easier production compared with solid blades.

The hollow blades were pressed over stubs on the turbine wheel and then brazed on. To allow cooling air to pass up into the blades, holes were drilled through the rim of the turbine wheel, the cooling air being forced out into the blades by centrifugal vanes attached to the upstream face of the wheel. This turbine wheel was lighter than that for solid blades and, owing to the lower stress, the pull of one blade on the wheel was about 3.4 tons. About 3 per cent of the total airflow through the engine was used for cooling purposes at the expense of engine efficiency, but, although cooling the turbine blades brought about a further slight decrease in efficiency, this effect was offset by advantages already mentioned. Unlike BMW, Junkers did not find an inner sleeve for their hollow blades necessary.

Specific data for the 109-004 B-4 are lacking but performance and external dimensions were generally as for the B-1, while the weight was slightly less. The complete engine could scarcely have been more economical in its use of scarce metals such as nickel, molybdenum and chromium, while no tungsten was used at all. A 109-004 B-4 with Tinidur turbine blades used 4.6kg (10.14lb) of chromium, 6.6kg (14.33lb) of nickel and 0.2kg (0.44lb) of molybdenum. The same engine with Cromadur turbine blades used 4.7kg (10.36lb) of chromium, 3.5kg (7.71lb) of nickel and 0.2kg (0.44lb) of molybdenum. These figures can be compared with those for the 109-004 B-1 (with solid Tinidur turbine blades), which had 6.35kg (14.0lb) of chromium, 9.85kg (21.72lb) of nickel and 0.205kg (0.45lb) of molybdenum.

CHAPTER TWENTY-TWO

Bayerische Motoren Werke A.G. (BMW)

Besides Junkers, the RLM selected BMW to develop another, but more advanced, first-generation turbojet, for which BMW was to be allowed more time. From 1933 BMW was working on large radial piston engines based on Pratt & Whitney Hornet and Wasp models from the USA. To deal separately with aero-engine development and production, the BMW Flugmotorenbau GmbH was formed at Munich in 1934. In 1938, BMW began sharing development with another aero-engine company, the Brandenburgische Motorenwerke GmbH (or Bramo) of Berlin-Spandau. BMW then absorbed Bramo in 1939 to form one of Germany's major aero-engine manufacturers.

In 1929, even before the merger, Hermann Oestrich, head of research at Bramo, had already made studies of jet propulsion but concluded that, for speeds then envisaged, jet propulsion was too inefficient to be practical. Thus, when the company was asked by Mauch and Schelp of the Technisches Amt in 1938 to study the field of jet propulsion, Oestrich turned his attention to ducted fans driven by piston engines (*Motorenluftstrahl* or ML units), which appeared more suitable for the speeds then in prospect.

P.3302 axial turbojet at Spandau

By the end of 1938, Bramo, under Oestrich's direction, had built its first ML unit, which consisted of a 160hp Bramo 314A seven-cylinder, air-cooled, radial engine driving a four-bladed airscrew at 2,200rpm within a short NACA annular duct or ring. The exhaust area of the ring could be altered to suit different speeds and altitudes. This unit was installed in a Focke-Wulf Fw 44J Steiglitz two-seat biplane in place of the normal engine and airscrew. Flight tests, conducted by the famous aviatrix Hanna Reitsch, indicated that the ducted-fan unit was worth pursuing and so an MLS unit, with provision for afterburning to further accelerate the airflow, was designed and built. The ducted fan of this unit was driven at 2,600rpm by a 1,200hp Bramo 323 radial engine. Unfortunately, very poor results were obtained from this MLS unit when it was tested in a wind tunnel at the AVA, Göttingen, partly owing to a low fan efficiency and low rpm.

Bramo did not pursue its MLS project far and abandoned it early in 1939 to concentrate on turbojet studies, which looked more promising as aircraft speeds increased. Even so, Bramo now believed that ML units and turbojets were theoretically equally efficient for high-speed flight, but that the ML unit could give the best take-off power and specific fuel consumption under partial load. Against this, the turbojet promised a less complicated construction and this, coupled with the disappointing performance of the MLS unit tested, appears to have influenced the decision to concentrate on the turbojet studies.

In these studies, which were begun at the end of 1938, the selection of the best type of compressor was considered and, with the parameters of both weight and size in mind, these were put in the order of favour as contra-rotating axial, straightforward axial and centrifugal. However, when the subject of simplicity and ease of development by a company without gas turbine experience was considered, it was decided to make a start on a turbojet engine with a straightforward axial compressor. Accordingly a turbojet was designed between December 1938 and April 1939; this later incorporated a single-stage turbine wheel, under development by BMW at Munich, when the two companies began pooling their development work. Other data for the turbojet design were gathered from Bramo experiments with combustion systems and other factors.

After its take-over by BMW in summer 1939, Bramo became known as the BMW Flugmotorenbau Entwicklungswerk Spandau and the Bramo turbojet design was given the designation of P.3302. In agreement with the RLM, this engine was at that time considered as a research engine to assist in the development of a more complicated turbojet with contra-rotating axial compressor and turbine, which it was believed could be made with a smaller diameter.

Weinrich's P.3304 (109-002) axial, contra-rotating turbojet

The contra-rotating engine was given the project designation of P.3304 and the official designation of 109-002. The layout and design of the 109-002 was the responsibility of the engineering consultant Helmut Weinrich of Chemnitz, who had been discovered belatedly by Mauch of the RLM and put in touch with BMW. In 1936 Weinrich had submitted plans to the RLM for a turboprop engine (PTL) using a contra-rotating compressor and turbine. Weinrich had followed this up with experiments but officialdom was apparently not ready to accept such radical powerplants, or else his proposals were never seen by the right people. Another effort in the field was Weinrich's patent (663,935) of August 1938 for a method of introducing fuel to combustion turbines. By 1940 intensive work was being performed on the 109-002, although only about twenty designers and engineers were available for this development at Spandau.

The P.3304 (109-002) featured four axial compressor stages on a drum rotor that was connected to another drum rotor mounting three axial turbine stages. Surrounding this compressor/turbine system was a shell rotor

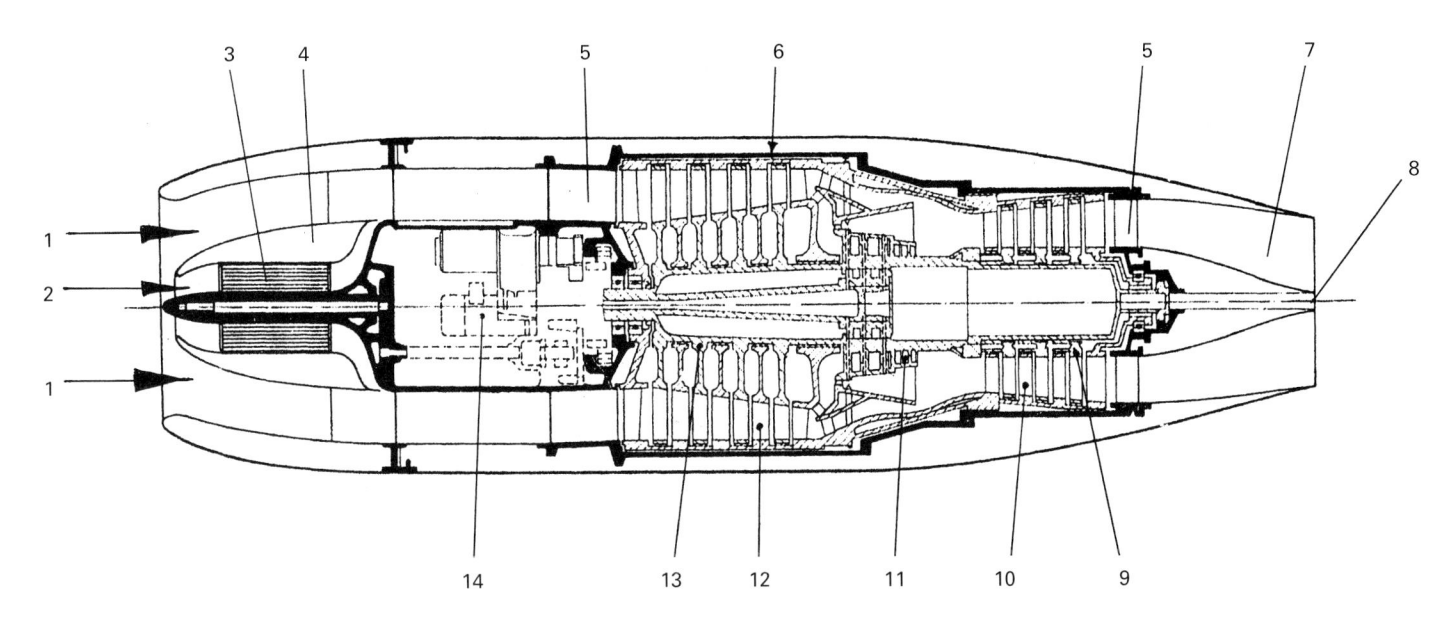

BMW P.3304 (109-002), Weinrich's contra-rotating turbojet: (1) air intake; (2) cooling air intake; (3) oil radiator; (4) oil tank; (5) support vanes; (6) rotating outer shell rotor; (7) exhaust nozzle; (8) cooling air exit; (9) 3-stage (inner) turbine; (10) 4-stage (outer) turbine; (11) labyrinth seal; (12) 5-stage (outer) axial compressor; (13) 4-stage (inner) axial compressor; (14) accessories.

that carried five intermediate compressor stages and four intermediate turbine stages, the whole rotating in the opposite direction to the inner drum rotor. Anti-friction bearings for the shell and drum rotors were provided in front of the compressor and at the rear of the turbines. An annular combustion chamber was provided with labyrinth seals to prevent hot gases escaping forward along the inner rotor surface. Cooling air was drawn in through a hollow intake centre fairing that also contained an oil cooling radiator. The cooling air passed through the inner, hollow drum rotor, cooled the turbine (roots?) and left through the centre of the fixed exhaust cone. Auxiliaries for the engine were contained in the space before the compressor. The support structure and fairing surrounding the whole engine was connected to the main bearing housings via fore and aft streamlined struts.

Individual stages of the contra-rotating compressor were built and tested, the forged aluminium alloy compressor blades being produced by WMF of Geislingen, and an annular combustion chamber was settled upon although not developed. As experience was gained, the realization dawned that the technical difficulties in developing even a straightforward axial compressor were very great indeed. Thus, by early 1942 Spandau had abandoned the 109-002 engine when it was decided to concentrate all BMW's resources on a straightforward axial turbojet, the P.3302.

Loehner's P.3303 centrifugal turbojet

In 1938, when BMW in Munich accepted the official jet propulsion study contract, the work was put under the supervison of its research head, Kurt Loehner. The decision was made that valuable experience could be gained by the actual construction and testing of an experimental turbojet engine, utilizing, where possible, BMW's previous experience with its exhaust-driven turbochargers for piston engines. This turbocharger work was, from 1937, largely supervised by Dr Max Adolf Mueller, already mentioned in connection with Junkers and Heinkel turbojet work. Mueller worked for BMW as an independent consultant. Under his name, the design of a hollow, air-cooled blade for gas turbines (especially in turbochargers) was filed for patent in December 1937; a patent (682,744) for this 'waste-gas turbine for aircraft' was granted in October 1939.

A typical arrangement of a BMW turbocharger consisted of a single-stage axial turbine driving a single-stage centrifugal impeller or compressor. By 1938 the hollow, air-cooled turbine blades in these turbochargers were allowing turbine wheels to run at working temperatures of up to 900°C. The elements chosen by Loehner, therefore, for the experimental turbojet engine included a single-stage axial turbine (with hollow, air-cooled blades) driving a two-stage centrifugal compressor and a combustion chamber with the then-high working temperature of 900°C. While this engine, which is believed to have received the project designation of P.3303, was being designed in more detail, experiments proceeded with the turbine wheel in which the first difficulties arose.

F.9225 axial turbojet project

Another project undertaken at Munich was for an axial turbojet, designated F.9225. This engine appears to have been influenced by Mueller. It had a seven-stage axial compressor (possibly with reaction blading designed by Rudolph Friedrich), an annular combustion chamber, a two-stage turbine and a fixed-area exhaust nozzle. The compressor was of disc construction, with anti-friction bearings at each end. It was connected via a flexible coupling to a hollow shaft that carried the turbines at its end and was itself provided with two anti-friction bearings. With this

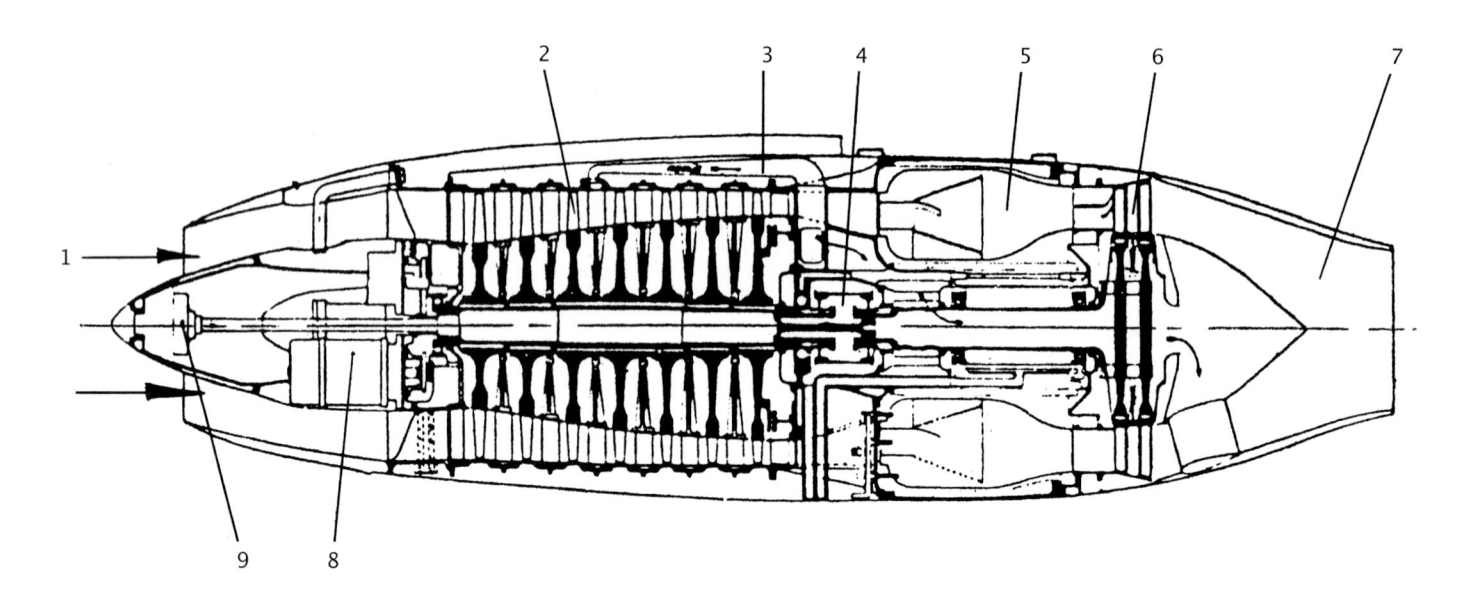

BMW F.9225 axial turbojet: (1) air intake; (2) 7-stage axial compressor; (3) cooling air pipe, tapped off from compressor; (4) flexible coupling; (5) annular combustion chamber; (6) 2-stage turbine; (7) fixed-area exhaust nozzle; (8) accessories; (9) starting clutch.

engine, the notion was hit upon to cool the turbine using compressor air instead of ram air entering a central intake. Accordingly, cooling air was tapped off from between the third and fourth stages of the compressor and led through the turbines' hollow shaft for cooling purposes.

With the outbreak of war in September 1939, the impetus was given for less experimental and more developmental work, so that it was agreed that BMW-Munich should drop the work on Loehner's centrifugal turbojet and Mueller's F.9225 and concentrate on the necessary production of piston engines for the Luftwaffe. Where possible, the experiences at BMW-Munich were to be used to assist in the development of the P.3302 axial turbojet at BMW-Spandau (formerly Bramo). This development was to evolve into the BMW 109-003, the company's principal turbojet offering during the Second World War.

At Spandau in 1939 only about forty engineers and designers were available for turbojet work, while construction and experiments had to be performed in the same workshops used for piston engines, which had priority. The position improved in 1942, when not only was the 109-002 turbojet dropped, as well as projects such as the MLS unit, but piston engine work was stopped at Spandau so that more personnel would be available to work on one turbojet project, namely the P.3302. From this point there were about 200 designers, scientists and test personnel available, plus about 700 people in the workshops. To a large extent, the plan to concentrate turbojet work at Spandau was followed but, owing to the dispersal of plants (because of the bombing of Berlin) and sub-contracting, the work was also conducted in various other places. A large dispersal was to caves at Wittringen (near Saarbrücken). Following the liberation of France in 1944, the work had to be removed from this border location and finally, after shorts stays at a couple of other sites, finished up in a disused salt mine at Neu Stassfurt (Shafts 6 and 7). This location was given the code name of Kalag and its twenty spacious, underground workshops were well equipped. Engine test-beds were located at the head of Shaft 6. By the end of the war in 1945, BMW had about 550 people on design and testing and about 1,000 in the workshops, largely by transferring personnel from piston engine to turbojet work.

The Managing Director of the whole organization was Fritz Hille. He was replaced about November 1944 by Dr Wilhelm Schaaf, while the Director in charge of all engine production was Dr Stoffregen, both men coming from the staff of Albert Speer's Armaments Ministry. From the Oberweisenfeld plant, near Munich, all BMW aircraft powerplant development was directed by Dr-Ing. Bruno Bruckmann, who had below him Dr Biefang (later succeeded by Dr Rolf Amman) directing piston engine development, Dr-Ing. Hermann Oestrich directing turbojet development and Dipl.-Ing. H. Zborowski directing rocket motor development. Bruckmann had been Technical Director in earlier days at Bramo.

Hermann Oestrich

Oestrich was born in Duisburg on 30 December 1903 and later studied at the Technisches Hochschulen in Hannover and Berlin. From 1928 he worked at the DVL's motor department in Berlin-Adlershof and it was there, around 1929, that he first became interested in theories about jet propulsion. In 1935 he worked at the Siemens motor works and gained experience with piston engines and high-altitude engines. From 1938 he was put in charge of the RLM turbojet study project at BMW-Spandau and then became, in 1943, the overall head of BMW turbojet development. Oestrich favoured an axial turbojet but without the complexities of a contra-rotating system that would need a long development period. From about 1942 Oestrich's turbojet staff included his assistant Dipl.-Ing. Hermann Hagen, who was involved in basic design and project calculations, and Dipl.-Ing. Hans Roskopf as his Chief Designer.

BMW, like Junkers, utilized in the first

stages of its turbojet development its existing piston engine facilities. Combustion test rigs had, at first, sufficient airflow to test only part of a combustion system so that BMW experimented with a segment of an annular combustion chamber. At least one test blower was driven by a 900hp Bramo 323 radial aero-engine. Gradually the facilities were improved and became more specialized, and in this respect the unique BMW high-altitude test plant is worthy of special mention. Designed by Dr Christoph Soestmeyer, construction of this plant began in May 1940 on the northern edge of the Oberweisenfeld factory site at Munich. Heavy bomb damage was sustained in May 1944, but by October 1944 the plant was repaired and remained in regular service from then until the end of the war. Both BMW 109-003 and Junkers 109-004 turbojets were tested in it. The engines could be tested at altitudes of up to 16,500m (54,100ft) with airflows of up to 25kg/sec (55lb/sec) and speeds up to 900km/h (559mph). Temperatures could be lowered down to –70°C.

Concentration on the P.3302 turbojet

During 1939 BMW-Spandau laid down the design of the P.3302 turbojet to produce a 600kp (1,323lb) thrust at 900km/h (559mph), or about 700kp (1,544lb) static thrust at sea level. An axial compressor driven by a single-stage turbine was planned, together with a very low combustion temperature of 600°C. When later in the year, however, Loehner's P.3303 centrifugal turbojet project was stopped at Munich, it was decided to incorporate the turbine wheel being developed for that engine into the P.3302. With this hollow-bladed turbine wheel, it was considered that the combustion temperature could be raised to the high figure of 900°C, while at the same time lowering the planned pressure ratio of the compressor. By this time the RLM had ordered ten experimental engines to be built.

Prof. Encke of the AVA designed the compressor form for the first experimental engine, the P.3302 V1. It was a six-stage axial type with a small percentage of reaction and gave a low pressure ratio of 2.77:1 at 9,000rpm. The rotor blades had AVA Göttingen high-speed profiles and test stand results indicated an 80 per cent efficiency for the complete compressor.

Testing the P.3302

Based on promising results from BMW-Spandau experiments in 1938 with annular combustion chamber segments, a combustion chamber of this type was chosen for the P.3302 that resulted in an engine diameter of only 0.670m (2ft 2⅜in), the RLM having asked for the minimum frontal area. Experiments with the turbine wheel led to the conclusion that the combustion temperature of 900°C was too high and so this was lowered to 750°C before the end of 1939. Unfortunately, when the V1 prototype first ran in August 1940, the result was disappointing and only 150kp (331lb) of thrust was produced at 8,000rpm on the test stand.

The causes of this poor result were several. Firstly, there was insufficient airflow for the combustion temperature, which had been lowered from that originally designed for. The lowering of the combustion temperature also lowered the velocity of the hot gases and therefore the power available from the turbine. There was as much as 370°C temperature difference around the combustion chamber, this resulting in distortion of the chamber inner liner and the turbine inlet nozzle diaphragm; this, in turn, created friction in the engine's bearings. Finally, there was a lot of fuel left unburned in the combustion chamber, lowering its efficiency down to 60–70 per cent, and high pressure losses were caused by the baffle plate system used. All in all, there were plenty of faults, but at least they were identified and work to eradicate them could begin immediately.

A new compressor, combustion chamber and turbine, plus innumerable detail improvements, were needed. This, of course, amounted to a complete redesign and the development of new components. For this work, the P.3302 V2 to V10 engines were used in addition to other research facilities. Outside assistance was obtained from the Brown Boveri, Brückner-Kanis and MAN companies in the development of a new turbine, while Henschel and the Technischen Akademie der Luftwaffe (TAL) assisted with the new combustion system. The branch of TAL responsible for combustion work was located at Eckertal (Harz) under a Dr Nagel.

BMW, however, was keen to tackle as much of the work as possible itself. Oestrich's team included Dipl.-Ing. H. Zobl and Dipl.-Ing. Kruppe, who headed combustion and fuel atomizer research, and Dr Sawert, who headed metallurgical research. Already in 1938 combustion work was started in Spandau to establish the best way of obtaining steady, efficient combustion without blow-out. It was decided that the right conditions could be provided by creating eddy regions in the airflow and so work began on the development of baffle plates inside an annular combustion chamber. Three main systems were tried. In the first, sixteen ceramic baffle plates were mounted in the combustion chamber and sprays from upstream fuel nozzles impinged on the forward faces of these. The total airflow passed around the baffles and was to carry the fuel into the lower-velocity eddy regions for combustion. However, not only were the blow-out characteristics unsatisfactory but the airflow around the baffles was not uniform around the annular chamber. The result was that temperature and velocity distribution was not even downstream and operation under full airflow conditions was not possible without a blow-out.

Modified from this system was a second system in which conical guide tubes were positioned in front of the ceramic baffles. These guide tubes permitted operation with full airflow without blow-out of the combustion, but temperature and velocity distribution still remained unacceptably poor. Also poor was the combustion efficiency, while the ceramic baffles frequently broke under thermal shock. Replacing these baffles with ones of steel gave an unlimited baffle life but did little for the other problems. In an attempt to obtain an even airflow around the baffle system, the individual circular baffles were abandoned and replaced by a continuous, perforated, steel plate annulus, but this gave no improvement and actually increased the pressure losses within the chamber.

It was probably the annular baffle plate that was used on the P.3302 V1 and a careful look at the poor results drew the conclusion that impinging the fuel sprays onto the baffle plate was wrong. Experiments whereby the fuel was injected upstream into the eddy regions were tried and immediately gave a 20 per cent improvement in combustion efficiency. There was also a 25 per cent improvement in the temperature and velocity distribution, although this was still far from good, and the total airflow still passed around the baffle plate and straight into the combustion chamber. Nevertheless, this improved baffle combustion system was

Two views of the BMW P.3302 axial turbojet (V2 to V10 series).

used for subsequent engines in the P.3302 series.

It was not until the next series that the baffle system was dropped and another combustion system, which was under parallel development and which used a divided airflow of primary and secondary streams, was introduced. The division of the airflow into primary for combustion and secondary for cooling and adding to the hot gases further along the combustion chamber was essential for correct combustion systems and was only gradually discovered by the various pioneering developers of turbojets.

By summer 1941 the P.3302 had been persuaded to give about 450kp (992lb) of thrust on the test stand, mainly by the combustion improvements mentioned above, and flight tests began from Berlin-Schönefeld, using a twin-engine Messerschmitt Bf 110 fighter with the turbojet suspended below. That autumn two of the P.3302 turbojets were delivered to Messerschmitt's Augsburg plant for fitting to the Me 262 V1 (PC+UA) prototype of the twin-engine fighter. Held up while waiting for turbojet engines, this aircraft had been flying since 18 April 1941 on the power of a 720hp Jumo 210G piston engine and airscrew in the nose. This piston engine was retained when, on 25 November 1941, an attempt was made to take off with the two P.3302 engines installed under the wings. This was a failure, however, since the turbines failed at take-off revolutions. Troubles had been experienced with the turbines for some time prior to this and failures had occurred at speeds as low as 8,000rpm. The type of turbine blade used in these early P.3302 engines was made by welding together two pieces of metal and the blades were attached to the turbine wheel by welding at the base. Cooling air entered the turbine wheel at the forward face only and passed by diagonally bored holes into the hollow blades and out through their tips. While this form of construction proved satisfactory in the smaller sizes used in BMW turbochargers, the larger size of the turbojet turbine proved too much and failures occurred in the welded joints through fatigue and in the blades through heat-induced brittleness. However, all the causes of failure took time to identify and this type of turbine construction appears to have been persevered with for most, if not all, of the first ten prototypes of the P.3302 series.

Early in 1942 two of the P.3302 V2–V10 engines were again delivered to Messerschmitt and fitted to the Me 262 V1. On 25 March 1942, with the nose-mounted piston engine and two turbojets running, Fritz Wendel cautiously prepared to take off in this aircraft. Improvements to the turbojets had raised the static thrust to about 500kp (1,103lb) each, but this was still some way below the planned thrust. According to the Messerschmitt flight test report 692/12 of 30 March 1942, following a long take-off run of between 800 and 900m (875–985 yards), the engines were throttled back after 20 seconds during the climb at 340km/h (211mph). At 1,000m (3,280ft) Wendel levelled off and, at a horizontal speed of 400 to 450km/h (248–280mph), despite further throttling, he could not get the turbojets below 7,000rpm, estimating that thrust was still over 200kp (441lb) per engine. (At that time, the BMW engine controls had no idling position.) When the fuel gauges showed 100ltr (22gal), a strong fluctuation in the fuel injection pressure was noticed and the left engine ran rough. On moving its throttle to the 'stop' position, the right engine throttle was inadvertently also moved so far back that it too cut out. The rpm indicators fell from 7,000 to below

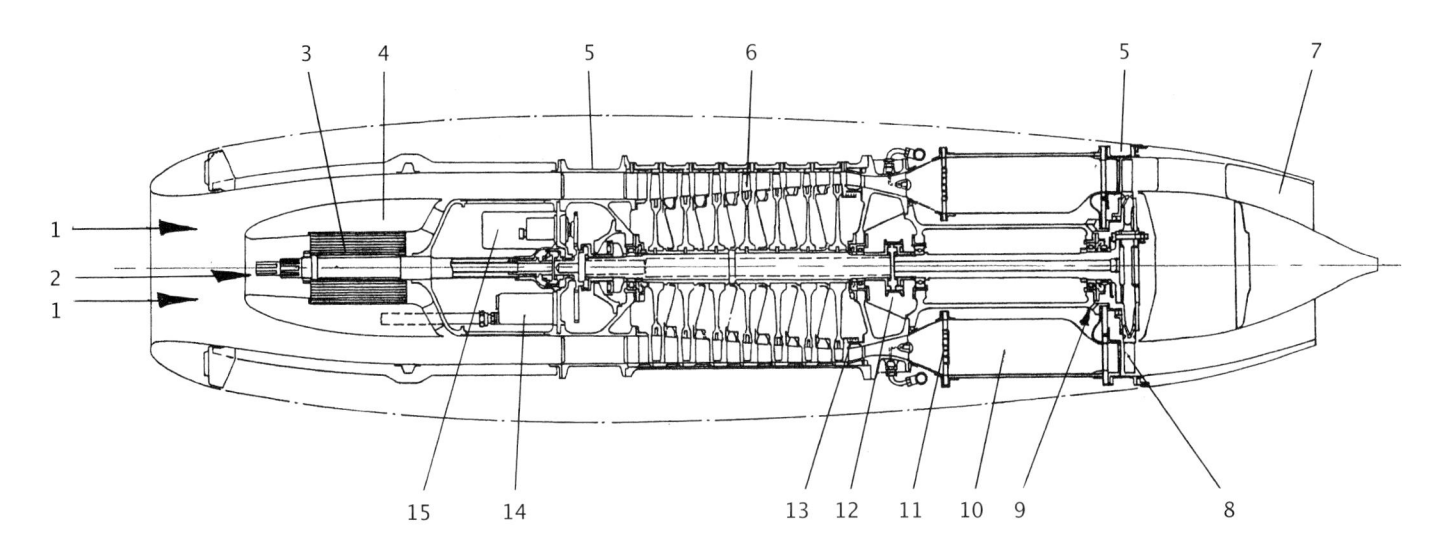

BMW P.3302 (V11 to V14 series) axial turbojet: (1) air intake; (2) cooling air intake; (3) oil radiator; (4) oil tank; (5) main supports; (6) 7-stage axial compressor; (7) fixed exhaust nozzle; (8) single-stage turbine, bolted on; (9) cooling air inlet; (10) annular combustion chamber; (11) baffle plate; (12) flexible coupling; (13) labyrinth seal; (14) oil pump; (15) accessories.

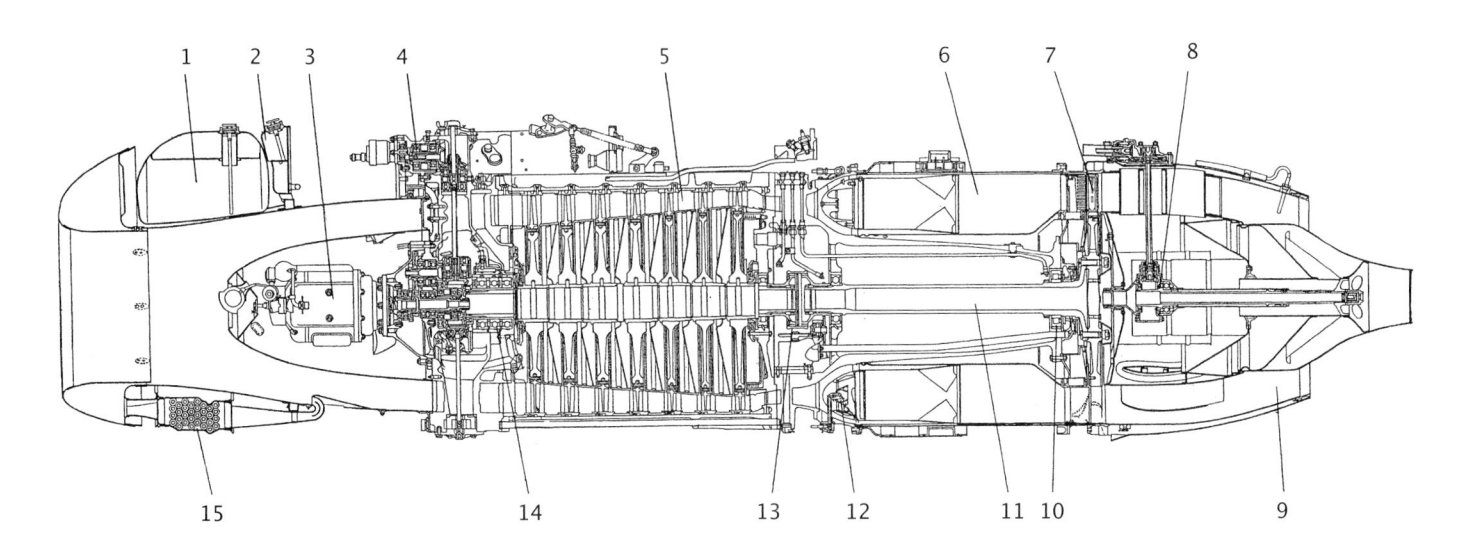

BMW 109-003 A-0 axial turbojet: (1) oil tank; (2) starter motor fuel tank; (3) Riedel AK-11 starter engine; (4) auxiliaries gearbox; (5) 7-stage axial compressor; (6) annular combustion chamber; (7) single-stage turbine; (8) exhaust cone drive gear; (9) variable-area exhaust nozzle; (10) rear roller bearing; (11) hollow rotor shaft; (12) fuel injection nozzle; (13) shaft coupling; (14) front ball bearings; (15) oil cooler.

2,000. Recommendation was that variable-area exhaust nozzles were needed before the next flight. Despite the high sinking speed during the landing approach at 240km/h (149mph), and with nose flaps fully extended, Wendel landed safely. After some two-thirds of the runway length, both steering mechanisms of the sprung undercarriage snapped off at their welding points for unknown reasons.

The Me 262 V1 was repaired and flew again on 2 March 1943, but this time used two Junkers 109-004 B-0 turbojets and practically all subsequent aircraft in the series were also powered with Junkers engines. Even so, up to about April 1942 both the Junkers and BMW experimental turbojets were running below their design thrust by about the same amount and both were beset with their own problems. Final data for the P.3302 V1-V10 series included a static thrust of 500kp (1,103lb) at 9,000rpm for a weight of 750kg (1,654lb). Using petrol, the specific fuel consumption was about 2.2 and the air mass flow was 15kg/sec (33lb/sec). With modifications, the diameter grew from 0.670m (2ft 2⅜in) to 0.690m (2ft 3¼in).

Following experiences with the prototypes from V1 to V10, BMW-Spandau

A sectioned BMW 109-003 A axial turbojet.

began construction of four more P.3302 prototypes (V11–V14) during summer 1942. For these a new compressor was designed that gave a 30 per cent greater air mass flow and an increased pressure ratio of 3.1:1 at a speed of between 9,000 and 9,500rpm. This new compressor was designed and built at Spandau and had seven instead of the previous six stages. The profiles of the rotor blades were changed to a thicker NACA type, probably on the advice of Dr Betz of the AVA, and 30 per cent of the pressure rise now occurred in the stator blades. Also, the lengths of the blades were decreased.

An improved turbine with greatly increased efficiency was also introduced, the rotor blades having a rounded leading edge and a longer, sharp trailing edge instead of the previous form with similar leading and trailing edges. To improve blade cooling, and therefore durability, more air was brought from both sides of the turbine wheel and into the roots of the hollow blades. The previous welding method of blade attachment was also abandoned and the new blades had longer roots, each being secured to the wheel by three axial pins that were buried half in the blade root and half in the wheel. These pins, which were arranged two on one side and one on the other side of the blade root, held the blades against centrifugal forces, while shoulders at the base of each blade countered radial or rocking forces.

Hollow vanes having profiles similar to the turbine rotor blades now replaced the previous, simple, curved metal vanes forming the turbine inlet nozzles. This not only gave greater rigidity and resistance to distortion for the inlet nozzle assembly but also allowed for air cooling. Cooling air entered the inlet nozzles at their bases and emerged through trailing edge slots. The combustion system, at this stage, was still of the annular baffle type with upstream injection of the fuel into the airstream's eddy area and without division of the airflow.

Test-bed running began with the P.3302 V11 towards the end of 1942, but no attempts were made to fly an aircraft with the new engine and flights were confined to those with the Bf 110 flying test-bed. At some stage a new intake cowling was introduced and the first tests with a variable-area exhaust nozzle began. The new intake cowling, of improved form more suitable for high-speed flight, was designed by Dr Küchemann of the AVA, aided by data obtained from wind-tunnel experiments at the Luftfahrtforschungsanstalt Hermann Göring (LFA; Hermann Göring Aeronautical Experimental Establishment). The variable-area exhaust nozzle used the so-called Ringspalt-Schubdüse, whereby a tapered ring could be moved axially inside the exhaust nozzle. Operation of this ring was carried out manually and both the hollow ring and its hollow supporting spokes were air cooled.

First runs with the new engine yielded at least 550kp (1,213lb) static thrust, but were disappointing since a number of deficiencies were revealed with the compressor, combustion system, turbine, exhaust nozzle and bearings. The first stage blades of the compressor fractured after a short time due to vibrations and starting characteristics were not as good as with the earlier compressor due to high drag. Serious problems remained with the baffle plate combustion chamber and its poor efficiency and temperature distribution could not be improved. Turbine blade failures had lessened but were still frequent. Distortion of the exhaust nozzle ring occurred and its operating mechanism overheated. Finally, the compressor axial thrust bearing overheated due to insufficient oil feed during starting and acceleration. Clearly, much development work still needed to be done. Salient data for the P.3302 V11-V14 series included a static thrust of 600kp (1,323lb) at 9,500rpm and a weight of 650kg (1,433lb). The pressure ratio was 3:1, air mass flow 18.8kg/sec (41.45lb/sec) and specific fuel consumption with petrol 1.6. The engine diameter, without auxiliaries, was 0.690m (2ft 3⅛in).

Development of the 109-003 axial turbojet

By the end of 1942 solutions had been found or were within sight for the most serious difficulties and a new engine had been designed and was under construction. This new engine, which was the final major redesign, was the responsibility of Hans Roskopf, now Chief Designer at Spandau. The P.3302 had by then received the official designation of 109-003, the pre-production models being designated 109-003 A-0. Some, if not all, of the new features designed to eliminate the previous faults were probably tested in the P.3302 V15 and V16 prototypes. However, improvements were still necessary at the pre-production stage of the 109-003 A-0 engines, so true pre-production engines were designated 109-003 A-00.

The first stage rotor blades of the new compressor used with the previous experimental engines had failed within one hour of running and this was obviously a serious problem requiring close scrutiny. Accordingly, a plastic window was fitted into the compressor casing facing the first stage of rotor blades. A small steel point or scriber was then fitted into the tip of one rotor blade at approximately the centre of gravity of its profile section. By this means, it was found that the 75mm (3in) long rotor blade took up a vibration motion on its natural frequency, the amplitude of which reached up to 2mm under the most critical conditions at approximately the rated speed of 9,500rpm. The blade frequency was, however, far too low to be excited by variation of aerodynamic pressures caused by the rotor blades passing through the wakes of the stationary vanes in front of the rotor, and it was found that the trouble emanated from further forward at the profiled ribs (or spider) supporting the front rotor bearings. These ribs generated a wake sufficient to create aerodynamic disturbances that were cut by the rotor blades. Unfortunately, when running at about rated rpm, the first-stage rotor blades cut the disturbances at a rate equal to the natural frequency of the blades, so that a resonant vibration was excited that led to blade fatigue and then failure. None of the rotor blades after the first stage ever gave trouble, since the aerodynamic disturbances grew less as they travelled rearwards, while the natural frequency of the blades in each stage progressively shortened.

Two modifications were introduced to cure the compressor trouble by creating as wide a gap as possible between the critical speed of the rotor and the rated speed of the engine. The 9,500rpm rated speed of the engine could not be readily altered and so other changes were made. The rotor blade profiles were made thicker, thereby increasing their natural frequencies and decreasing centrifugal and bending stresses on them. Secondly, a new support spider was designed for the front rotor bearing and the number of ribs was decreased, thereby reducing the frequency of the excitation transmitted in the rib wakes. Also, to break the sequence of the impulses, the ribs were positioned with unequal spacings.

These modifications were entirely successful and no further compressor blade failures were suffered, but the compressor efficiency was reduced from 80 per cent to 78 per cent. BMW also tried, at one stage, compressor blades with a degree of looseness in the rotor. Running with such blades was quite possible, centrifugal forces keeping the blades in a stable position, but the critical vibrations were worse and so firmly fastened blades were retained.

Despite the considerable improvement of the combustion chamber, results had remained poor and it was decided that the baffle plate system was basically wrong. Another type of combustion system was therefore put under development and was ready for introduction into the 109-003 A-0 engine. A fundamental difference of the new system was the division of the airflow into primary and secondary streams. The primary stream (60–70 per cent of the total) passed to burner cones to support combustion, while the secondary stream (30–40 per cent of the total) was introduced downstream to mix with the hot gases. Each of the sixteen burner cones had its own primary air supply, the downstream edge of the cone producing toroidal eddies into which the fuel was directly sprayed. Further downstream there were eighty hollow fingers that protruded into the annular combustion chamber from the inner and outer walls, the secondary air flowing from these fingers into the gas stream. A later

The world's first four-jet aircraft. The Arado Ar 234 V8 prototype (c/n: 130008, GK+IY) was fitted with four BMW 109-003 A-0 turbojets as part of the trials to ascertain the best arrangement of engines for the more-powerful Ar 234C. Comparisons were made with the Ar 234 V6 (c/n: 130006, GK+IW), which had four BMW 109-003 A engines in separate nacelles, and it was found that the combined nacelles (V8) were superior.
V8, Philip Jarrett

improvement admitted a small amount of air around the actual fuel nozzles, this reducing the temperature at the centres of the the flames and cooling the nozzles. This new combustion system was a success. Combustion efficiency was raised up to 90 per cent or more, pressure losses were greatly reduced and the temperature and velocity distribution was greatly improved. At this stage, petrol was still being used as fuel.

Improvements to the turbine section of the engine concentrated on improving the cooling and extending the life of the rotor blades. To replace the previous turbine blades made from two welded halves, blades forged from tubular stock by the Leistvitz company were tried; although these gave a longer life, they proved unsuitable for mass production in the envisaged quantities. BMW then developed a new blade made by rolling sheet stock, folding and welding down the trailing edge. Inside the hollow blade was a sheet-metal insert that restricted the cooling-air passage to a small volume around the inner wall of the turbine blade. Adequate cooling was thus achieved using the minimum of air, the cooling amounting to about 2 per cent of the total engine airflow in the 109-003 A-0. Not only did the cooling-air insert (*Kühleinsatz*) afford outstanding economy in cooling the turbine blades, but it contributed largely to the subsequent freedom from blade failures; friction between the insert and the blade occurred every time the blade flexed and thereby gave a strong damping action. Since the new blades had a less solid foot than before, a new method of attachment to the wheel was needed. A pin was welded to the blade and cooling insert feet and the whole assembly was fitted with two side pins into an appropriately shaped cut-out in the turbine wheel rim. A shaped circular plate on each side of the wheel prevented longitudinal movement of the blades and pins, and also served to duct cooling air to the blades. This cooling air was tapped from the fourth stage of the compressor. Although the blades and their inserts were still made from heat-resisting steel, the amount used was less.

Results with the new turbine wheel were good and its life was considerably increased. Furthermore, the 109-003 A-0 was designed so that the turbine wheel could be replaced easily without removing the engine from the aircraft, an advantage not enjoyed by the Junkers 109-004. When turbine blades did fail (initially after about twenty hours), the fracture occurred at the blade neck, just above the foot, and was caused by hardening of the metal.

Since the movable exhaust nozzle ring had proved deficient, a new method of varying the exhaust nozzle exit area was devised. The new device was known as the mushroom exhaust nozzle (*Pilz-Schubdüse*) and in principal was the same as the movable bullet type of device already described for certain Heinkel-Hirth and Junkers engines. The power to move the bullet axially, and thereby alter the exit area, was derived from an electric motor and transmitted via a shaft to bevel gears, a radial shaft to the engine axis and other gears that rotated a nut. The nut rotated on a threaded shaft connected to the bullet. Air was used to cool part of the drive mechanism and the sheet metal exhaust structure. The new system was less susceptible to distortion.

The final principal modifications introduced with the 109-003 A-0 were a new and extensive lubrication system and the use of anti-friction (ball and roller) bearings at all the main points. Previously some bearings had been of the plain, bushed type. All auxiliaries were now fitted, including a Riedel starter motor.

With the appearance of the 109-003 A-0, the BMW flight-testing section began using a twin-engined Junkers Ju 88 as a test-bed. The turbojet under test was mounted beneath the aircraft's fuselage, with its fuel tank in the bomb bay, and the first flights with this Ju 88 were made in October 1943. The new features introduced with the 109-003 A-0 proved to be on the correct lines and, with systematic development, the engine reached its new rated static thrust of 800kp (1,764lb) at 9,500rpm. Trouble-free runs of at least twenty hours were then obtained. The engine's weight was 570kg (1,257lb), its diameter 0.690m (2ft 3⅛in) and length 3.565m (11ft 8⅜in) with the exhaust bullet extended. The specific fuel consumption,

Preparing the Arado Ar 234 V6 for flight. On early prototypes, such as this, starting of the BMW 109-003 A turbojets was by means of a piston engine on a trolley, a shaft and a clutch – a method going back to the early days of piston aero-engines and the Hucks starter. E.J. Creek

using petrol, was between 1.35 and 1.4 and the air mass flow was 19.3kg/sec (42.55lb/sec) for a pressure ratio of 3.1:1.

The year of 1943 saw the end of competition between the He 280 and Me 262 jet fighters, since the He 280 was abandoned for reasons already mentioned, and plans to fly the He 280 with BMW 109-003 engines were also abandoned. The 109-003 A-0 was next tested to power experimental prototypes of the Arado Ar 234 A jet bomber. This had a straight, shoulder-mounted wing, conventional tail empennage and a cockpit glazed flush with the nose and fuselage in unbroken lines. Initially, the prototypes took off on a jettisonable trolley and landed on a skid, pending the later development of a retractable, tricycle undercarriage. The maiden flight of the Ar 234 V1 (TG+KB) prototypes had taken place on 30 July 1943, piloted by Selle. This and other early prototypes were powered by two underwing Junkers 109-004 turbojets but, since this engine was largely earmarked for the Me 262 fighter, the RLM was keen for the Ar 234 bomber to use BMW 109-003 engines.

With the Ar 234 V8 (GK+IY, c/n: 130 008), four 109-003 A-0s, arranged in pairs, were fitted in order to give sufficient thrust. In this form the V8 first flew on 4 February 1944, but soon the turbines needed to be changed. Also, severe shock waves were generated between the paired nacelles and the fuselage. Therefore, the Ar 234 V6 (GK+IW, c/n: 130 006) was fitted with four 109-003 A-0 engines in spaced-out nacelles beneath the wings. The V6 first flew on 25 April 1944. The pilot, Ubbo Janssen, became quite fatigued, however, owing to the necessity of constantly juggling with the engine throttles in attempts to equalize the thrusts of the engines, this problem being compounded by the lack of automatic speed governors. On the seventh flight, on 1 June 1944, all four engines failed one after the other, but Janssen succeeded in making a good emergency landing. An engine control and governor system was then adapted from the Junkers 109-004 design. Equipped with the new control system and governors, four 109-003 A-1 engines were fitted to the Ar 234 V13 (PH+SU, c/n: 130 023), once more in the paired but improved arrangement since this was considered the most efficient. The V13 first flew on 6 September 1944, the pilot again being Janssen, but three of the engines failed and the aircraft was badly damaged in the emergency landing that followed. From then on, all Ar 234 Blitz bombers (as they were nicknamed) were flown with BMW 109-003 engines.

By August 1944 a hundred 109-003 A-0 engines had been built, of which sixty were intended for flight, and more than 4,000 hours of running had been performed on test stands for experimental and developmental purposes alone. At this time, however, BMW's engine was lagging behind the Junkers 109-004 in performance and further delays were to come in adapting the engine to run on J2 fuel instead of petrol. Even so, the RLM opinion was that the BMW engine had great development potential and it was hoped to produce it more rapidly and economically than the Junkers engine. The annular combustion chamber, superior accessibility, smaller frontal area and weight of the 109-003 were all looked upon with official favour.

The first examples of the mass-production model, the 109-003 A-1, began coming off the assembly lines around October 1944, although a few examples were released months earlier for testing. The differences between the A-0 and A-1 models were mainly in detail design to suit mass production. Whereas performance was hardly altered, apart from the engine's life, three serious limitations had to be taken into account. These limitations, officially requested, were that the engine had to be produced for about 500 man-hours per unit, that each engine should use no more than about 0.6kg (1.323lb) of nickel (far less than for the Junkers 109-004 B) and that the engine should be broken down into the maximum number of easily made components. Another requirement, easy accessibility, was already largely met in the 109-003 A-0 model.

Therefore, before the engine was committed to mass production, the design was turned over to a Dr Fattler and his staff to simplify ruthlessly wherever possible, without falling below the performance of the A-0 model. Dr Fattler had experience with mass-production techniques in the American automotive industry before the war and he substituted sheet-metal parts for many of the cast and machined ones. The enthusiastic substitution of such parts would have been taken further than it was, were it not for long and acrimonious meetings between the staffs of Dr Fattler and Dr Oestrich finally thrashed out a compromise between the purely mass-produced and the technically good engine. The final engine did not meet the exact requirements of production time and material limitations but came reasonably close.

To increase the life of the engine, most attention was given to improving the turbine section. An improvement to the turbine blade fixing was made, whereby a wedge was positioned on each side of the blade. These wedges were forced tight by centrifugal action and prevented the tangential blade movement that wear or bad fitting had previously allowed. This modification was actually tried on the last few A-0 models.

Other small but effective improvements eventually allowed the 109-003 A-1 to run up to fifty hours or more without trouble, the final limit being set by the turbine. Of course, this time between overhaul (TBO) was low compared with British engines but was the best that could be done given the limitations in available metals. By careful metering, a reduction in cooling air was found possible and air for the turbine was tapped off after the last compressor stage instead of at the fourth stage, as in the A-0 model.

Changes made in the 109-003 A-2 engines were almost exclusively to ease production and not all the latest modifications to improve performance had been incorporated on the production lines by the end of the war. Thus, while experimental versions of the A-2 model are reputed to have produced static thrusts of up to 1,200kp (2,646lb), the general performance of the production models remained similar to the A-0 and A-1 models. The weight, however, increased by 30 to 40kg (66–88lb) because of the simplification and strengthening of various parts. While the number of turbine rotor blades was decreased to sixty-six (compared with seventy-seven for the A-1), the thickness of the blade metal was increased from the previous metric gauge of 2.15 up to 2.65. Two small fins, which were fitted to the exhaust cone of the A-2 and late examples of the A-1, reduced vibration.

A setback for the production engines came in 1944 when an official decree required that all turbojets should run on the more readily available J2 fuel, the use of which had been fully proven by Junkers turbojets. A BMW fuel system was not ready and so a system similar to that for the Junkers 109-004 B was adapted.

However, this required a considerable amount of time to accomplish. The most obvious change was the replacement of the Henschel centrifugal pump with a Barmag gear pump in order to reach the pressure of 60 atmospheres required at sea level. Changing to J2 fuel initially gave some sooting problems, but these were overcome by correction of the fuel nozzle spray angles. When the burners were finally matched to the new fuel system, the heat distribution was found to be better than with the petrol system. When hot spots did occasionally develop in the combustion system, they were sometimes visible on the exhaust cone.

109-003 A-1 turbojet

The essential layout of the 109-003 A-1 turbojet, the main production model, still followed the general lines of the original P.3302 prototypes, thereby confirming the correct thinking of Dr Oestrich and his team. The most obvious changes from the P.3302 were the removal of the oil cooler tank, cooler and auxiliaries from the central inlet fairing to make room for the Riedel starter unit, the vastly improved annular combustion chamber and the provision of a variable-area exhaust nozzle. Thus, clustered around the intake duct, behind the streamlined cowling, were the oil tank, oil cooler and a small petrol/oil tank for the Riedel two-stroke starter engine and a petrol tank used to initiate starting of the turbojet prior to switching over to J2 fuel. The Riedel starter was housed in the central cowling, well back inside the intake duct.

The compressor casing was cast in Elektron alloy and was not split longitudinally, as in the Junkers turbojet. Most of the auxiliaries and drive gears were attached to this casing. The intake duct was attached to its forward end, while, attached to its rear end, there was a cast ring supporting the compressor rear bearing, with eight circumferential slots through which air from the compressor passed to the annular combustion chamber.

The compressor was of the seven-stage axial type and used a small amount of reaction, the blades having almost symmetrical profiles. Each rotor blade was secured to its disc groove by a single pin through a tongue in its root. The rotor discs were forced onto a shaft, which was supported by three ball races at the front and a single roller race at the rear. Eight rows of stator blades were secured between inner and outer shroud rings. The hot air tapped off from the last compressor stage was still suitable for turbine cooling. No surging problems were experienced with this compressor.

From here on, the engine was largely of sheet metal construction. The annular combustion chamber had forty inner and forty outer hollow fingers to feed in secondary air after the sixteen burner cones. These injected J2 fuel downstream and imparted some swirl to the fuel to encourage atomization. At the top of the combustion chamber were two petrol injectors and two sparking plugs to initiate starting. Because the hollow fingers experienced temperatures up to 800°C, they were made from heat-resisting steel but the rest of the combustion chamber was constructed from aluminized mild steel. The combustion chamber had a life of about 200 hours.

The single-stage turbine wheel was bolted to a flange on the hollow rotor shaft that was supported at each end in a single roller race. A splined coupling connected the turbine shaft to the rear of the compressor rotor. A life of about fifty hours was expected for the turbine, with an inspection every ten hours. The turbine inlet nozzles and rotor blades were made from heat-resisting steel and were of hollow construction to permit air cooling. Each hollow rotor blade had an insert to reduce the internal air space and also to give friction damping when the blades flexed. Attachment of the blades to the turbine wheel was by means of pins and wedges. After about fifty hours of running the turbine blades were apt to harden and crack, but it was possible to change a turbine wheel in about two hours.

Variation of the exhaust nozzle area was by means of a movable bullet or tail cone. Hollow walls and spaces were used in the exhaust system to allow air cooling. Movement of the bullet was by means of an electric motor driving through bevel gears, shafts and a threaded sleeve. An electric switch in the cockpit allowed the setting of the exhaust nozzle area at maximum for starting and idling, minimum for take-off and climbing and intermediate for horizontal flying. In the exhaust system only the hollow vanes supporting the tail cone structure and bullet final drive mechanism were made from heat-resisting steel. All other sheet metal was of aluminized mild steel.

The fuel system and centrifugal all-speed governor, which regulated the fuel flow and engine speed, was derived from that used by the Junkers 109-004 B engine. Hauptmann Bispink of BMW's flight test section flew one of Arado's Ar 234 bomber prototypes with 109-003 A-1 engines with the all-speed governors up to 13,000m (42,600ft) altitude in September 1944. Flame-outs would occur if less than full power was used at such a height and re-starting was only possible at or below 3,500m (11,500ft). For more efficient operation at all altitudes, a duplex fuel injector was under development but not ready before the war ended.

The 109-003 A-1 and A-2 turbojets were designed for underwing installation but were produced in modified forms as the 109-003 E-1 and E-2 to allow installation above a fuselage. To produce the E models mainly involved the siting of attachment points below the engine, moving most auxiliaries to the top of the engine and some re-routing of pipes and cables. Re-positioning of the auxiliaries gave the cowling of the engine its characteristic hump-backed appearance.

The static thrust of the 109-003 A-1 was 800kp (1,764lb), or 705kp (1,554lb) at 900km/h (559mph) at sea level, at a speed of 9,500rpm. Its net dry weight was 570kg (1,257lb), its diameter was 0.690m (2ft 3⅛in) and length 3.565m (11ft 8⅜in) with the exhaust bullet fully extended. The air mass flow was 19.3kg/sec (42.55lb/sec) and compression ratio 3.1:1.

CHAPTER TWENTY-THREE

Flight-Testing and Production

Junkers 109-004 B turbojet

Junkers typically used the Ju 88 medium bomber to flight test its 109-004 B turbojet, the engine under test being suspended from the bomb shackle points between the port engine and the fuselage. Examples of such aircraft are GH+FQ and DE+DK. Thrust in flight was ascertained by measuring the pressures before and after the turbine and at the exhaust nozzle while fuel consumption was measured using flow meters.

Official tests, both static and in-flight, were also carried out on both experimental and production engines at E-Stelle Rechlin on a regular basis. For flight tests, a Junkers Ju 52/3m tri-motor transport aircraft was used: the turbojet was suspended beneath the fuselage in a parallel-motion frame and the pressure on a load cell gave an electrical reading equating to the thrust. Later, various Me 262 A fighters with 109-004 B-1 engines were flight-tested at Rechlin. The Me 262 flight tests showed that, for a 15°C difference in air temperature, a difference of 30 per cent in thrust and 25 per cent in fuel consumption resulted. Consequently, the average level speed of the Me 262 A fighter at 7,000m (22,960ft) varied from 868km/h (536mph) in winter to 820km/h (508mph) in summer.

Towards the end of 1943 the RLM was ready to put jet aircraft into production and 109-004 B-1 turbojets were being produced in quantity. The first production engines were made at Dessau, where all the experimental engines were also built, but from August 1944 production was taken over by Junkers factories at Köthen, 25km southwest of Dessau, and at Muldenstein, to the east of Dessau across the Mulde river; the last of these eventually achieved the highest total production. The vast underground factory known as Mittelwerke GmbH in the Harz mountains near Nordhausen joined in the production programme in October 1944. This was followed by a factory at Zittau (about 75km east-southeast of Dresden) the following month. The final production centre in Germany, known as Mitteldeutsche, was located at Leipzig, but production of the 109-004 B had barely begun when the war ended. One other plant outside Germany had also just begun producing engines: this was near Prague, possibly at the Letov factory. Between them these factories produced a total of 6,010 completed 109-004 B-1/B-4 and experimental engines. Production ceased in March 1945, two months before the end of the war in Europe.

The time taken to produce one 109-004 B was about 700 man-hours, which compares with about 600 man-hours for the BMW 109-003 turbojet. In some cases, of course, engines took longer than this to produce. At the Mittelwerke, where the Junkers section occupied the rear portion of a tunnel about 3km long northwest of the town, the programme was rather disorganized because of the heavy load of work in the machine shops on parts for V-weapons and other aero-engines. The large number of measuring gauges there was indicative of the lack of jigs for production, at least at the Mittelwerke. Nevertheless, it seems that the later total production rate for the 109-004 B was close to the scheduled rate.

Other production centres were being set up apart from those already mentioned, but were either too late or were abandoned to advancing Allied forces. The largest such centre was probably situated at Strasbourg in France, where, from June 1940, Junkers had taken over the Matford plant, which had formerly built Ford cars. Renamed as the Junkers-Meinauwerke, this plant was re-equipped to produce 250 new piston engines (for example, Jumo 211) and repair 500 piston engines per month. Production of the 109-004 B at Strasbourg was scheduled at twenty engines in September 1944, increasing to a maximum of a thousand per month by July 1945, but bombing of the area on 27 May 1944 set the programme back. Dispersal was then begun around the general Strasbourg area to make use of the existing facilities. However, no complete 109-004 engines were turned out in France before the liberation. Nevertheless, it can be seen that, had the war continued a little longer, a great number of turbojets would probably have been produced in Germany.

BMW 109-003 turbojet

The finest of BMW's test facilities was its high-altitude test chamber at Oberweisenfeld, Munich, in which pressure and temperature conditions could be maintained equivalent to an altitude of up to 16,500m (54,000ft). Because this plant required a power load of about 10,000kW, only night-time operation under Munich off-peak hours was possible. Besides this plant, there were, of course, many other test cells and stands for ground tests. A great deal of BMW's test work was also performed by its flight test section, which at one time operated from Berlin-Schönefeld under Flugkapitän Staege. In July 1944 Dr Denkmeir was made responsible for the sub-section devoted to turbojet flight tests. Towards the end of the war, however, when long-range research became pointless, Dipl.-Ing. Peter G. Kappus was put in charge of flight testing and was primarily concerned with readying the 109-003 E for the Heinkel He 162 fighter. Previously Kappus had headed project studies (department EZD) but was far from being desk-bound, since extensive flight training with the DVL before the war enabled him to later fly jet aircraft.

The first aircraft used for 109-003 flight tests by BMW was a twin-engine Ju 88, which began flying with an underslung A-0 turbojet in October 1943. Further flight test installations for turbojets were considered at a meeting held by BMW on 2 May 1944 and requests for more aircraft were made. This resulted, in the latter half of 1944, in two more aircraft being taken on as flying test-beds, these being the

Arado Ar 234 V15 and V17. They were originally Ar 234 B Blitz bomber prototypes and each was fitted with two 109-003 A-1 engines.

The main work involved in the flight-testing was exploring the altitude flame-out and restarting limits with the fixed orifice fuel injectors, testing the variable-area exhaust nozzle and testing the J2 fuel system, which had been introduced at a late date. Another flight programme covered the testing of a new surface oil cooler that was to be introduced on late 109-003 A-2 engines. This cooler was integral with the intake cowling to help combat compressor icing and was easier to make than the tubular cooler. Work on the fuel system, however, became the most urgent priority: by 1945 the BMW flight test section had acquired a further Ar 234 B prototype (possibly the V18), a four-engine Ar 234 C prototype, another Ju 88 and two Messerschmitt Me 262s to facilitate the work. Nevertheless, because of the acute shortage of fuel, flying time was progressively and drastically reduced as each main problem was solved.

Towards the end of the war the BMW aircraft were moved about to different bases for safety from the advancing Allied forces. The first such move in 1944 was to Oranienburg and this was followed by a move to Burg (near Magdeburg). According to Peter Kappus:

> Towards the end of the war, I was assigned the responsibility of flight-testing the BMW 003 engine. At that time, a decision had been made to mass produce the so-called People's Fighter plane (He 162), which was to use our engine, and it was necessary for the engine development to be completed in a very short time. The flight-testing took place over a period of several months at Oranienburg, near Berlin, and was then transferred to near Magdeburg. At BMW, we initially used the ... Ju 88s. We then switched to the twin turbojet reconnaissance bomber, the Arado Ar 234, which had its engines mounted [on] the wings and was a fairly advanced-looking, high-speed modern plane.

> The critical turbojet problem at the time was high-altitude flame-out. We had a fuel system and injection system that was not very reliable at high altitudes. We then had to dive down to a relatively low altitude without power and fire up the engine(s) again.

> Another problem was to make use of the very crude fuels that were available. We had started using gasoline in turbojets but, because of the destructive moonlight attacks on the refineries, all we could get was a crude heavy oil. We adapted the Junkers fuel control system to the BMW engine and managed to operate the engines with this heavy fuel but we still needed gasoline to get the engine started.

Orders then came to evacuate Burg, where it is believed the two 109-003-powered Me 262s were abandoned since they were not ready for testing. BMW flights continued from the Autobahn south of Munich, but it is doubtful if much work was possible there at that late stage of the war. In any event, Kappus himself flew one of the Ar 234 B prototypes from Burg to Lechfeld (south of Augsburg) and then on to Neubiberg (south of Munich). There, BMW finally blew up its equipment as the US 42nd Division approached Munich. The only known figure for the total of 109-003 flight tests is the vague one of 500 to 1,000 hours.

Another flying test-bed for the 109-003 was the He 219 twin-engine night fighter that the Heinkel company had at Vienna. The single turbojet fitted below the fuselage was for the purpose of gaining data on the spread of the exhaust stream in the neighbourhood of the fuselage and not for testing the engine. This data was to assist in the design of aircraft, such as the He 162, where the turbojet was fuselage mounted.

As early as 1937 BMW had requested, with much foresight, that its production plants should be moved underground, but this was not officially sanctioned. The first real bombing damage to its plants occurred in February 1944; the following month BMW production began moving underground and to dispersed sites. Manufacture of the 109-003 A and E production turbojets did not begin until this dispersal was complete by October/November 1944. Only the manufacture of pre-production 109-003 A-0 and A-00 engines was carried out earlier at BMW's Spandau and Basdorff plants, with final assembly at Eisenach (50km west of Erfurt), the latter plant being codenamed Town-Werke.

Following the dispersal, Eisenach was supplied with parts from salt mine factories at Heiligenroda, Abterroda, Bad Salzungen and Springen in the Eisenach area and also from salt mine factories at Plömnitz and Egeln, south of Magdeburg. There were other dispersed plants making complete engines, notably in one of the tunnels at the Mittelwerke at Nordhausen in the Harz mountains. Dispersed production was facilitated by the engine being designed with a maximum number of easily made components.

Dr Fattler had modified the 109-003 to be built in 500 man-hours but, in practice, 600 man-hours per engine was the best that could be achieved. The optimistic production figures were never achieved either: by the war's end a total of 559 engines, mostly A-1 and E-1 variants, had been built. Setbacks were caused by delayed development, bombing and the inaccessibility of the factories located in salt mines and disused mines. These mines usually had small, old lifts that limited the size of machinery that could be installed and so impaired efficiency. Humidity in the lifts and vertical shafts caused corrosion problems. Where there was humidity in the salt mine shafts, concrete pouring to make improvements was impaired due to the need to limit the escape of moisture, which would dissolve the walls.

The first production 109-003 A-1 turbojets were delivered in October 1944 from Zühlsdorf, near Oranienburg, to where production from Spandau had been moved. Zühlsdorf, in fact, completed most of BMW's turbojets by the end of the war.

CHAPTER TWENTY-FOUR

German Turbojets Go to War

Contrary to what is generally published, the British, as we have seen, were the first to get their jet aircraft operational. However, they found little to do since the Luftwaffe was all but swept from the skies of coastal Europe by the time RAF Meteor fighters arrived. On the other hand, there was more than enough for the Luftwaffe to do in tackling the armadas of conventional fighters and bombers that were hammering Germany and the introduction of their jets was eagerly awaited.

Although many aircraft were planned, the two main jet types that went into operational service with the Luftwaffe were the Me 262 fighter and Ar 234 bomber.

Towards the very end of the war the He 162 fighter was also, but barely, operational. From July 1944 the 1st Staffel of the Versuchsverband Ob.d.L, followed by the 3rd Staffel, began using Me 262s and Ar 234s on experimental reconnaissance missions. Servicing of the new aircraft was initially performed with the assistance of

This newly built Me 262 A-1a is in the final stages of having its Junkers 109-004 B turbojets fitted before flight testing and delivery to an operational unit. E.J. Creek

A line-up of Messerschmitt Me 262 A-1a Schwalbe (Swallow) fighters of Erprobungskommando 262 at Lechfeld in September 1944. Nearest is 'White 2' (c/n: 170071), followed by 'White 3' (c/n: 170087) and 'White 5' (c/n: 170045). E.J. Creek

personnel from Messerschmitt, Arado and Junkers. The aircraft were found to be excellent in these trials, being easier to fly and much faster than piston-engine fighters, although they were less manoeuvrable and required long take-off and landing runs. Their Junkers 109-004 B-1 engines, however, were prone to premature failure and would stand no mishandling from the pilot. In particular, failures occurred in the compressor, turbine inlet nozzles and the turbine wheel, and new engines were given a general overhaul before being put into service. Jet aircraft types being prepared for mass production and service were the Me 262 A-1a Schwalbe ('Swallow') fighter, the Me 262 A-2a Sturmvogel ('Stormbird') bomber, which Hitler had ordered, and the Ar 234 B reconnaissance and bomber aircraft. In addition, various sub-variants were to follow for specialized roles such as night-fighting.

Messerschmitt Me 262

The Me 262 stemmed from Messerschmitt's P.1065 project of June 1939; based on this, the design was worked out primarily by Rudolf Seitz. This design phase ended in mid-May 1940 and resulted in the two engines being planned for underwing mounting, instead of inside the wings, to simplify wing spar design. The airframe was considerably cleaner and had less drag than the British Meteor, although there was little to choose as regards the drag of the engine nacelles of the two types. The design of the Me 262, although radical in nature, was quite conservative in its aims in order to ensure success, and it was estimated that the requisite performance could be obtained with engines of 680kp (1,496lb) static thrust each. A stressed-skin, all-metal structure was used with formers and stringers for the fuselage, ribs and stringers for all flying surfaces and sheet-metal skin overall. The moderately swept-back wing, at 18 degrees 32 minutes on the leading edge, was low mounted and had long-span ailerons, plain flaps inboard of the engines and typical Messerschmitt full-length automatic leading edge slots. A triangular fin carried at about mid-height the tailplane and a two-piece elevator. Trim tabs were provided for ailerons, rudder and tailplane. The fuselage, which had a rounded-off

GERMAN TURBOJETS GO TO WAR

Me 262 A-1a Schwalbe fighter		
Span	12.5m	(41ft 0⅜in)
Length	10.605m	(34ft 9½in)
Wing area	21.68sq m	(233.3sq ft)
Height	3.830m	(12ft 6¾in)
Empty weight	4,000kg	(8,820lb)
Loaded weight	6,775kg	(14,938lb)
Maximum level speed at 7,000m (22,900ft)	868km/h	(536mph)
Limiting Mach number	0.86	
Initial climb rate	1,200m/min	(3,936ft/min)
Landing speed	175km/h	(109mph)
Approx. service ceiling	11,000m	(36,000ft)
Range at 9,000m (29,560ft)	1,050km	(652 miles)

with about twenty Me 262 A-1a fighters. These first operations, against USAAF bomber formations, proved disappointing and a number of Me 262s were shot down when they reduced speed because of the inability of some pilots to attack with accuracy at high speed. Other aircraft were lost in accidents, of which 34 per cent were due to undercarriage failures, 33 per cent to turbojet failures and 10 per cent to tailplane failures. In the latter case, resonance from the turbojet exhausts set up oscillation leading to structural failure.

In September 1944, Me 262 A-2a bomber units began forming; these were known as Kommando Schenk and Kommando

triangular cross-section, carried the fuel tanks fore and aft of the cockpit, which was provided with a framed canopy having a hinged centre section for access. Front views of the fuselage had a shark-like appearance. Four MK 108 cannon and ammunition were carried in the nose. In addition, in the case of the bomber variant, either one 1,000kg or two 500kg bombs were carried below the nose. Ing. Josef Helmschrott was in charge of all constructional problems and the high degree of accuracy required in production was ensured by special attention to detail manufacture: wing ribs, for example, were produced from metal dies. Nevertheless, although the Me 262 was cheap to build, its quality of construction varied due to the dispersed production.

By September 1944 the first semi-operational Luftwaffe jet fighter unit, Erprobungskommando 262, had finished evaluating the Me 262. Also, at that time, a special Ar 234 test unit, Sonderkommando Götz, was formed. The first fully operational Luftwaffe jet unit, Kommando Nowotny, began operations on 3 October

On 24 February 1945 a USAAF Republic P-47 Thunderbolt fighter shot down an operational Arado Ar 234 jet bomber of III./KG 76. The aircraft crashed in woods near the village of Segelsdorf and became, after a skirmish, the first Ar 234 to be captured. It was dismantled and shipped to England for detailed examination. Of special interest were its two Junkers 109-004 B-1 turbojets (c/ns: 1163 and 2006). One of these is shown in the woods, ready for removal, and the other in a German workshop awaiting shipment. Parts of the Arado's wings are seen also.

Edelweis. From November 1944 two further small Ar 234 reconnaissance units were set up, but were replaced in the new year by 1.(F)/100, 1.(F)/123 and 1.(F)/33, which were able to carry out reconnaissance over the British Isles and other areas with impunity. Another unit, Sonderkommando Sommer, performed similar work over northern Italy. A new fighter wing, JG 7, was established in November 1944 and attempts were made to solve the problems of high-speed attack, chiefly by fitting rocket armament to the Me 262 fighters. In order to familiarize pilots more safely with the characteristics of jet aircraft, training units with two-seat versions of the Me 262 were set up for both fighter and bomber units. Training for the first Ar 234 bomber unit began in November 1944 at Alt Lönnewitz, where the main Ar 234 production centre was located. During the Ardennes offensive of December 1944 to January 1945, the Ar 234s of 6./KG 76 made the first jet bomber attacks in history when they attacked Allied positions. Soon other units of KG 76 became operational with the Ar 234 bomber. Although operations were limited by the severe fuel shortage, sorties by KG 76 increased from early March, often supported by Me 262s of I. and II./KG 51 bomber units. Some of the most desperate, and often suicidal, of these operations were against bridges, such as that at Remagen spanning the Rhine. By the end of March 1945, however, Ar 234 bomber operations had dwindled to a low level and had almost petered out by the end of the war.

In the meantime, the Me 262 was performing in its most successful role, a short-range reconnaissance aircraft, being virtually immune to interception when flown fast and without deviation. For such work, Sonderkommando Brauegg was formed in December 1944, equipped with Me 262 A-1/U3s. On New Years Day, 1945, the Luftwaffe made its last big effort by attacking Allied airfields in Europe with all available aircraft. Following this operation, in which the Luftwaffe lost many aircraft, a rapid decline followed and soon the jet aircraft were among the few Luftwaffe aircraft still flying. April saw a few jets being used in the night-fighting role, Kommando Welter using the Me 262 and Kommando Bonow using the Ar 234. On 10 April the Luftwaffe carried out its

Another view of the Me 262 A-1a Schwalbe fighters of Erprobungskommando 262 at Lechfeld. These eight aircraft had recently arrived from the factory. They carry the unit's yellow recognition bands on the fuselage and white tactical numerals on their noses. Everyone looks very pleased with the new fighters.
E.J. Creek

Me 262 A-2a Sturmvogel (Stormbird) fighter bombers of I./KG 51 at Rheine airfield late in 1944. A Kettenrad is towing out one aircraft (9K+BH, c/n: 170096) to prepare it for another sortie over the Western Front. E.J. Creek

last mission over the British Isles, when an Ar 234 of 1.(F)/33 flew from Stavanger in Norway on a reconnaissance mission over the north of Scotland.

Although the new year saw various former conventional bomber units preparing to begin operations with the Me 262 fighter, of these only I.KG(J)54 became operational. By the end of the war, only III./JG 7 and Jägdverband 44 were still operational as Me 262 fighter units. The famous Jägdverband 44 (JV 44) was formed by the ace pilot Gen. Lt Adolf Galland as a last-ditch effort to concentrate the cream of the Luftwaffe's fighter pilots with the best fighter aircraft into a single unit. As one of Germany's final attempts to hit back at the Allied bomber streams that had already crushed her, it showed what the new jet fighters could do once the tactics for their employment were worked out. JV 44, formed on 10 February 1945, began operations on 31 March from München-Riem with about twenty-five Me 262s and fifty pilots. A surfeit of almost 100 jet fighters was soon collected from other, disintegrated units, although only about six Me 262s were kept serviceable at a time. By the time JV 44 surrendered on 3 May 1945, the unit had destroyed some forty-five enemy aircraft.

Heinkel He 162 fighter

Whereas the Junkers 109-004 B turbojet saw considerable wartime service, largely in the Me 262, the BMW 109-003 engine was only just entering service by the end of the war because its development had fallen behind the Junkers engine by about nine months. The chief use for the BMW engine was in powering the He 162 Volksjäger fighter and particular versions of the Ar 234 C with four 109-003 engines. While development of the Ar 234 C was proceeding, albeit slowly because of disruption of the Arado plant and the number of sub-variants involved, the Volksjäger or People's Fighter was being planned. On 8 September 1944 the specification for this fighter was issued as the outcome of a conference of industrial leaders and the Speer Ministry in late summer 1944. The specification, which arose largely from the ideas of Karl Otto Saur, by then chief of the Technisches Amt, called for an aircraft to make use of existing aircraft components and the barest essentials in the way of equipment. The power was to be supplied by the BMW 109-003. Aircraft gross weight was to be no more than 2,000kg (4,410lb) and the wing loading not more than 200kg/sq m (41lb/sq ft). Its top speed was specified as 750km/h (466mph) and its endurance was specified as 20 minutes at sea level. The fighter was to be regarded as 'a piece of consumer goods' and to be ready by 1 January 1945.

Great pressure was put on the five aircraft companies willing to submit designs for this specification and Heinkel's P.1073 project was finally selected on 30 September and given the designation He 162 Spatz (sparrow). A month later, day and night work had produced detailed drawings and the first prototype (He 162 M1, c/n: 200 001) flew on 6 December with Flugkapitän Dipl.-Ing. Gotthold Peter at the controls. This was a truly amazing achievement by the Heinkel company. Although the He 162 M1 prototype crashed four days after its maiden flight, killing Peter, work went ahead at top speed to solve the major structural and aerodynamic defects and to pursue general development with thirty-two prototypes (He 162 M2 to M33). Simultaneously, production went ahead with the He 162 A-1 and then the A-2, the aircraft by then being known as the Salamander.

The design of the He 162 was under the chief project designer Siegfried Günter and the Chief Designer Karl Schwärzler. The most noteworthy features of this very small fighter were its low span/length ratio, the mounting of the turbojet above the fuselage and the pronounced dihedral of the tailplane with its twin, rectangular fins. The mounting of the engine outside the fuselage was in line with Heinkel policy to avoid problems with jet intake and tailpipe ducting. The He 162 had a semi-monocoque fuselage of duraluminium construction, but with the nosewheel doors and radio compartment of plywood. The cockpit had various standard fittings, a minimum of instruments and a simple, cartridge-operated ejector seat. Cockpit visibility was generally good except aft, where the turbojet rendered visibility zero. The shoulder-mounted wings were of wooden construction throughout, except for aluminium alloy tips that were given anhedral. The tips were also removable to permit internal servicing with the aid of special tools and a mirror because the wing skin was permanently bonded on without access panels. The complete wing was attached to the fuselage by four vertical bolts, and there were three other connections for the turbojet. The wing auxiliary spar carried the ailerons and flaps, the latter being hydraulically operated, as was the undercarriage during retraction, the hydraulic pump being driven from the engine. For longitudinal trimming the entire tailplane incidence could be altered about the hinge at its forward end, while the twin rudders could be offset for directional trimming. These trim changes could be made in flight, but the trim tabs on the ailerons could only be adjusted on the ground. The normal fuel capacity of 695ltr (153gal) was contained in a fuselage tank and could be increased to a maximum of 765ltr (168gal) and further augmented by carrying another 180ltr (39.6gal) in the space between the main and auxiliary wing spars, the wood being suitably sealed. Throughout the design, every effort was made to simplify and use materials readily available.

With only a small amount of development, the He 162 had no major flaws and was a stable gun platform. However, it was a fighter suitable only for experienced pilots and the Nazi idea of using lots of hastily trained Hitler Youth pilots was really a non-starter. It seems likely that, had time allowed, the target of producing a thousand He 162s per month would have been reached by June 1945, but obtaining the necessary fuel and pilots was a problem without solution.

Modifying the 109-003 A engine into the E model for over-fuselage attachment presented little difficulty, but the same was not true of engine production and considerable writing down of the original schedules was necessary. Since the He 162 was regarded as 'a piece of consumer goods',

A BMW 109-003 E-1 turbojet modified to permit mounting above a fuselage, as for the Heinkel He 162 fighter.

He 162 A-2 fighter		
Span	7.20m	(23ft 7¾in)
Length	9m	(29ft 8½in)
Wing area	11.15sq m	(120sq ft)
Height	2.55m	(8ft 4⅜in)
Empty weight	1,750kg	(3,859lb)
Normal loaded weight	2,490kg	(5,490lb)
Maximum loaded weight with extra fuel	2,700kg	(5,953lb)
Maximum level speed at 6,000m (19,680ft)	835km/h	(522mph)
Limiting Mach number	0.75	
Initial rate of climb	1,290m/min	(4,231ft/min)
Landing speed	165km/h	(103mph)
Approx. service ceiling	11,000m	(36,080ft)
Endurance at sea level	20min	
at service ceiling	57min	
Maximum range with maximum fuel at service ceiling	1,000km	(620 miles)

only about 20 per cent of engine spares were planned, compared with 40 to 50 per cent spares for the 109-003 A-1 (for the Ar 234).

Faith in the *Wunderwaffen*

Despite the difficulties, the Nazis, who were largely responsible for conceiving the He 162 and were by then controlling almost everything, were determined to get the jet fighter into service at a time when the Me 262 could not be produced in sufficient quantities to combat the Allied air onslaught on Germany effectively. During 1944, with the war going inexorably against Germany, increasing faith (real or feigned in propaganda) was put in the so-called *Wunderwaffen* (wonder weapons) being developed by German industry. These included rockets, flying bombs and guided missiles, but high on the list were jet and rocket aircraft.

Conferences were held regularly with high-ranking personnel from the RLM, Luftwaffe, SS and the aircraft industry to plan future developments, production and deployment of jet aircraft as if there was plenty of time available. In this, many were

out of touch with reality. The SS in particular put their hopes in the *Wunderwaffen*. Following the assassination attempt on Hitler on 20 July 1944, the SS moved into a position of much greater, indeed omnipotent, power, since Hitler distrusted virtually everyone after the attempt on his life. With the appointment of SS Obergruppenführer Max Jüttner as Chief of Army Weapons and Commander of the Reserves, aeronautical matters and much else came increasingly under the influence of the SS.

The SS dreamt of having its own aviation branch or Flieger-SS that would be better equipped than the Luftwaffe. From the ranks of the Waffen-SS, Hans Kammler was promoted by Hitler to a position of very great power. On 27 March 1945, when the war had less than two months to run, Hitler sent Kammler a letter of authority in which he belatedly gave jet aircraft 'priority over all other military and economic matters' and instructions to achieve the goal of high production of these aircraft 'by combining all available resources of the Reich under a strict leadership'. Kammler had at his disposal 'all command centres, departments and facilities of the Wehrmacht, the Party and the Reich, who are to follow his directives'. Albert Speer, Reich Minister for Armaments and War Production, had done a good job of making the best of available resources, but he was now subordinate to Kammler, whose new authority made him the most powerful man in the Reich, after Hitler.

Dr-Ing. Hans Kammler held the military rank of General in the Waffen-SS, the fighting arm of the SS. He combined a high technological ability with the ruthlessness of an enthusiastic Nazi and he had pushed through in short order many projects, including the underground rocket, missile and jet aircraft factories near Nordhausen. Ruthless use of slave labour was one of the keys to his success. Of the many new aircraft projects pushed by Kammler, the one that came nearest to success was the Heinkel He 162 jet fighter. Every possible facility was pressed into service in a well-planned production programme

Operational (barely) Heinkel He 162 A-2 Salamanders of I./JG 1. E.J. Creek

(at least for the airframes) and by January 1945 the operational proving of the He 162 had begun. This work was performed by Erprobungskommando 162 (alias Einsatzkommando Bär), based at E-Stelle Rechlin and later at München-Riem. On 6 February 1945 the unit I./JG 1 transferred from the Eastern Front for He 162 training at Parchim. The following month, II./JG 1 moved to Warnemünde for similar training. Encroaching Allied forces disrupted the numerically extensive training programme planned. By 4 May 1945 units under training had been merged into one unit at Leck in Schleswig-Holstein. This unit was designated I./(Einsatz)Gruppe/JG 1 and had about fifty He 162s.

Combat with the enemy was, however, forbidden since more training was needed to reach operational readiness and on 8 May 1945 the unit at Leck surrendered with all its aircraft intact to British forces. Five days previously, Adolf Galland's Me 262 unit, JV 44, which had been joined by the He 162s of Einsatzkommando Bär, had also been forced to surrender. Thus, while the He 162 was the aircraft that most nearly took the BMW 109-003 E turbojet into full operational service, all the great effort and ingenuity expended on the Volksjäger programme came to nothing in military terms. By the end of the war about 275 He 162s had been completed and about 800 nearly completed. A large proportion never received engines.

Heinkel He 162 A-2 Salamander (c/n: 120222) of JG 1.

Although the bulk of 109-003 engines were earmarked for the He 162 programme by September 1944, other aircraft continued to be planned for the BMW engine, as well as the Ar 234 C. March 1945 saw the construction of the Henschel Hs 132 V1 prototype of a dive-bomber of very similar layout to the He 162 but with a transparent nose for a prone-lying pilot. A single 109-003 E-2 engine was to provide power and the first flight was expected in June 1945, but before then the V1 and two other partially built prototypes had been captured by Soviet forces at Schönefeld.

Junkers Ju 287

The Junkers Ju 287 low-speed test aircraft for a bomber with swept-forward wings was first flown on 8 August 1944 on the power of four Junkers 109-004 B-1 turbojets. Although official backing for this development was soon dropped because of the overwhelming need for fighters to defend the Reich, backing was given again in March 1945 on the orders of Hans Kammler. Work went ahead on the first true prototypes and pre-production examples of the Ju 287, each of which was to be powered by six 109-003 A-1 turbojets. These engines were mounted one on each side of the fuselage nose and two in a paired nacelle forward of each wing main spar. The unfinished Ju 287 V1 and V2 prototypes were captured by Soviet forces and the programme was continued by the Soviets after the war (*see* the Soviet section in Volume 2). As for Kammler, he was last known to be heading for Pilsen at the war's end and all trace of him was lost. Many other aircraft were designed to use the turbojet in Germany during the war but they remained on the drawing board and space precludes their mention here.

Flying on the Junkers 109-004 B

Of course, the other Axis powers were interested in German turbojet work and Japanese delegations made regular visits to the Dessau works to inspect the turbojet work there, the last such visit being as late

A Heinkel He 162 A-2 Salamander in flight, powered by a BMW 109-003 E turbojet. This small fighter handled well in flight but its endurance was very short and it was not for new pilots. Some He 162s had larger fuel tanks fitted. Philip Jarrett

Arado Ar 234 Ar 234 C-3 multi-purpose bomber Four BMW 109-003 B-2 turbojets			
Span		14.44m	(46ft 3½in)
Length		12.65m	(41ft 5½in)
Wing area		27.3sq m	(284.167sq ft)
Height		4.29m	(14ft 1½in)
Empty weight		6,500kg	(14,400lb)
Loaded weight		11,000kg	(24,250lb)
Maximum speed	at sea level	800km/h	(496mph)
	at 6,000m (19,680ft)	852km/h	(530mph)
Climb to 10,000m (32,800ft)		16.7min	
Service ceiling		11,000m	(36,091ft)
Maximum range		1,215km	(765 miles)

as 5 March 1945, when all the latest developments were seen. (The Japanese work is recorded in its own section in Volume 2.) By the end of the war, hundreds of German aircraft had flown on the power of the Junkers 109-004 B turbojet. Most of the aircraft used twin, underwing-mounted engines and the most prolific type was the Me 262, of which 1,433 were built. However, fewer than 200 or so reached operational units.

In service the throttle of the 109-004 B turbojet had to be handled with care. If the throttle lever was moved too quickly to increase power (pilots were used to a rapid throttle response from piston engines), the turbojet would either cut out or fail to reach full speed. Attempts were made to introduce a time delay in the governing system to avoid this problem, but the governor gave trouble at high altitude and so the pilot had to be relied on to a great extent for successful operation.

Owing to the slow throttle response with these early turbojet engines, a pilot was committed to land his aircraft once on final approach. At that stage he was highly vulnerable to an attacking enemy aircraft, so Allied fighters would seek out Me 262 landing grounds (or autobahns) as virtually the only place where they could shoot the jets down. The approach path of the Me 262s, however, was often guarded by piston-engined fighters and lined with anti-aircraft guns.

An incident connected with the 109-004 B movable exhaust bullet is of interest and concerns several Me 262s that crashed, in each case following the failure of one engine. At the time no explanation was forthcoming as to why a crash should follow immediately after one of the engines stopped. The answer was found after the war when the German test pilot Ludwig Hoffman was ferrying an Me 262 for the Americans. This flight was planned from Lechfeld to Cherbourg, but an engine cut out *en route* and the aircraft immediately went into a dive that Hoffman was unable to counter. Fortunately the pilot was able to bale out and his evidence enabled the phenomenon to be explained. Apparently the movable bullet in one of the engines had become detached and blown back, blocking the airflow through the engine. The resulting severe effect on the airflow then caused the aircraft to sideslip and dive, following which the tail controls became ineffective due to the bulk of the aircraft preventing airflow over them. The first Me 262 to which this sequence of events occurred was the Me 262 V3 (PC+UC), which crashed after it had been rebuilt following a previous crash, killing the pilot.

Towards the end of the war, following the destruction of the refineries, German turbojets in service, specifically the 109-004 B, were being run on unrefined oil, which was only cleaned centrifugally. The oil was heated before filling the aircraft tanks and no trouble was experienced, apart from sooting up of the spark plugs. The lack of trouble was very largely because Junkers had developed the 109-004 from the start to run on J2 or diesel-type oil. This choice of fuel was made on the grounds of safety, economy and other technical reasons, and not because of any envisaged shortage of regular aviation gasoline.

Preparing the Junkers Ju 287 (RS+RA) for flight. This low-speed test aircraft had swept-forward wings and was powered by four Junkers 109-004 B-1 turbojets, two of which were mounted below the wings. Where possible, this aircraft was made up from existing aircraft parts to save time. This programme, to produce a jet bomber, was continued by the Soviets after the war. E.J. Creek

CHAPTER TWENTY-FIVE

Heinkel-Hirth's 109-011

On 18 July 1941 the Technisches Amt issued a specification for a bomber to be powered by a new turboprop engine. The turboprop layout was planned by Helmut Schelp's department and consisted of two gas turbines (with their own compressors and combustion chambers) that acted as hot gas producers to feed a third power turbine driving a variable-pitch airscrew. Heinkel's investigation of this scheme suggested that it would be necessary to first build a turbojet engine that could later be employed as one of the gas producers of the turboprop unit. The first turbojet layout considered for development by von Ohain's team at Marienehe is believed to have been the HeS 9 (see Chapter 20). Before the end of 1941, however, Heinkel had been persuaded by Schelp to begin work on another turbojet that incorporated the latter's ideas on compressors. The compressor of the new engine consisted of a diagonal or oblique flow stage (a supposed compromise between axial and cetrifugal flow), followed by three axial-flow stages. The efficiency of the diagonal stage could not be high and its use meant an increase in engine diameter, but it was claimed that, having robust blades, it would be less susceptible to damage from dirt and stones and it would be less likely to ice up. Also it was thought to be easier to start and less likely to stall than an axial compressor.

Von Ohain's team was not enthusiastic about the proposed compressor system and a similar system had indeed been studied and rejected some years earlier when Mueller was at Junkers. Nevertheless, Ernst Heinkel accepted the job of developing it because it was tied up with the acquisition of the Hirth Motoren works. Heinkel hoped to use these works for the full development of the promising HeS 30 engine, although this aim was never realized.

Prototypes and development of the HeS 011 turbojet

While the proposed dual turboprop unit never materialized, work went ahead on the turbojet engine. This was designated the HeS 011, or officially the 109-011, and Schelp had become keen to have it developed into a Class II engine in its own right. This was considered all the more important by the end of 1942, when the Class I 109-004 turbojet being developed by Junkers was making rapid progress and its mass production seemed imminent. An official contract for the 109-011 was awarded in January 1942 and a meeting was held at the RLM on 16 July 1942 to finalize details of the engine. Von Ohain's ideas were discussed with specialists from the Technisches Amt and the AVA, and full support was given to the project by Schelp.

By the end of 1942, experimental work for the 109-011 had begun, preliminary studies having been completed in September that year, and official orders were given for Heinkel to concentrate on that engine

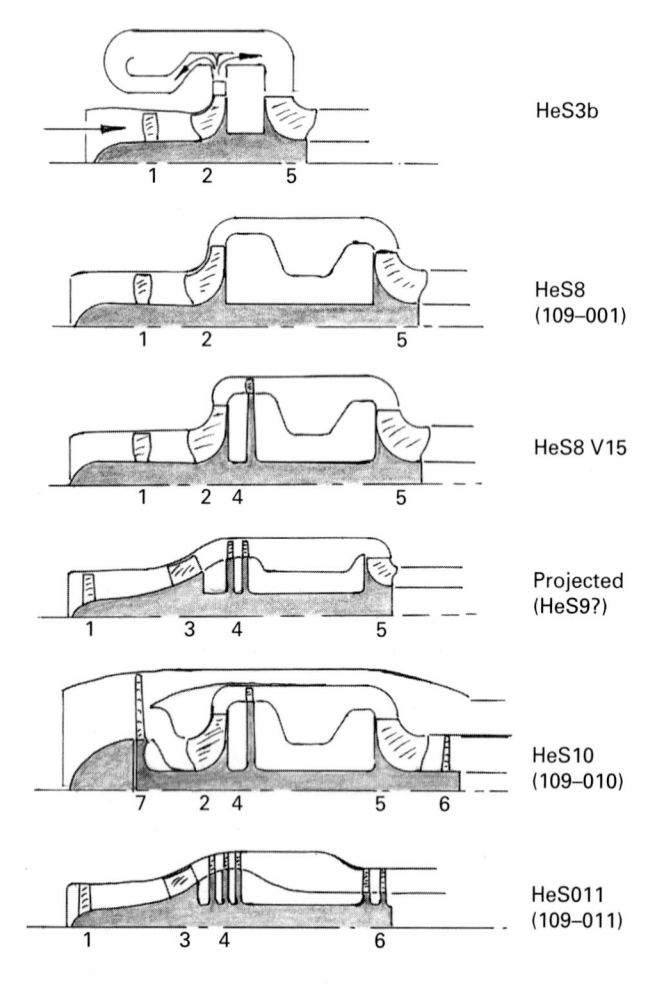

Flow path development of Heinkel turbojet engines leading to the Heinkel-Hirth 109-011 engine: (1) axial inducer; (2) centrifugal compressor; (3) diagonal compressor; (4) axial compressor; (5) radial inflow turbine; (6) axial turbine; (7) ducted fan. HeS 3b has reverse flow combustion, all others straight flow.

alone. The requirements laid down for the engine were a static thrust of 1,300kp (2,866lb), a cycle temperature of 750°C, a pressure ratio of 4.4:1 and an air mass flow of 30kg/sec (66.15lb/sec). Von Ohain's team at Marienehe was eventually moved and consolidated with Mueller's team to work together on the 109-011 at the former Hirth works at Zuffenhausen, Stuttgart. There, about 150 engineers, designers and research workers were concentrated on the development. The two teams did not work well together and so, although von Ohain remained as Chief Engineer in charge of design and construction, Harald Wolff was brought from BMW and sent to the Zuffenhausen works as technical controller of Heinkel-Hirth turbojet development. By March 1943, on Erhard Milch's orders, Wolff was in complete charge of the whole works but, despite his competence, he never welded the personnel into a unified team and was often in dispute with Heinkel himself.

Development of the 109-011 up to the production stage was planned as follows:

- Prototypes V1 to V5, differing in major details, were to be built to establish the basic form of the engine.
- Prototype engines V6 to V25 were to be used for flight tests.
- Prototype engines V26 to V85 were for experiments spread over a long period and were to be mostly modified from production engines.
- The first pre-production engines were to be designated 109-011 A-0.

Before building the first experimental engine, it was necessary to establish the dimensions and form of the diagonal-flow compressor more accurately than could, at that time, be calculated. Tests were therefore carried out with three different axial inducers in combination with three different diagonal-flow compressors. The final combination selected was then tested with each of the three axial compressor stages. This work was carried out at speeds below the design point on the inadequate 1,600kW test rig at Zuffenhausen.

When the first experimental engine, the 109-011 V1, was built the compressor system failed after less than one hour's running time. The diagonal stage of the compressor, which had been machined from a solid cheese, broke through centrifugal stresses and poor material, but the failure of the axial stage was attributed to fatigue caused by vibration. The solution to these problems was entrusted to Dr Max Bentele, whose activities included the aerodynamic testing of compressors and special stressing work. His first report on the 109-011 V1 was issued in October 1942. He recommended various methods of stress calculation, checked when possible with evidence from the engine itself and supplemented with frequency tests. Unfortunately, the first vibration failure caught Heinkel-Hirth unprepared, so these frequency checks were carried out using a piano and detecting by ear. (Later, a set of specially calibrated tuning forks was acquired and, later still, electronic and pneumatic equipment was borrowed from the FKFS). From the calculations and tests it was found that the four-armed bearing support spider following the axial compressor stage (in the same plane as the inlet guide vanes to the combustion chamber) was the cause of the failure.

Dr Bentele was not in a position to redesign the compressor but he redesigned the support spider with spacings of 60–120–60–120 degrees between the four arms, instead of the previous equal spacings. This altered the harmonic value of the engine speed and increased the running time to four or five hours without failure. This improvement was but a start, however, and various experiments were made on compressor blades for later engines. To reduce root-bending stresses on the diagonal compressor blades, each blade was fitted with a bulb root to allow it to rock freely and take up a position where the air load was balanced by the centrifugal force. Friction damping plates were fitted between the blades. Also tested was the rolling of short grooves into the tips of the axial compressor hollow blades to stiffen the trailing edge.

The 109-011 V1 was built up from steel castings and sheet metal, and no attempt was made to save weight or meet aircraft design standards. The compressor set, drum-type rotor and turbine set were connected by a long tie rod to take out axial imbalance, and the whole assembly was carried by three sets of anti-friction bearings. The milling out of the diagonal compressor from a solid cheese of aluminium alloy proved to be extremely difficult, but was finally achieved by J.M. Voith GmbH of Heidenheim. An incredible 3,000 man-hours were required on the milling machine to produce one diagonal compressor. This compressor was bolted to the front of the bolted assembly of axial rotor blade discs and these discs, in turn, were bolted to the front of the drum rotor. A similar construction was used for the turbine at the other end of the rotor. The two-stage, axial flow turbine used solid blades that were held in their discs by means of pinned V-roots. A fuel and combustion system was provided similar to that of the HeS 8 V15 engine and some secondary air was introduced into the annular combustion chamber through radial fingers or sandwich mixers. A two-position tail cone, operated by a hydraulic cylinder, was used to vary the exhaust nozzle area. Since this experimental engine was not for flight, no accessories were fitted apart from an electric starter motor fitted above the front support ring. A drive shaft led from the starter gearbox through one arm of the support spider. The same arms were set at an angle to act as guide vanes for the air flowing from the axial inducer to the diagonal compressor. The 109-011 V1 prototype turbojet gave a static thrust of 1,115kp (2,459lb) at 9,920rpm. Its cycle temperature was a low 660°C and its air mass flow was 21.65kg/sec (47.7lb/sec).

According to the specification laid down in September 1942, a considerable advance in performance was to be aimed at by increasing the cycle temperature and air mass flow. The compressor set was not tested up to its full speed of 11,000rpm until December 1943, or a little later, when a DVL test bed at the Augsburg works of MAN was used. The first attempt to replace the expensive turbine blades made from solid heat-resisting steel was in the V5 prototype, which had a DVL-designed, single-stage turbine with hollow, air-cooled blades. This work was carried out under Dr Fritz and A.F. Schmidt.

However, Dr Bentele was not satisfied with the DVL's work because it seemed that too much emphasis had been placed on the cooling system and not enough on mechanical and manufacturing considerations. Bentele therefore studied a dozen concepts of cooled turbine blades before conceiving the *Faltschaufel*, or folded blade, made entirely from sheet metal. Then, from two interested manufacturers, the famous Württembergische Metallwaren-Fabrik (WMF) at Geislingen/Steige was chosen to produce blades along these lines. After further development, WMF produced the turbine blade known as the *Topschaufel* (tubular or bootstrap blade) and this became very successful in turbochargers and, finally, the 109-011 turbojet. Experimental methods of construction for the compressor set included forging and casting the axial inducer, fabricating the diagonal compressor and forging and sheet-metal-forming the axial stator blades. The five engines in this series were completed somewhere between the end of 1943 and early 1944 but the running hours accumulated are unknown.

Work on the 109-011 V6 to V25 series of engines, following the previous broadly experimental engines, was aimed at refining the design and developing it to suit aircraft. This phase was started at Zuffenhausen early in 1944, but the threat from enemy bombing necessitated the dispersal of factories in the Stuttgart area and the continuation of work at safer locations. One such location was the salt mine known as Staatliche Saline Friedrichshall at Kochendorf. The Organisation Todt, using prisoners of the SS, started preparations for this mine, code-named Eisbak, in spring 1944. A quarter of the installation was allocated to Heinkel-Hirth and was known as the Ernst Werk. Heinkel-Hirth began the move there in June 1944 and was ready to resume work on the 109-011 engine by August/September 1944. By this time almost two years had been spent on the protracted difficulties with the first five engines. The compressor, in particular, was still giving trouble. Increasing demands were made to speed up development and as early as the beginning of 1943 the specification of the 109-011 engine had been issued to aircraft companies to begin designing new, badly needed, jet aircraft.

The major layout changes introduced with the V6/V25 series of engines included a shortening of the engine by redesigning the combustion system and a reduction of the rotor support bearing sets from three down to two. The compressor set and turbine were now bolted to the ends of a much simplified drum rotor. Now unbroken in the middle for a centre bearing, the new rotor had only a bearing set at each end. Lubrication and cooling air were fed to the roller race behind the turbine through the arms of its support spider.

Composite construction was now used for the diagonal-flow compressor. Its twelve blades were machined from aluminium alloy forgings and each was held in the steel hub by a bulb root. Damping blocks between the blades at their bases defined the inner wall of the air passage. Curvature of the blades was in the direction of rotation at the inlet and against the direction of flow at the outlet. At the rear of the hub was a flange to which were bolted succeeding stages of the axial compressor, and at the front was a stub shaft for the front bearings. This shaft was internally splined for connection of a quill shaft leading to the auxiliary drive gears and axial inducer at the front.

Mounting of the rotor blades of the three-stage axial compressor was by U-shaped feet, the first and second stages being in aluminium alloy and the last stage in steel. Work was performed in both rotor and stator blades of the compressor, but the amount of reaction is unknown. The blades followed a modified aerofoil section with a parabolic camber line. Although the axial compressor was still a three-stage type, the stator blades were reduced from four to three rows, the last stage of the rotor blades being followed immediately by a single row of steel inlet guide vanes leading to the combustion chamber. The stator blades were held in the outer aluminium alloy casing, with mechanical means of adjusting their incidence. Each stator blade had a shaft, located axially in steel split rings keyed into the casing, the shaft having a boss with a slot for adjustment. A circular base to each blade fitted into a ring and labyrinth seal.

A complete change was made in the combustion system. In place of the former evaporative fuel system with 128 nozzles, there were 16 downstream injection nozzles. These fuel nozzles were set inside an annular shrouding ring that formed the front end of a primary air chamber. Secondary air was introduced through sandwich mixers set just aft of the primary chamber. In addition to the attempt to establish a primary combustion zone, an attempt was also made to manufacture combustion chamber parts of aluminised mild steel instead of heat-resisting steel. This was an essential change in view of the acute shortage of metals for heat-resisting steel. At the exit from the combustion chamber were fitted nozzle guide vanes that were adjustable for angle, a similar feature having been used on some of the previous experimental engines. These guide vanes were hollow to permit air cooling, the cooling air being carried to them from annular ducts surrounding the combustion chamber to holes at the top and bottom of each blade, the air leaving through slotted trailing edges.

Air-cooled turbine blades were apparently not tried on the 109-011 V6. For obscure reasons, the former variable-area exhaust nozzle was dropped and a simple fixed nozzle was used instead. Inside this nozzle was a large, domed fairing at the centre of the rear bearing support spider. Since the V6 engine, and those that followed, was intended to be suitable for flight tests, its external shape had to be made aerodynamically acceptable. Inlet ducting was attached to the front flange of the front annular casting. A sheet-metal housing led back from the intake duct to enclose the accessories at the top of the annular casting and an oil tank at the bottom.

Little is known about the engines subsequent to the 109-011 V6, but only this and three others in the V6/V25 series accumulated any running time. At least one engine had a cast main casing with external bracing ribs in a large diamond pattern. The four engines tested in this series were mostly run on the ground but one or two flights were made in January 1945, using a Junkers Ju 88 as a test-bed. (This Ju 88 was one of several flying test-beds kept at the Lufthansa servicing facilities at Berlin-Staaken for use by engine manufacturers.) At this stage, the Junkers speed governor and thrust regulator had been adapted for use but the design static thrust of 1,300kp (2,866lb) was still not being achieved. No engine flew under its own power. Altogether, the four engines had accumulated 184 hours of running time by January 1945. Of these, 154 hours were at thrusts below 800kp (1,764lb) and only 3 hours of the remainder were at thrusts in excess of 1,100kp (2,425lb).

109-011 A production turbojet

By the end of 1944 Heinkel-Hirth was forced to turn its attention prematurely towards the first pre-production engine, the 109-011 A-0, despite the fact that its speeded-up efforts had still not produced a satisfactory experimental engine. As we have seen, the experimental engines were permutations of various layouts and features, making the final choice very difficult at an early stage in development. The compressor assembly was still imperfect and probably presented the major problem. Dr Encke of the AVA conducted a test on the compressor assembly and pronounced its performance poor. He expected, in general, an 85 per cent stage efficiency from an axial compressor, but the flow efficiency of the 109-011 compressor was only 80 per cent owing to the lowering effect of the diagonal stage. Trouble was also being given by the combustion system, there being burning of the metal devices, but there were no carbonizing troubles. Runs were being made with both B4 (paraffinic gasoline) and the cruder J2 fuel, the essential switch to the latter having been accompanied by some starting problems.

A major redesign was undertaken for the production engine, partly as a result of the previous experiments and partly to adapt the engine for mass production. This was scheduled to begin in March 1945 at a former cotton spinning mill at Kolbermoor, Bavaria, which was already being used as a dispersal factory for BMW-Munich. Components were to be made at various other locations. The compressor/turbine set for the 109-011 A-0 pre-production engine was similar in layout to the V6/V25 series but simplified by the omission of the forward bearing roller race, leaving just one ball race. Eleven blades were provided for the axial inducer on the quill shaft and there were twelve inserted blades for the diagonal compressor. All blades for the three-stage axial compressor were now of hollow, sheet-steel construction, the stator blades being riveted into

A Heinkel-Hirth 109-011 A-0 turbojet at the Cranfield Institute of Technology, Bedford, March 1970.

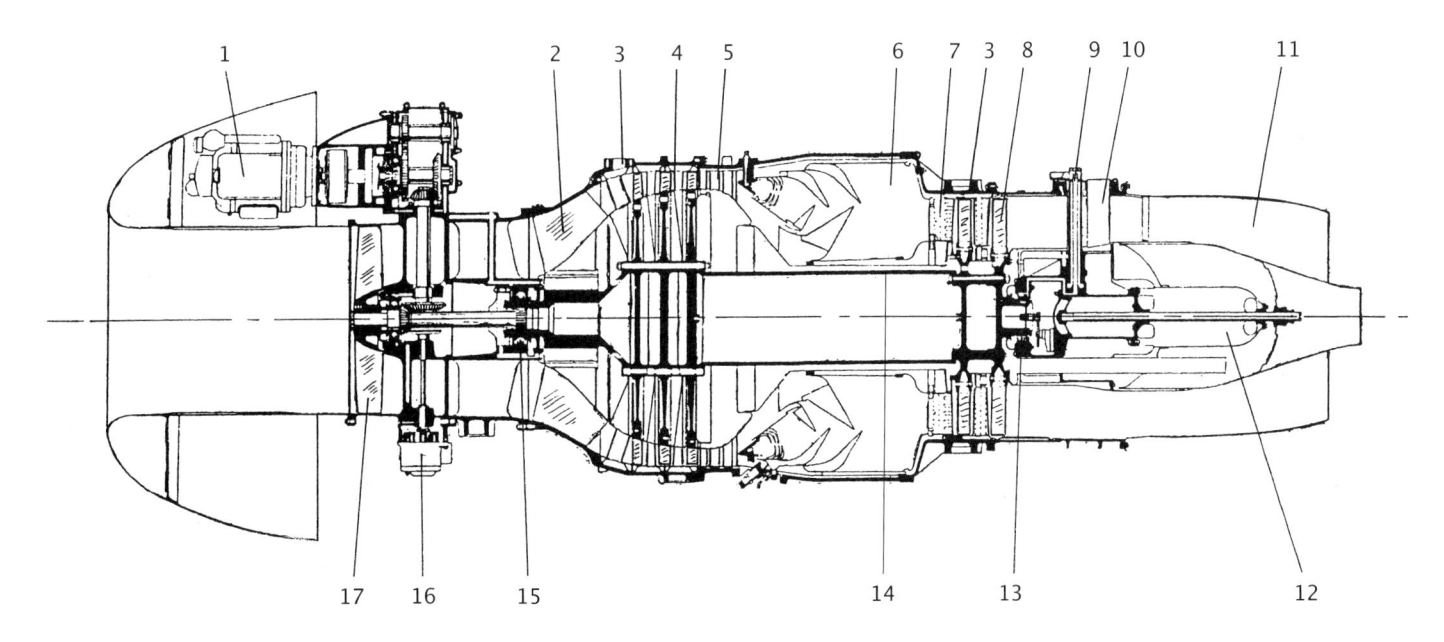

Heinkel-Hirth 109-011 A-0 turbojet: (1) Riedel AK-11 starter motor; (2) diagonal compressor; (3) attachment points; (4) 3-stage axial compressor; (5) three rows of inlet guide vanes; (6) annular combustion chamber; (7) turbine inlet nozzle guide vanes; (8) 2-stage turbine; (9) oil inlet; (10) ventilation tube; (11) variable-area exhaust nozzle; (12) hydraulic system for tail-cone adjustment; (13) rear roller bearing; (14) rotor drum; (15) front ball bearing; (16) oil pump; (17) axial inducer.

Air-cooled turbine blades of the Heinkel-Hirth 109-011.

sheet-metal shroudings and the rotor blades being riveted into their discs. Where previous engines had used a single row of inlet guide vanes, following the last row of the rotor blades, there were now three rows of inlet guide vanes leading into the combustion chamber.

Air-cooled, hollow blades of Bentele/WMF design (*Topfschaufel*) were used for the two-stage turbine. Each blade was held by its hollow root by a single pin in the turbine disc and was therefore able to rock to avoid root stresses in a similar manner to the diagonal compressor blades. Cooling air for the turbine inlet nozzle guide vanes, turbine rotor and stator blades was supplied from the space surrounding the annular combustion chamber. The working temperature in the region of the turbine was about 750°C, while the maximum temperature in the blades was kept below 600°C by cooling.

A considerable redesign of the combustion system was made for the 109-011 A-0. Ideas on a primary combustion zone had advanced further. Sixteen downstream fuel nozzles were provided, the fairing for each nozzle acting as supporting spokes for a V-sectioned ring shrouding the nozzles in a similar manner to that used in the V6. Primary air fed around this shrouding ring and mixed with the fuel with the assistance of six rows of small mixing fingers set aft of the primary chamber. For the introduction of secondary air into the combustion chamber, a return was made to the radial finger type of sandwich mixer used for the first experimental engines, but now four rows of fingers were used. These fingers scooped air from the annular spaces surrounding the combustion chamber. As always, the aim was to mix the air with minimum flow losses. Quite good mixing was achieved with the fingers for a pressure loss of about half an atmosphere. However, oxidation of the mild steel sheet metal was rapid.

Production engines were to be turned out fitted with sixteen fuel nozzles of the duplex type. This was an important feature and of special interest since no German turbojet was so equipped, those turbojet engines that went into service having single-nozzle fuel injectors. Since a fuel nozzle not only sprays but also meters the fuel, a single nozzle cannot supply the correct amount of fuel under all conditions. Thus, when the engine is running slowly or is operating at high altitudes, that is when the combustion chamber pressure is low, the fuel leaves the injectors in a coarser spray (much of it ejected in unburnt droplets) and combustion efficiency is lowered. For the duplex nozzle, two fuel manifold inlets were used, one to supply the primary stages of the atomizer and the other the secondary stages. Two fuel pumps were provided, one a service pump to draw fuel from the main tanks and to feed the inlet of the second pump, which delivered fuel under pressure to the two manifolds of each injector. A loaded relief valve permitted fuel to pass to the primary stage manifold at pressures of about 1.55kg/sq cm (22lb/sq in), above which pressure both manifolds were fed with fuel.

A variable-area exhaust nozzle was again introduced, movement of the open-ended tail cone being by means of a hydraulic cylinder. An increase in the overall length of the engine occurred as a result of a rearrangement of the forward end. Auxiliary drive shafts now passed through their own two hollow fairings, which were positioned in front of the support spider and guide vanes. The axial inducer was in front of the two hollow fairings. The quill shaft, from the rotor, drove the auxiliary drive shafts via bevel gearing and these shafts drove the accessory gearbox and oil pump. A Riedel two-stroke petrol engine, which could be started by electric motor or by hand, fitted inside the forward engine cowling and spooled the engine up through a clutch and bevel gearing.

The 109-011 A-0 was designed to give a static thrust of 1,300kp (2,866lb) at 11,000rpm. Its weight was 950kg (2,095lb) but steps were being taken to lighten this to 865kg (1,907lb). The hoped-for fuel consumption was 1.31. Its dimensions were a length of 3.455m (11ft 4in), width of 0.864m (2ft 10 in) and a height of 1.08m (3ft 6½in).

The sole prototype of the Messerschmitt P.1101 experimental fighter. The swept-back angle of its wings could be altered on the ground. It was awaiting a Heinkel-Hirth 109-011 A turbojet when captured. Here an Allison J35 axial turbojet has been fitted at the Bell Aircraft Corporation's plant for study prior to building the Bell X-5 research aircraft in 1951.

Planned applications of the 109-011 A turbojet

Most of the aircraft planned to use the 109-011 A turbojet were fighters, but one of the first designed for it was the Heinkel He 343 D, a four-engined reconnaissance version of the fairly conventional, straight-wing He 343 bomber project. When details of the engine were released in 1943, aircraft companies, especially Messerschmitt and Focke-Wulf, began planning single-engined jet fighters that could more efficiently equal or surpass the performance of twin-engined jet aircraft such as the Me 262. Officialdom, however, was sufficiently satisfied to continue with existing jet aircraft until near the end of the war, when the problem of unreliable operation of Class I turbojets above about 11,000m (36,000ft) was recognized as a serious shortcoming. This was all the more serious when it was realized that even the German speed advantage could be challenged with the imminent appearance of Allied jet aircraft. How much the Germans knew about the British Meteor fighter is unknown. It was hoped that the 109-011 A turbojet would operate at ceilings in excess of Allied aircraft, especially if the duplex fuel injectors (already described) came up to expectations. At the end of 1944, therefore, the Oberkommando der Luftwaffe (OKL) was demanding a new jet fighter to defend a reeling Germany, and so an emergency fighter competition was instigated by Col Siegfried Knemeyer (Chef/TLR).

The specification issued to all the main aircraft companies called for the new fighter to be powered by a single 109-011 A turbojet, have a level speed of 1,000km/h (621mph) at 7,000m (23,000 ft), operate at altitudes up to 14,000m (46,000 ft), carry about 1,500ltr (330gal) of fuel and be armed with four 30mm MK 108 cannon. In response to this specification, eight proposals had been received from five companies (Blohm und Voss, Heinkel, Junkers, Messerschmitt and Focke-Wulf) by February 1945 and were considered at an official meeting on 27 and 28 February. From these proposals, Focke-Wulf's Ta 183/I was chosen for immediate development and production. The first prototypes were to be powered by Junkers 109-004 turbojets, pending delivery of the first 109-011 A turbojet, but, despite prodigious efforts, even detail design had not been completed when all Focke-Wulf factories suitable for production were overrun by Allied forces in April 1945. To enhance its interception performance, the Ta 183 was to have a 1,000kp (2,205lb) thrust rocket motor fitted beneath the turbojet, rocket fuel for a 200-second burn being carried in underwing drop-tanks.

Space precludes mention of the many other projects planned to use the 109-011 A turbojet. In July 1944, however, well before the competition to find an emergency fighter, Messerschmitt designed its P.1101 and began construction of a prototype on its own initiative. This prototype could have the angle of its swept-back wings altered on the ground and its construction

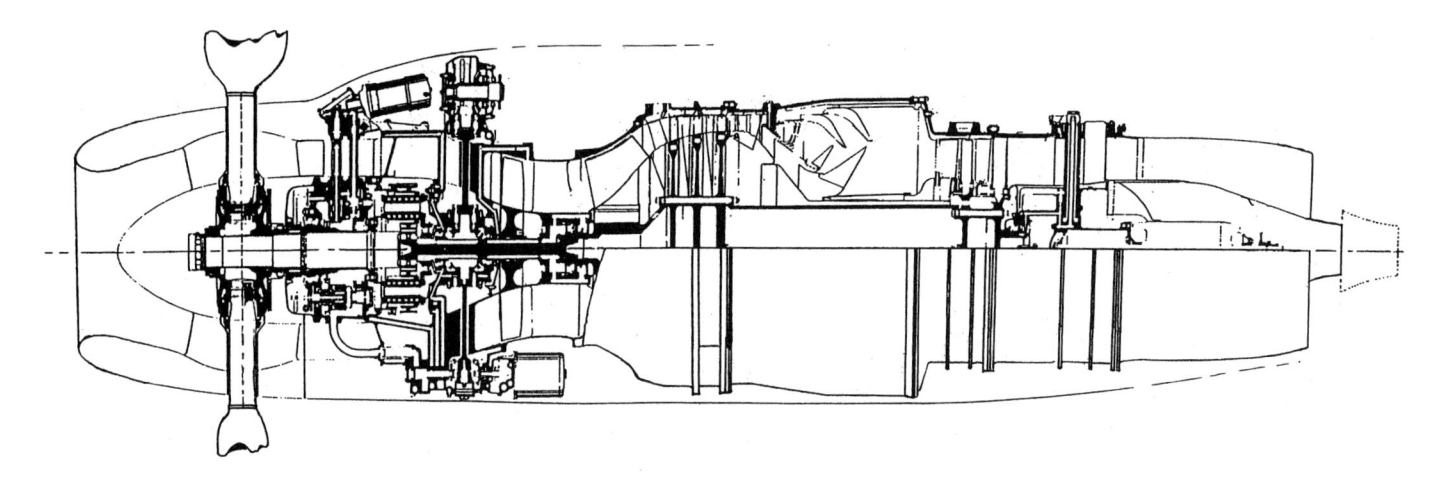

The projected Heinkel-Hirth 109-021 (PTL) turboprop, based on the H-H 109-011 A turbojet with reduction gears added to drive an airscrew. This project was later taken over by Daimler-Benz and a third turbine was added, plus the gearing, to drive contra-rotating airscrews.

was about 80 per cent complete when it was captured by the Americans. Only a mock-up of the 109-011 A engine was fitted, however, and the P.1101 had been scheduled to begin flight tests with a Junkers 109-004 engine. There was an awareness, therefore, that the 109-011 A turbojet still had a long way to go before it was available in a reliable form. Finally, as late as March 1945, when the war was virtually over, the RLM instructed BMW to develop a replacement for the 109-011 A. This was the axial BMW P.3306, which would, it was hoped, be in production by summer 1946 (see p. 237).

The Tuttlingen engine

By the end of 1944, when development of the 109-011 A turbojet was more or less frozen, von Ohain and Harald Wolff appear to have departed from Kolbermoor and were working on a new engine at a dispersed site at Tuttlingen on the river Danube, 100km from Stuttgart. This engine, named after the town, was intended to improve on the Holzwarth constant-volume combustion turbine by avoiding the intermittent combustion of the cycle with its associated mechanical difficulties. The essential feature of the Tuttlingen engine was a single rotating element that performed the function of a compressor and turbine. There was also fixed valving arranged around the rotor element to ensure an intermittent hot and cold gas stream. The scheme had many variations proposed. According to von Ohain, a small test model of 60hp was built and run at Tuttlingen, but it either blew up or was dismantled and the parts dispersed. In addition, detailed drawings were lost and little evidence remained at the end of the war.

The 109-021 turboprop

In January 1945, a turboprop (PTL) version of the 109-011 A turbojet was projected, designated 109-021. This was based as far as possible on the 109-011 A-0, but with the rotor shaft extended forward to drive a single airscrew via planetary gearing, but no extra turbine stages were added for this. There was a short ducted spinner forming the air intake. Little else is known but it was the simplest way of adapting the 109-011 turbojet to give airscrew power. Strangely, the project was also given to Daimler-Benz, which projected a more complex version of the 109-021.

CHAPTER TWENTY-SIX

Further Junkers Developments

By the end of 1943, once the urgent task of getting the 109-004 B turbojet ready for production was completed, further development of this turbojet, and also of a new, larger engine, was begun.

109-004 turbojet developments, models C to H

The 109-004 C was a redesign of the 109-004 B-4 engine with detail refinements to give an increased static thrust of 1,000kp (2,205lb) while slightly reducing the size. Its diameter was 0.755m (2ft 5¾in) and its length was 3.830m (12ft 6¾in). However, a Junkers document of January 1945 states that this model was not being proceeded with and information previously issued was to be destroyed.

Modification of the combustion system of the 109-004 B-4 permitted hotter combustion and increased thrust at some expense to engine life. This appears to have been merely an interim thrust uprating and the engine incorporating the modification, the 109-004 D-4, was not for large-scale production. The static thrust of this engine was 1,050kp (2,315lb) and it was to fit the same nacelle as the engines in use.

The 109-004 D-4 engine was to be introduced with afterburning as the 109-004 E in July 1945. A 20 per cent increase in thrust was expected: to test this, a 109-004 was fitted with a long exhaust section, but a 13 to 14 per cent thrust increase was the most obtained without excessive exhaust section temperature and deterioration of the structural work. Two fuel injection methods were tried. First, injection at turbine inlet nozzle level was tried, this giving perfect stability but badly affecting the turbine. Next, upstream injection immediately behind the turbine, without baffling, was tried; this did not give perfect stability but was the preferred method since it did not increase the turbine temperature. These tests were performed on the test-bed only and no flight tests were made. One source states that about a hundred hours of running were performed on experimental engines, both with and without the afterburner section. The 109-004 E was planned for a static thrust of 1,200kp (2,646lb) with afterburning, but, since only about 1,140kp (2,514lb) was practical, the type was not worth pursuing in the light of other developments. Nothing is known of a 109-004 F engine, but it may have been intended as a turbojet with water injection since Junkers carried out experiments with water injection, both before and after the turbine, to increase thrust.

The projected 109-004 G was based on the C model and was designed to have an eleven-stage compressor, eight combustion chambers and the usual single-stage turbine. With its air mass flow increased, the engine was expected to give a Class II static thrust of 1,700kp (3,749lb) but was not built.

The projected 109-004 H was a complete redesign and virtually a new engine. It was planned to give double the thrust of the

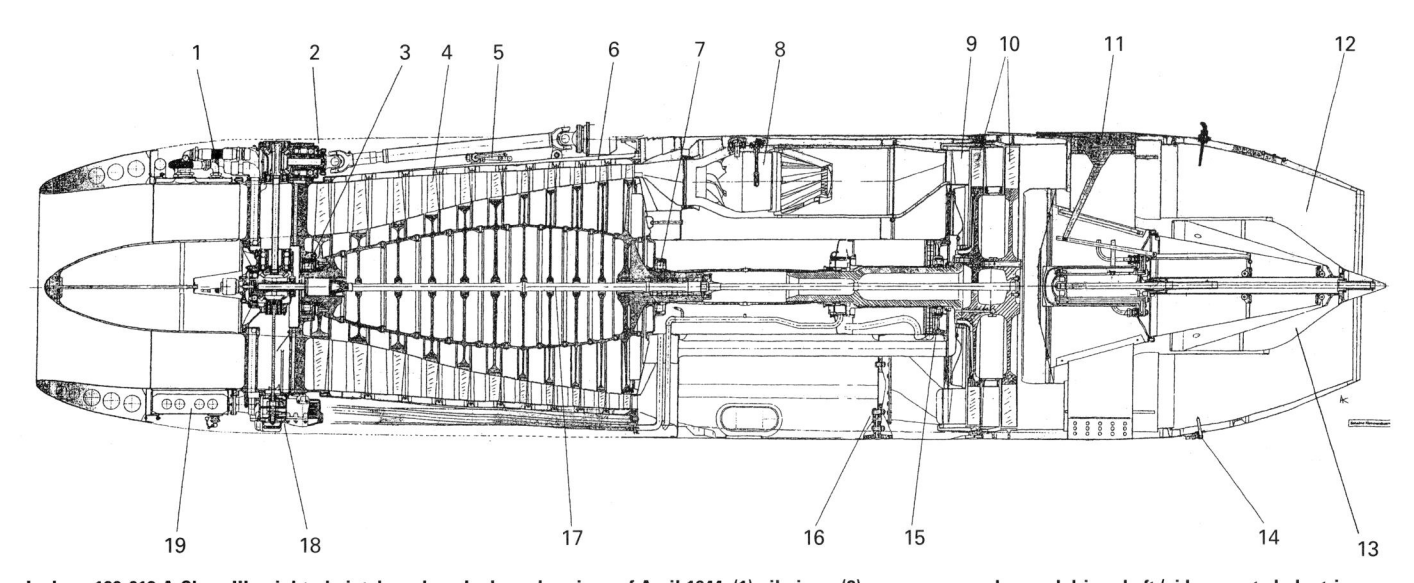

Junkers 109-012 A Class III axial turbojet, based on Junkers drawings of April 1944: (1) oil pipes; (2) accessory gearbox and drive shaft (side-mounted electric starter motor not shown); (3) front roller bearings; (4) 11-stage axial compressor; (5) bleed air pipes; (6) mild steel sheet fabricated casing; (7) compressor rear roller bearings; (8) eight combustion chambers with swirl-plate intakes; (9) turbine inlet nozzles; (10) 2-stage, air-cooled turbine; (11) cable and linkage to actuate exhaust cone cylinder; (12) variable-area exhaust nozzle; (13) movable cone; (14) exhaust gas temperature pyrometer; (15) turbine roller bearings; (16) support for outer casing; (17) compressor tie rod; (18) oil pump and filter; (19) annular oil tank. Author

FURTHER JUNKERS DEVELOPMENTS

B model, with only a slight increase in size. It was to have an eleven-stage compressor and a two-stage turbine and fell into Class II. By the end of the war, however, no parts had been made and the engine was still in the final drawing stage because the RLM had allotted a higher priority to the 109-012 turbojet (*see below*). The 109-004 H was designed to have a static thrust of 1,800kp (3,970lb) at 6,600rpm for a weight of 1,200kg (2,646lb). The compressor's presure ratio was 5.0:1 and was to give 29.5kg/sec (65.05lb/sec). The engine's diameter was 0.860m (2ft 9⅞in) and its length 4.0m (13ft 1½in).

Rare pictures of Junkers turbojets. TOP: The 109-004 C, which had detail refinements over the B-4 model to give more thrust. Production was just beginning in 1945. BOTTOM: The 109-012, partially completed mock-up. Designed to have three times the thrust of the 109-004 B, its size dictated a sheet-steel, fabricated compressor casing to save weight.

109-012 turbojet

Once the Class I 109-004 B turbojet was in production and its principal shortcomings had been remedied, the Technisches Amt urged Junkers to develop a Class II turbojet. However, the 109-004 G and H designs at first put forward were not considered powerful enough and so an entirely new, larger engine, the 109-012, was begun for Class III. The design of this engine was led by Dr Stein but was hampered by a lack of facilities and the fact that the work was dispersed from Dessau to Brandis, Aken and Mosigkau.

In layout the 109-012 followed previous practice to a large extent. The compressor was an eleven-stage axial type with solid blades, all enclosed in a neatly designed casing fabricated from mild steel sheet with band and rib strengtheners. Some initial difficulty was expected with distortion of this casing, but a cast casing was ruled out for weight reasons and also because difficulties were being experienced in obtaining castings in quantity. The compressor blades, designed by Encke of the AVA with a critical Mach number of 0.8, were thinner than those for the 109-004 and had much sharper leading edges. Their maximum thickness occurred at approximately the 55 per cent chord point. Provision was made for an air bleed after the fifth or sixth stage to provide, if required, an auxiliary compressed air supply. A fabricated compressor exit assembly led into eight combustion chambers that were housed inside a sheet metal casing. The turbine was of the two-stage axial type with hollow, air-cooled blades and a movable cone was used to vary the exhaust nozzle area.

Ten 109-012 turbojets were ordered for initial development. Some parts for the first three were made at Dessau, where nine exhaust cones were also ready, the cones having been made elsewhere (probably by the firm of Opel). None of the parts was tested, however, because the RLM ordered a stop to the development in December 1944 for unknown reasons. The 109-012 was expected to give a static thrust of 2,780kp (6,130lb) at 5,300rpm for a weight of 2,000kg (4,410lb). A pressure ratio of 6.0:1 and an air mass flow of 50kg/sec (110lb/sec) was planned. Its data included a diameter of 1.063m (3ft 5⅞in) and a length of 4.862m (15ft 11⅜in), excluding the tail cone. The projected Junkers Ju 287 B-2 (with swept-forward wings) and Heinkel He 343 bombers were each planned to be powered by two 109-012 turbojets.

109-022 turboprop

In line with the official turbojet development programme, turboprop versions were to be developed from the new turbojets of Class II and larger so that the performance of large aircraft could be considerably improved. The projected turboprop version of the 109-012 turbojet was the 109-022, which Junkers was to develop as a turboprop intermediate in size between the smaller Daimler-Benz 109-021 and the larger BMW 109-028 (see Chapter 27).

Basically, the Junkers turboprop was to consist of the 109-012 turbojet with an extra turbine stage added so that 3.50m (11ft 5¾in) diameter contra-rotating, variable-pitch, three-bladed airscrews could be driven through reduction gearing. Between 40 and 60 per cent of the turboprop's power was to be delivered to the airscrews, depending on speed. By the end of the war, the design of the 109-022 was not complete and the problems associated with the design of the airscrew reduction gearing were still being wrestled with. In any event, work probably stopped in December 1944 when the turbojet was shelved. The power rating of the 109-022 turboprop at sea level and a speed of 800km/h (497mph) was expected to be 9,500eshp. With the airscrew rotating at 1,200rpm this power was to be shared between a jet thrust of 1,200kp (2,646lb) and a shaft power of 3,500hp. With a weight of 2,600kg (5,733lb), the diameter was 1.063m (3ft 5⅞in) and the length 5.640m (18ft 6in), including 1.605m (5ft 3⅛in) for the airscrew spinner. Continuation of Junkers work in the Soviet Union is detailed in Volume 2.

CHAPTER TWENTY-SEVEN

Further BMW Developments

Despite the fact that the development of the basic BMW 109-003 turbojet up to production status fell a good way behind schedule, the company found that they had sufficient staff to consider and work on various improvements and advances for the engine. Some of these developments reached the test stage, while others remained at the project stage, depending on their priority and time of conception.

109-003 C turbojet

Much was directed at making the 109-003 both capable and suitable for operation at the higher altitudes being demanded as a tactical necessity. Many Allied piston-engined aircraft had operational ceilings greater than those imposed on German turbojets. Problems brought with high-altitude flight included intake and compressor icing, combustion flame-out and, of course, the need for cockpit pressurization. Although a new type of surface oil cooler was under test and planned for late 109-003 A-2 and E-2 engines, other means of combating icing were being investigated, such as electrical heating of the first-stage compressor stator blades. Work was also about to commence on tapping off compressor air for cockpit pressurization.

While the existing fixed-orifice fuel injectors were designed for operation at altitudes up to 12,000m (39,500ft), an engine flame-out could easily occur at altitudes above 10,000m (33,000 ft) if the throttle was touched. In any case, even the designed altitude had to be improved upon and so investigation into duplex types of fuel injector, with a separate spray nozzle for high altitudes, was proceeding. There was also interest in a Bosch fuel system that used injection pumps.

Redesigns to increase the turbojet thrust were made. One 109-003 B with a static thrust of 900kp (1,985lb) was built, but details of it are lacking. Experiments were conducted in applying water injection and afterburning to the basic engine to give temporary increases in thrust. Ready for testing, using an Ar 234, was an automatic control for continuously variable operation of the exhaust nozzle exit area, but time ran out before flight tests could be made. A complicated control system, under Hermann Hagen's direction, was intended to give a constant flight Mach number for a constant throttle setting, but this was only in the design stage at the war's end.

Following a suggestion of the Technisches Amt in 1941, the industrial turbine firm of Brown Boveri & Cie (BBC) at Mannheim was given the job of designing and developing a new axial compressor for the 109-003 A-0. This engine was then being designed with the aim of eliminating faults in the previous P.3302 experimental engines, major troubles being experienced with the compressor. Although BMW was working on resolving the compressor problems, the BBC compressor, if successful, was to be an interchangeable alternative.

The work at BBC was designated the Hermso project and was the responsibility of Dipl.-Ing. Hermann Reuter, whose team included Hermann Schneider, Joseph Krauss and Karl Waldmann. A major difference in the BBC compressor, compared with the AVA design, was that 50 per cent reaction blading was used, which shared the pressure rise equally between rotor and stator blades. The aim was to achieve greater efficiency, air mass flow and pressure ratio while still being interchangeable with the AVA compressor. This aim was achieved and the seven-stage Hermso I compressor reached an efficiency of 84 per cent and gave an air mass flow of 20kg/sec (44.1lb/sec) with a pressure ratio of 3.4:1. These results were gained in runs with the compressor on its own, it being housed in a specially built casing with multiple pressure-reading points, a vertical outlet duct and a rear extension shaft for an external drive.

Unfortunately this work took a considerable time, so that the BMW turbojet intended to receive the Hermso I compressor, the 109-003 C, was not constructed by the end of the war. BBC also built another compressor intended for the 109-003 engine. This was the ten-stage Hermso II axial compressor, which achieved an efficiency of 90 per cent, a higher figure than for any other German compressor.

A review of the 109-003 C engine was made in the first week of March 1945 and decisions were made as to how the project should be developed. By this time, design work on the engine was the responsibility of H. Dabruck. The intake of the engine was to be improved by elimination of the enclosing hoods and cooling conduits of the nose-tank cooler. Improvement was to be made to the casing of the Riedel starter to obtain better cooling for it. Manufacture of the new BBC compressor was to be carried out by the VDM airscrew company and it was to be delivered as a whole unit from BBC, which would test the compressor at Oberhausen. Improvements to the combustion chamber, in accordance with tests made by the LFA at Braunschweig, were to include extended-length mixing fingers in order to obtain a more even temperature distribution in front of the turbine. Also, smoother inner walls before the turbine were to be striven for in order to avoid flow separation ahead of the turbine. Because of the greater air mass flow of the new compressor, the turbine duct diameter was to be enlarged without increasing the engine diameter. It was to have the same turbine as the 109-003 A-2, but with new cooling inserts for the rotor blades. It was planned that work on the new parts would be concluded by July.

The planned static thrust of the 109-003 C was 900kp (1,985lb) at 9,800rpm, the weight of the engine being 610kg (1,345lb). Its specific fuel consumption was expected to be between 1.27 and 1.3, using J2 fuel. The diameter of the engine was 0.690m (2ft 3⅛in) and its length 3.415m (11ft 2⅜in) with the exhaust bullet extended.

109-003 D turbojet

Another projected improvement of the 109-003 A turbojet was the 109-003 D model. Although it was planned to use, as far as possible, such items as the combustion system, variable area exhaust nozzle and structural design of the 109-003 A, the D model was, in fact, a new engine. It had a new eight-stage axial compressor designed to give 30 per cent greater air mass flow, a higher pressure ratio and improved efficiency. In a second version of the engine, the turbine was to be of the two-stage type.

Development of the new compressor was begun by the Brückner-Kanis company in cooperation with BMW; the BBC Hermso II compressor, however, was to be used instead if it turned out to be more efficient. At Brückner-Kanis the compressor work was in the hands of Rudolph Friedrich who, years earlier, had designed the compressor for the advanced Heinkel HeS 30 (109-006) turbojet (see Chapter 20). As with the 109-006 compressor, Friedrich chose 50 per cent reaction blading for the new BMW engine.

By the end of the war no prototype of the 109-003 D had been built, although some documents state 'in construction'. Considerable headway, however, was made with the initial compressor development. A test compressor, without the full number of stages, was running at an efficiency of 89 per cent to give a pressure ratio of 3.2:1, and the indications were that the requisite pressure ratio of 4.95:1 and an efficiency of 85 per cent could be achieved, but the compressor with the full number of stages was not completed in time. It was anticipated that a more powerful Riedel starter engine of about 20hp would be required. For short-duration increase in thrust, water injection into either the compressor, combustion chamber or turbine blades was under consideration. The central hub in the intake was to be configured for de-icing. The performance laid down for the 109-003 D was a static thrust of 1,050kp to 1,100kp (2,315–2,426lb) at 10,000rpm for a weight of 620kg (1,367lb). A specific fuel consumption of 1.1 and an air mass flow of 25kg/sec (55lb/sec) were anticipated. The engine's diameter was 0.70m (2ft 3½in) and its length 3.656m (11ft 8⅜in) with exhaust bullet extended.

109-003 R turbojet/rocket engine

During 1943, department EZS under Dipl.-Ing. Peter Kappus identified the combination turbojet/rocket (TLR) as a superior concept to those of water injection and afterburning, which received a lower priority. With such a TLR unit it was hoped to obtain the advantages of cruising and fast flight with reasonable fuel economy on the turbojet, while having in reserve the rocket unit for rapid acceleration and fast climbing, without much penalty in weight. The rocket could also be used for short-time increases in altitude. Since BMW was already working on rocket engines for missiles and aircraft, the

A BMW 109-003R turbojet with BMW 109-718 rocket motor with its propellant pumps driven by a shaft from the turbojet. This method of thrust boosting for combat became of general interest after the war, until afterburning was sufficiently developed. E.J. Creek

Technisches Amt accepted its TLR proposal and a development contract was issued in September 1943. Thus, work began on the world's first combination jet power unit, which received the official designation of 109-003 R. This unit was to consist of a 109-003 turbojet modified to drive the fuel pumps of a liquid-propellant rocket motor mounted on top of the engine.

The initial objective, therefore, was the development of the rocket engine, which received the BMW project number of P.3395 and the RLM designation of 109-718. Previously, BMW rocket work had begun in 1938 to supplement similar work being performed by other German organizations. In autumn 1939 Dipl.-Ing. Helmut Philip von Zborowski was asked to act as technical consultant during the planning and construction of a BMW rocket research and testing facility at the Zühlsdorf factory. In the course of later research, led there by Zborowski, nitric acid (SV-Stoff or Salbei) was adopted as an oxidant despite the fact that this dangerous, though untried, liquid was viewed with considerable disfavour by rocket engineers. For fuel, methanol (M-Stoff or Mell) was originally used because it gave safe ignition with the nitric acid when a pyrotechnic or electrical ignition system was used. Later, methanol was replaced with liquid hydrocarbon fuels because of their higher density and, finally, the standard aviation fuels were employed to simplify the supply in service use.

The desire to simplify and increase the safety of rocket engines, by dispensing with auxiliary ignition systems, led to the search for fuels that would spontaneously ignite when brought into contact with the nitric acid oxidant. At first, organic metal compounds, such as zinc ethyl, were tried but these had the disadvantage of ageing in atmospheric oxygen, thereby degrading their properties. Initial tests with turpentine fuel, which was accidentally found to work when copper was added, led to a search for a self-igniting fuel among the amines that finally resulted in the range of fuels collectively termed R-Stoff or Tonka by BMW. Tonka compositions were many and complicated but were finally reduced to three, namely Tonka 93, 250 and 500, all of which contained raw xylidine.

By the time the 109-718 rocket motor for the TLR unit began development, Tonka self-igniting fuel and nitric acid oxidant were well-established BMW propellants. A thrust of 1,250kp (2,756lb) for a maximum three-minute period was aimed at, the thrust to be turned on and off at will. From the start of the development, high combustion temperatures generated by the self-igniting agents gave trouble, burning and buckling even the externally cooled SAS-2 (chrome-nickel steel) combustion chamber liner used in the first fifteen test units. An improved cooling system was developed whereby a proportion of Tonka fuel flow was diverted direct from the head of the injection nozzle in such a manner as to form a film of cooling liquid around the inner wall of the combustion chamber. Thus, the chamber then had two cooling systems, namely a nitric acid external cooling jacket and a Tonka fuel internal cooling film.

Results with the new rocket combustion chamber were excellent and the immediate improvement was such that an hour's continuous running was made without trouble. Further improvements reduced fuel consumption, raised the thrust, raised the exhaust velocity and reduced the combustion chamber weight. Largely because of the success of the new cooling system, it became possible to redesign the combustion chamber using almost exclusively light alloy, thereby halving its weight to 12.77kg (28.16lb). This almost dispensed with expensive and scarce heat-resisting steel components and made the chamber very much easier to machine. Altogether, this was a remarkable development.

The 109-178 combustion chamber was attached above the turbojet combustion chamber section. The rocket propellant pumps were mounted on a gearbox at the forward end of the rocket, the 200hp required to drive them being taken from the turbojet main auxiliary gearbox to the rocket pump gearbox via an extension shaft (with two universal joints) and an electrohydraulic coupling. The pumps were of the centrifugal type and were only started up when the turbojet was running at its full speed of 9,500rpm when the rocket fuel pump ran at 16,300rpm and the oxidant pump at 20,000rpm. Other equipment included the necessary piping and electrohydraulic shut-off valves, while the rocket propellants were carried inside the aircraft.

Operation of the 109-003 R was as follows. When the pilot operated a tumbler switch, an electrohydraulic coupling or clutch engaged the shaft taking the drive to the rocket pumps. Once the necessary pressure had built up for both propellants, pressure valves opened and admitted them to the combustion chamber where spontaneous ignition occurred. If the requisite propellant pressure was not reached within three seconds, the drive coupling automatically disengaged. For twin-engine installations, one rocket unit was automatically cut out if the other one failed. The propellant feed pressure into the combustion chamber increased gradually in order to give an increasing thrust up to the fixed maximum, but throttling of the thrust was not possible.

Once the development of the 109-003 R was completed, flight tests and official type- testing were arranged. For the type-test, the 109-718 rocket was put through about fifty runs, each of three minutes' duration. A meeting was held on 19 November 1943 to discuss the installation of two 109-003 Rs in the Me 262 test aircraft. The order was then given to proceed with the manufacture of an experimental batch of these engines and to begin the installation in a test aircraft at Messerschmitt's Augsburg factory. However, while development of the engine had proceeded remarkably rapidly, experimental manufacture of the 109-718 rocket components at BMW's Allach and Bruckmühl plants fell a good year behind schedule, due to bombing and dispersal.

Installation of two 109-003 Rs into a prototype of the Me 262 C-2b was begun at Augsburg by November 1944, but a rocket unit exploded during the first ground test of the aircraft at Lechfeld, apparently because of careless handling. The extent of damage to the aircraft is unknown but it was not until early in the following year that the same or another aircraft (c/n: 170 078) was ready for testing. On 28 March 1945 Flugkapitän Karl Baur flew the aircraft from Lechfeld on its maiden flight, the rockets units being started once the aircraft was in the air. Only two flights were made before a fault developed in one of the turbojets and an engine change could not be affected before the war ended. Although these flights were reported as spectacular, performance figures are unknown. However, by turning the rockets on at 700km/h (435mph) at sea level, the Me 262 C-2b was expected to achieve an initial climb rate of 5,100m/min (16,728 ft/min), reach an altitude of 10,000m (33,000ft) in 1 minute 55 seconds and an altitude of 13,000m (42,600ft) in 2 minutes

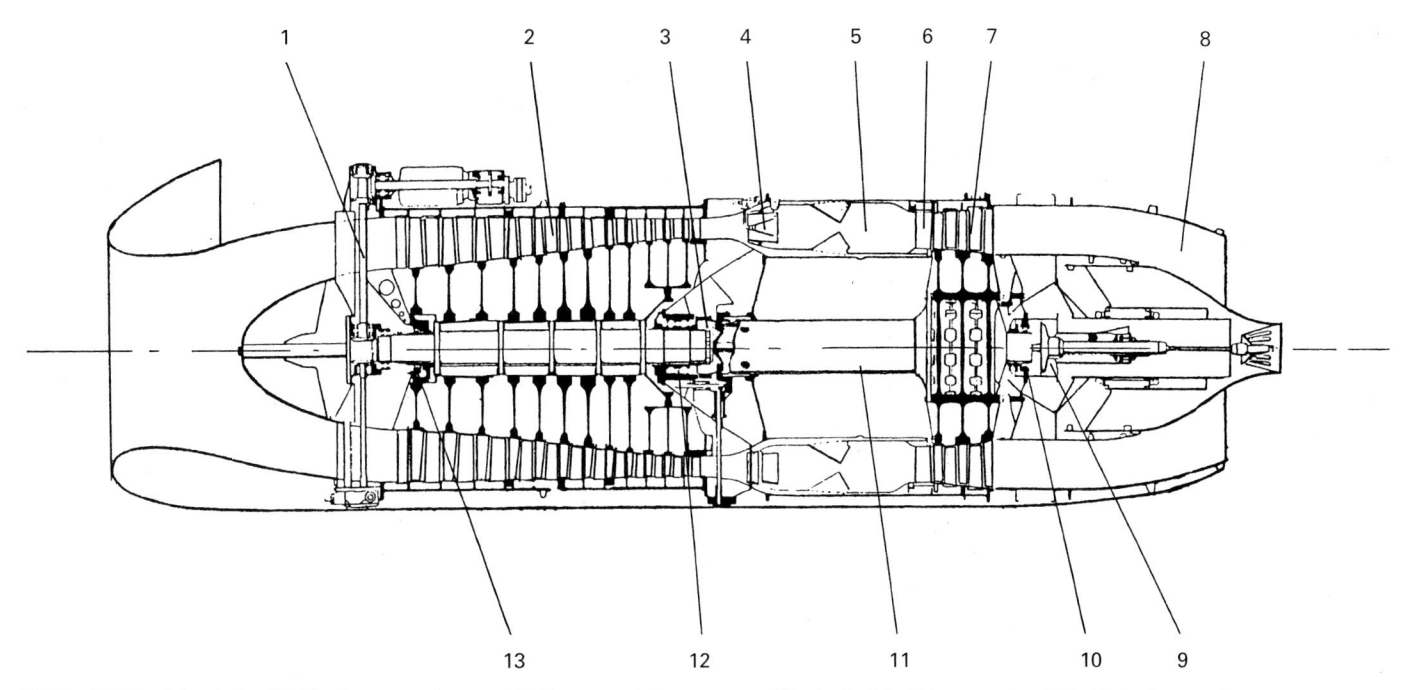

BMW 109-018 axial turbojet: (1) drive for external starter; (2) 12-stage axial compressor; (3) spherical flexible connector; (4) fuel injection nozzle; (5) annular combustion chamber; (6) turbine inlet nozzle; (7) 3-stage, air-cooled axial turbine; (8) variable-area exhaust nozzle; (9) hydraulics for movable tail cone; (10) rear roller bearings; (11) hollow shaft; (12) compressor rear ball bearings; (13) front roller bearings.

20 seconds. Maximum level speed expected was 900km/h (559mph) at 9,000m (29,500ft) altitude. The best altitude could be obtained by switching on the rockets at the normal service ceiling when, if enough propellant was carried, a theoretical ceiling of 18,000m (59,000ft) was possible. On the other hand, the best range could be obtained by switching on the rockets immediately after take-off to push rapidly through the denser atmosphere and then fly relatively lighter at the most efficient altitude for the turbojets. By this method, an Me 262 with a drop tank faired into the fuselage could achieve a range of 1,700km (1,056 miles). Competition for the 109-003 R included simpler schemes whereby a jettisonable ATO rocket was fitted beneath, for example, the Me 262 C-3a, but all these schemes were too late to see operational service.

109-018 turbojet

In addition to developing the 109-003, BMW worked on larger turbojets. In accordance with the broad turbojet development programme envisaged by Schelp of the Technisches Amt, once the Class I turbojets were developed and/or in production, more powerful engines were to begin development. Heinkel-Hirth worked on the Class II turbojet, the 109-011, and Junkers worked on the Class III turbojet, the 109-012. BMW was chosen to work on a larger still turbojet in Class IV and was given the order to proceed with such an engine in 1942, despite the fact that its 109-003 turbojet was far from perfected and certainly not ready for series production. The new engine, designated 109-018, was aimed at giving a static thrust of 3,500kp (7,717lb) and was intended for operation at altitudes up to between 15,000m (49,200ft) and 18,000m (59,000 ft) and for bringing short-range bombers up to the critical speed of Mach 0.82.

The main features and layout of the 109-018 were similar to the 109-003, but with appropriate scaling up and additional compressor and turbine stages. The chief design difficulty was that the size of the engine necessitated the use of fabricated instead of cast structural parts. Although such parts were to be normalized and stress-relieved, the distortion under running conditions was likely to give rise to some initial difficulties. The same concern was felt for the Junkers 109-012. Owing to difficulties with the development of the 109-003, which did not go into full production until the latter part of 1944, the design of the 109-018 was not completed until the end of 1943, or later, when work began on the construction of three developmental prototypes. Further delays were caused by the experimental group having to move four times because of bombing and the liberation of France.

Development of the 109-018 was officially stopped at some time during November or December 1944 (about the time when the Junkers 109-012 was stopped) because it was regarded as a long-term development for which no airframe was available anyway. However, work continued beyond this time on BMW's own initiative. One complete compressor was built but, to avoid its capture, it was destroyed at Stassfurt before it could be tested. The tests were to be done at either Dresden or Oberhausen. By the end of the war, combustion tests were about to commence to determine the best angles for the secondary air sandwich mixers, and the three experimental engines were partially complete at Stassfurt. The plan decided upon in March 1945 was that these three engines should be completed at the former cotton-spinning mill at Kolbermoor, the dispersed factory in Bavaria that was shared with other companies, before testing and development at Oberwiesenfeld. Most of the special tools, including those for blade manufacture, were moved to Kolbermoor

A BMW 109-018 Class II axial turbojet, partially completed. The largest turbojet worked on by the Germans by 1945, it was remarkable for its time, being designed to operate above 15,000m (49,200ft) and power bombers up to Mach 0.82. The bellmouth intake on the right is for static testing.

but, with defeat imminent, all engine components appear to have been destroyed.

The compressor of the 109-018 was of the twelve-stage axial type, designed for an efficiency of 79 per cent, and was the responsibility of two engineers named Loffler and Karl Fickert. A light construction was used, the first five discs being of duraluminium and the remaining seven, hotter ones, of steel. For the first seven stages, the rotor blades were of duraluminium and were attached to the discs by hollow rivets, while the remaining, high-pressure rotor blades were of steel. Much of the strength of the sheet-metal compressor casing was derived from the substantial webs, or channels, formed on the outside of the compressor stator supporting rings, which, in turn, were covered by a steel skin. The compressor shaft was hollow (although not a drum type) and was supported by two large ball races at the rear and a smaller, roller race at the front.

The annular combustion chamber was of the same design as that for the 109-003, although naturally larger, and probably had even the same number of inner and outer hollow fingers for directing secondary air. However, there were twenty-four main burner cones and eight auxiliary fuel injectors, the latter being for starting purposes. A mean gas temperature of 800°C was planned. The altitude aims for this turbojet were the most ambitious planned by the Germans.

A three-stage, air-cooled axial turbine was employed, the discs of which were held together with long bolts, these bolts being the only items in the engine that were subject to selective assembly. Departing previous practice, the turbine had only one bearing, this being of the roller type and at the rear of the turbine. For the forward

Another BMW 109-018, seen from the intake end. This engine was captured by the Soviets and taken to the USSR.

234

support of the turbine, a large, hollow shaft was provided that was located at its forward end inside a spherical seating attached to the compressor shaft, an offset stud transmitting power from the turbine to the compressor shaft. This spherical seating arrangement permitted a certain amount of axial misalignment that structural distortions could produce and thereby avoided serious stresses from this cause. In front of the spherical seating, a long tie-rod was attached to the turbine shaft and transmitted the thrust load on the turbine to the forward end of the compressor. This tie-rod was screw-threaded to assist in the expulsion of the turbine during dismantling of the engine. The turbine was expected to have an efficiency of 75 per cent.

Cooling air for the turbine was tapped off after the fifth stage of the compressor and was fed to the turbine blades via the hollow turbine shaft and hollow drum formed by the turbine discs. For cooling the inlet guide vanes, some of the secondary air flowed past the combustion chamber into the guide vanes and out through their trailing edges. For the cooling of the exhaust nozzle and tail cone assembly, air scooped from the atmosphere was used in a similar manner to that for the 109-003, but the design of the movable bullet mechanism was different. To vary the exhaust nozzle exit area, the bullet or cone moved longitudinally on rollers fitted around a central tube. Partly inside this tube was a hydraulic cylinder to move the bullet, the internal actuating valve being operated by means of a Bowden-type cable. Initially, however, the bullet was to be operated by electromechanical means, pending development of the hydraulics that were to be integrated into a fully automatic system giving constant variation of the exit area.

For starting the engine, a 50hp Riedel two-stroke petrol engine was to be used since this was already available, but later a small gas turbine, to be developed by Heinkel-Hirth, was to be substituted. The starter, of whatever type, was mounted on the compressor casing and connected to the turbojet rotor by means of bevel gears and a shaft that passed through one of the hollow arms of the front support spider. This arrangement was very much simpler than that for the 109-003. Other auxiliaries were grouped around the front of the engine but had their own gearbox driven by a separate auxiliary shaft.

The 109-018 turbojet was expected to give a static thrust of 3,400kp (7,497lb) at 5,200rpm for a weight of less than 2,200kg (4,851lb). Experimental engines were to weigh more, at 2,500kg (5,512lb). Using J2 fuel, the specific fuel consumption at static thrust was expected to be 1.1 to 1.15 and the air mass flow 83kg/sec (183lb/sec). The diameter of the engine was 1.250m (4ft 1¼in) and its length was 4.950m (16ft 2⅞in).

Pursuing the 109-018 design further, a design study was made of a TLR version (109-018 R) on similar lines to the 109-003 R layout. It is believed that the rocket unit was to be a development of the 109-718 rocket, giving a thrust of 2,000kp (4,410lb) for between 60 and 90 seconds. BMW considered the rocket best designed as a tri-fuel unit in the sense that it would run on J2 fuel and Salbei (nitric acid or SV-Stoff) oxidizer, having been started with a small amount of an igniting agent such as Tonka (R-Stoff). Not only did this propellant system, desirable for all TLR designs, give increased safety but also, if the rocket was not needed on a sortie, its allocation of J2 fuel could be used by the turbojet engine for an increase of endurance. There was, however, no plan to build a 109-018 R since the TLR type of engine was considered suitable for fighters only, and a large enough fighter was not then envisaged. However, at least one high-altitude fighter was projected for the engine without the rocket booster. Naturally, bomber projects were the most numerous planned for the large 109-018. An adaption of this turbojet for use in a twin turboprop layout is described below.

In addition to the aircraft industry, BMW also put forward aircraft projects of its own designed around the 109-018 turbojet. A BMW group was set up to make such aircraft studies in collaboration with the aircraft industry and other interested parties; this, along with similar work by Daimler-Benz, represents an early attempt at 'systems engineering', whereby airframe, engine and equipment were considered and integrated from the beginning. The 109-018 turbojet was the last BMW design to get past the drawing-board stage before the end of the war. The following BMW wartime designs described, therefore, remained in the project stage.

109-028 turboprop

As early as 1940 BMW's project department made studies for a large turboprop engine (PTL) and by early 1941 the development of such an engine had the backing of the Technisches Amt. Although the design was never finalized, the engine resulting from the development was to be designated 109-028 and, by July 1942, was planned to develop 8,000eshp at 800km/h (497mph) at 8,000m (26,240ft) altitude. At about this time the decision to develop the 109-018 turbojet was taken so that, for a given engine size, BMW's turboprop work preceded its turbojet work instead of vice versa, as other companies chose to operate, for example in the development of the Junkers 109-022 PTL from the 109-012 TL.

Nevertheless, the first designs for the 109-028 bore many features seen in the 109-018 turbojet, such as a twelve-stage axial compressor of disc construction, similar combustion and exhaust nozzle assemblies and the same method of supporting and connecting the compressor and turbine shafts. Although the turbine was also of the same construction as for the 109-018, another stage was added to extract power for the airscrews. Therefore, the 109-028 had a four-stage, axial turbine using about 30 per cent reaction. The turbine stage added for airscrew power was not independent of the other three stages but all four were bolted together. Power was transmitted to the contra-rotating airscrews by a shaft from the front of the compressor driving through planetary gearing within the airscrew spinner.

For this first design of the turboprop engine, the airscrew spinner was enclosed in a duct or shroud that began in front of the spinner nose and led up flush to the engine air intake. The airscrew drive and intake section were chiefly the responsibility of Dipl.-Ing. Wilhelm Stoeckicht of BMW and Ebert of the VDM airscrew company and this part of the turboprop posed the biggest problems. Not only was the mechanical design made difficult by the space and weight limitations but also the aerodynamic problem of ensuring a good airflow into a compressor preceded by airscrews was considerable.

Unless a suitable design was arrived at, the compressor would suffer a loss of intake ram air owing to the aerodynamic disturbances caused by the airscrews and their centrifugal effect on the airflow near the spinner. The latter effect was to be minimized by using two large airscrews revolving at the minimum speed without sacrificing efficiency, but other intake disturbances were not so readily avoided. The reasoning

FURTHER BMW DEVELOPMENTS

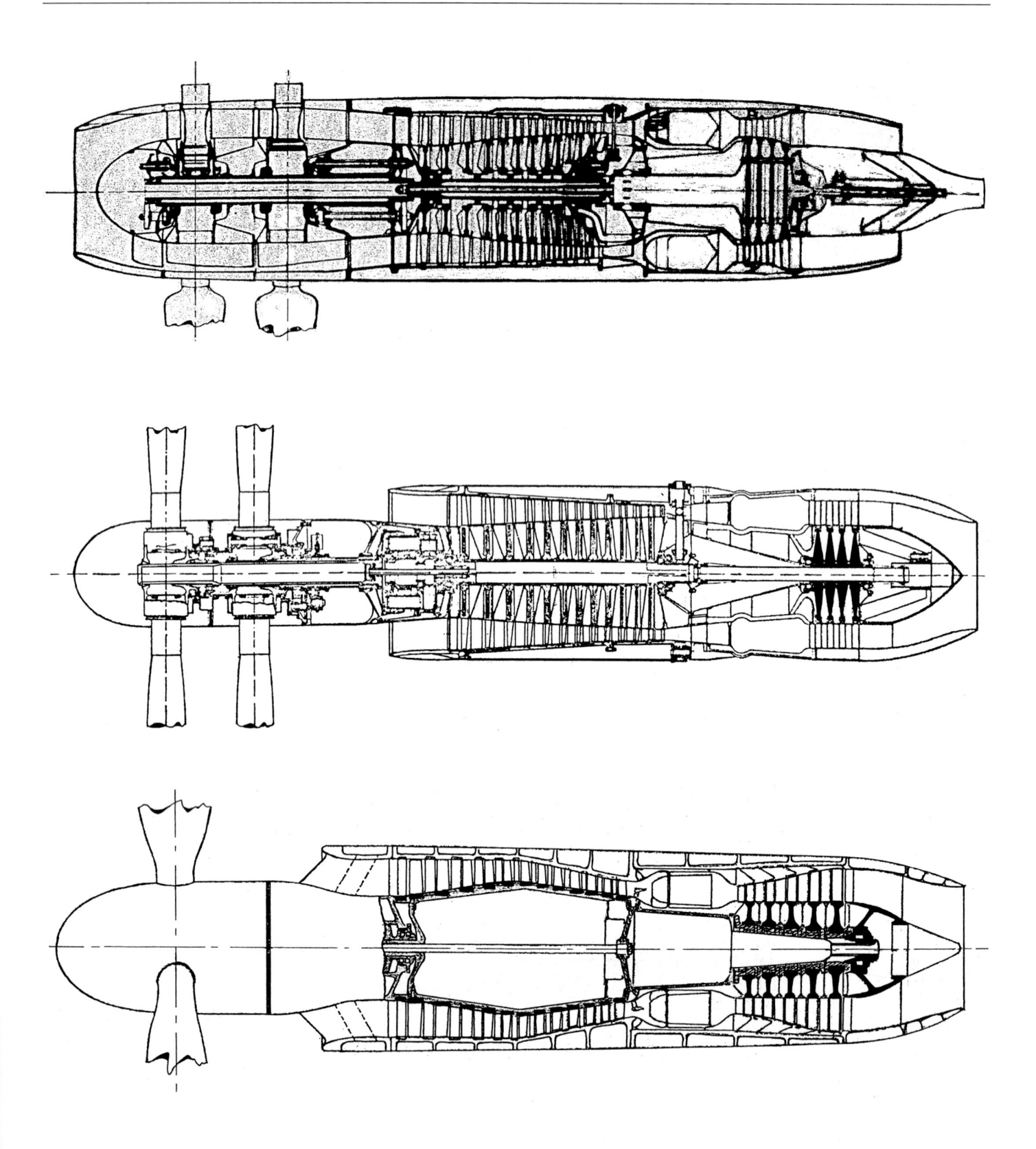

Three layouts for the projected BMW 109-028 (PTL) turboprop engine: (top) with contra-rotating airscrews, ducted spinner housing planetary gearing, 12-stage compressor and a 4-stage turbine; (centre) with contra-rotating airscrews and gearing moved aft, shortened intake, 12-stage compressor and 4-stage turbine; (bottom) with a single airscrew, short intake, 12-stage compressor and 5-stage turbine.

behind the ducted airscrews spinner was that air would be drawn in before the major airscrew disturbance, but a disadvantage here was that the annular space between the duct and the spinner was considerably cluttered with guide and support vanes in addition to the airscrew blade roots.

Although no airflow tests were made for this or any other intake design, another one was proposed without a duct or shroud around the airscrew boss. In this second design, the aim was to position the airscrews as far forward as possible from the engine intake. A long airscrew spinner therefore resulted, together with the shortest possible compressor intake, the latter having sharp, slightly outward-turned lips. Also, in order to minimize the cantilever load on the projecting airscrew section, its main gear drives were moved back from inside the spinner to a position just in front of the compressor. Further refinements were also introduced in this design, such as a modified combustion chamber and a new exhaust nozzle bullet that did not project beyond the end of the exhaust nozzle. The nozzle area was to be adjusted in accordance with altitude and airspeed, the associated control cam profile being designed so that the airscrews would not receive more than the 7,000hp for which their reduction gears were designed.

Other details known for the engine, which was still in a fluid state of design, are as follows. The contra-rotating airscrews were each to have three or four blades, depending on the diameter decided upon (3.5–4.0 m). They were to have variable negative-pitch range, permitting reverse-thrust braking upon landing. For the hydraulic operation of the airscrew pitch, an all-speed governor was designed that was set by a cam, the profile of which was calculated to give good fuel consumption at part load. Engine accessories were to be driven by a radial shaft between the compressor and the combustion chamber, up to 300hp at full speed being available for this purpose. Hydraulic operation was planned for the variable area exhaust bullet, and the turbine and exhaust sections were to be air cooled. Efficiencies aimed for were 70 per cent for the turbine and 80 per cent for the compressor, together with an average combustion temperature of 770°C before the turbine. The exhaust temperature was to be limited by a valve that measured air mass flow (in terms of temperature and pressure upstream of the compressor) and suitably metered the fuel flow.

Finally, a complete redesign of the projected 109-028 turboprop was undertaken to introduce simplifications and thereby enhance the chances of early success in development. The contra-rotating airscrews were replaced with a single airscrew to eliminate the complex gearing, a wise decision in view of the later problems experienced in the development of large turboprops in the Soviet Union. An all-new twelve-stage axial compressor with a more refined intake was designed, together with a five-stage turbine.

A plan existed to conduct the first flight tests by fitting two 109-028 turboprops into a Heinkel He 177 Greif bomber, which normally used two large airscrews driven by four piston engines coupled in pairs. However, since the detail design of the turboprop engine was not completed, not even the building of a prototype was begun. The 109-028 was planned to give a static jet thrust at sea level of 2,200kp (4,815lb) plus a shaft power of 4,700hp, a total equivalent of 6,570shp. At 800km/h (497mph) at sea level, the jet thrust was expected to decrease and the shaft power to increase, the total equivalent being 12,600shp. The weight was estimated to be around 3,500kg (7,717lb). Its specific fuel consumption was expected to be as low as 1.06 at 12,000m altitude. The diameter of the engine was 1.250m (4ft 1¼in) and its length 6.0m (19ft 8¼in).

Apart from the plan to convert an He 177 for flight tests, there were few schemes for the utilization of the 109-028. Calculations were made to fit two of them to the Messerschmitt Me 264, which was originally designed to fly non-stop to the USA and back, albeit with a small bomb load owing to the enormous fuel load needed. BMW also put forward a bomber project of its own.

Because of the difficulties expected in trying to obtain a good intake with the 109-028 turboprop, a fresh look was taken at the whole question of the layout of large turboprop engines. Until then, the basic idea was to take a turbojet engine and add one or two extra turbine stages to divert a major part of the power to an airscrew at the front, thereby giving a larger mass of slower-moving air more suitable for larger, slower aircraft. On the advice of the independent consultant Dr Alfred Müller, however, the Technisches Amt directed BMW to make a study of a new turboprop layout. This, in essence, comprised a 109-018 turbojet with an enlarged

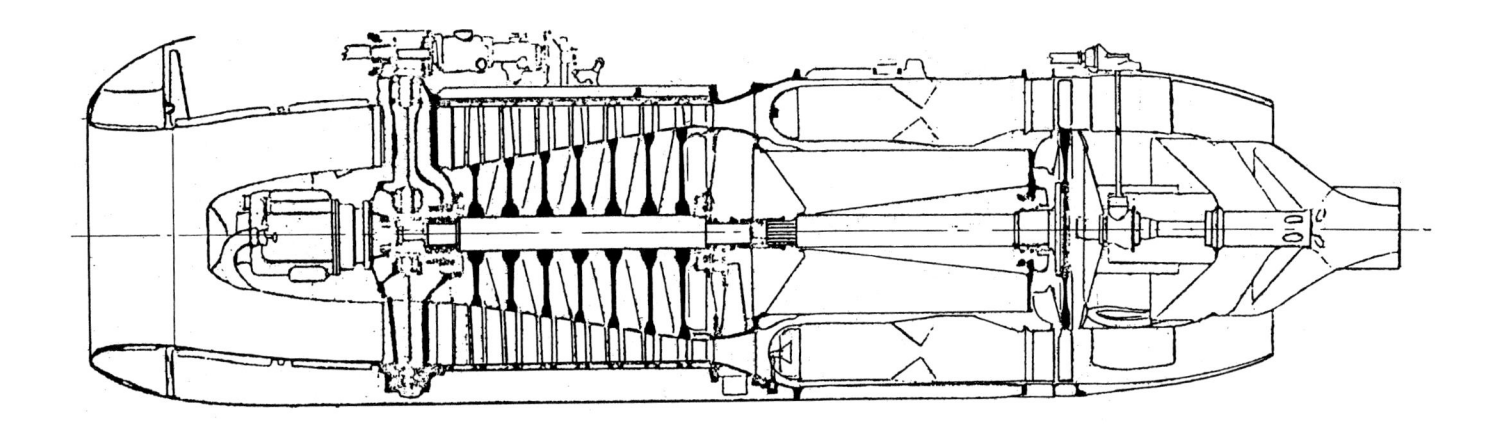

The projected BMW P.3306 turbojet was to be interchangeable with the Heinkel-Hirth 109-011 and was to introduce simplifications for more rapid production. It was hoped to have it running by summer 1946.

compressor from which air could be tapped off to supply two independent power units in the aircraft wings. Each power unit consisted of a combustion chamber supplying hot gases to a turbine driving an airscrew, the turbine operating at the then high temperature of between 900 and 1,000°C. Assuming this interesting project was pursued, no headway could be made with it until the basic 109-018 turbojet had undergone some development. Nevertheless, this scheme makes an interesting comparison with the one looked into by Heinkel whereby two smaller turbojets were to supply hot gases to an independent turbine driving an airscrew. For both schemes, the companies concerned gained the investigation incentive from officialdom.

P.3306 turbojet

During a conference in March 1945 the Technisches Amt sought BMW's opinion on the Heinkel-Hirth 109-011 turbojet. This engine was supposed to be in production by then but was still beset with various technical deficiencies. Although plans were being made for BMW to give assistance to Heinkel-Hirth for the production of the 109-011, Dr Bruno Bruckmann, Director of BMW powerplant development, thought that an official order was about to be issued to his company for the development of a new turbojet engine. This was to be interchangeable with the 109-011 but with a 1,700kp (3,749lb) instead of a 1,300kp static thrust. An operational ceiling of 15,000m (49,200ft) was specified. In view of the obvious impending collapse of Germany and the fact that there had been considerable official pruning of projects and developments, Bruckmann must have had some concrete evidence that an order was imminent because he ordered planning work to begin on a new engine. Given the BMW project number P.3306, the project was more than just an exercise to keep its considerable design staff busy and it seems that there was every intention of pushing ahead with the full development of the turbojet after making the first runs in summer 1946. It was thought that, once a prototype was running, full official support would be forthcoming.

The layout of the P.3306 was simplified and showed no remarkably novel features but largely followed the layout and practices used in the 109-003 while incorporating all the improvements and lessons learned during the development of that engine. A seven-stage axial compressor employing disc construction was used, and was to give a pressure ratio of 4.2:1 and an air mass flow of 40kg/sec (88.2lb/sec). It was to be driven by a single-stage axial turbine. Sheet-metal construction was used to a greater extent than in the 109-003, and castings were either dispensed with or reduced to a minimum size. The usual combustion chamber with secondary air fingers was used but with the average combustion temperature raised to 750°C, and there was a mechanically driven, open-ended bullet to vary the exhaust nozzle area. A Riedel starter motor was mounted inside the intake central fairing and the usual auxiliaries, oil pumps and such were mounted externally and driven from the rotor by radial shafts. For the mounting of the main auxiliaries (on top of the compressor casing) and for the attachment of the engine, a reinforced section was fitted to the sheet metalwork.

The planned static thrust of 1,700kp (3,749lb) for the P.3306 was to be attained at a speed of 8,700rpm but this could be increased to 1,820kp (4,013lb) for a short period for take-off. Its weight was calculated as 900kg (1,985lb) and its specific fuel consumption as 1.2. Dimensions included a diameter of 0.850m (2ft 9½in) and a length of 3.20m (10ft 6in).

Expendable P.3307 turbojet

Another BMW project was for an expendable powerplant for missiles. This, the P.3307 turbojet, was designed to use a maximum of sheet metal and to have a production time of only 100 man-hours. Unfortunately, nothing is known of the background of this project or if it was to go forward. It seems unlikely that it was a direct competitor of the Porsche 109-005 (see Chapter 30), which, although planned for the same duty, bore little other resemblance in its specification. From the weight and size points of view and, no doubt, from its general layout also, the P.3307 was similar to the 109-003. The compressor was of the axial type but different sources give its number of stages as five, six or seven. This possibly indicates that the design was at a very early stage.

The P.3307 was to avoid the use of scarce materials. Therefore, the turbine wheel was to be made of unalloyed steel but the engine's maximum temperature was to be between 600°C and 650°C. A fixed-area exhaust nozzle was to be used and starting of the engine was to be by external means only. For testing, a useful life of ten hours was to be aimed at, this easily covering the mere two hours running in operational use. No exacting requirements were specified for the engine's weight. The P.3307 was designed for a static thrust of 500kp (1,102lb), for a weight of 650kg (1,433lb) and a specific fuel consumption of 1.5. Its diameter was 0.690m (2ft 3⅛in) and its length 2.850m (9ft 4¼in).

CHAPTER TWENTY-EIGHT

Daimler-Benz A.G.

After the economic difficulties following the First World War, the Daimler engine company merged with the Benz vehicle company to form Daimler-Benz A.G. (DB) in summer 1926. In the aero-engine field DB concentrated on liquid-cooled engines under the direction of Dipl.-Ing. Fritz Nallinger. The Second World War saw DB as one of Germany's major aero-engine companies, with eleven main factories making its aero-engines alone. The facilities in Stuttgart-Untertürkheim formed the main research, design and development centre for all DB products (including vehicles), while pre-production engine work (O-series) was carried out at the Berlin-Marienefeld plant known as Werk 90. As a further illustration of the company's size, there were at the end of 1944, for example, 63,500 personnel (including just over 40 per cent foreigners) employed by DB in all its branches and factories.

Evolution of the turbojet programme

Given its reputation for pioneering and first-class engineering products, one would have expected DB to be in the forefront of turbojet development, but this was not the case, due to an initial conservatism. Before the Daimler-Benz merger, however, the Daimler Motoren Gesellschaft of Berlin did make some studies of the gas turbine. This is illustrated by its patent (426,009), granted on 3 March 1926, for a combustion-turbine with auxiliary liquid, whereby a petrol-air mixture was to be exploded in a chamber to compress a volume of water, which formed a jet to impinge upon the buckets of a turbine. Very little heat was passed to the turbine and the curious, though interesting, proposal was a far cry from the pure gas turbine.

When DB was approached by Mauch and Schelp of the Technisches Amt in autumn 1938 in connection with turbojet development, this approach proved fruitless. Largely this was because of the scepticism of Fritz Nallinger, head of development at DB, who completely refused to become involved with turbojets or jet engines. As we have seen, by 1939 all the other major German aero-engine, and some airframe, companies were either officially or privately involving themselves with jet engine work. Then, even though the company was very heavily committed to the development and production of its piston engines, the decision was made to devote a small team to jet propulsion study. The official view was that the DB effort would be best directed along a new line, unlike those of the other companies, and the suggestion of a turbine-driven ducted fan engine was taken up.

Fortunately, a first-class engineer had arrived at DB in 1939 with five years' previous experience at the DVL working on exhaust-driven turbochargers and the study of the general theories of jet propulsion by turbojet and ramjet engines. This was Dr-Ing. Karl Leist, under whose direction the DB turbojet project was begun at Stuttgart-Untertürkheim. Leist began his research by studying a ducted fan driven by a piston engine. A DB 604 24-cylinder engine of 2,500hp was to drive the ducted fan and a compressor feeding an afterburning chamber to provide extra thrust. This engine, designated DB 670 (and also known as the ZTL 5000) was expected to give a thrust of 600kp (1,323lb) at a speed of 900km/h (559mph) at an altitude of 6,000m (19,700ft). However, this scheme was doomed due to its estimated weight of 1,700kg (3,749lb)! Leist then briefly looked at the possibilities of the pulsejet before turning to the turbojet project.

This project proved to be one of Germany's most complicated turbojets because it involved, as well as a ducted fan, a contra-rotating central compressor, both being driven from the same turbine. It is not known if this ambitious scheme originated with Schelp and his department alone, or if Leist also held a brief for it. As previously related, other German companies considered or tried turbojets with either a ducted fan or a contra-rotating axial compressor, but not both together. (The Heinkel 109-010, a projected ducted-fan version of the HeS 8 turbojet, was abandoned when the HeS 8 was overtaken. The contra-rotating compressor was considered for the Junkers 109-004 turbojet but not adopted, while Bramo's 109-002, which did use a contra-rotating compressor, had to be abandoned by early 1942 because of great developmental difficulties.)

Yet another novel feature marked out the DB turbojet from other German engines: the method of cooling the turbine. Whereas the general German practice eventually involved the use of hollow turbine blades through which cooling air passed, Leist proposed that his turbine should be only partly immersed in the hot gas stream, leaving about one third of its circumference available for immersion in a cooling air stream. Of course, this meant that less of the turbine would be available for producing power, but this was to be balanced by using an abnormally high gas temperature, as high as 1,100°C being thought possible.

Leist's ideas on such a turbine went back at least to his days with the DVL. From Aachen, on 23 February 1934, he filed a patent for his partially impinged combustion or gas turbine (767,078), the specification also listing earlier patents on a similar theme. In this scheme, a large combustion chamber supplied hot gases to drive an axial turbine that drove a compressor and also, via gearing, an airscrew. Air from the compressor passed to the combustion chamber via a pre-heater fitted inside the exhaust tube and nozzle, carrying the hot gases away from the turbine. This pre-heater appears to foreshadow later attempts (not at DB) to improve the thermal efficiency of turbojet and gas turbine engines by developing heat exchangers. The most important aspect of Leist's scheme was that hot gases were to be admitted to the turbine over only about 36 per cent of its circumference. For the

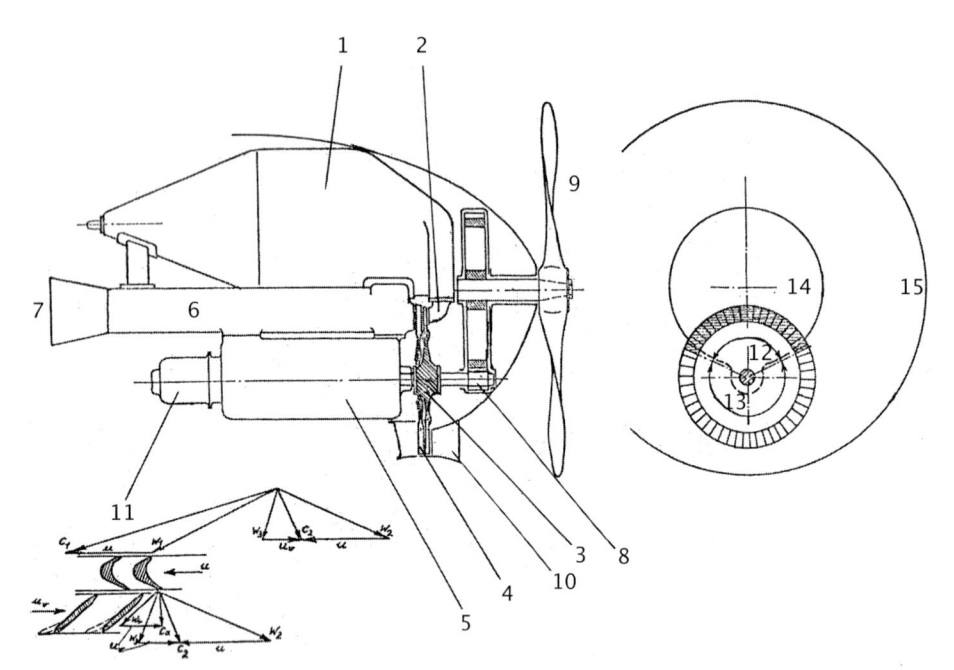

Karl Leist's scheme for a partially impinged combustion or gas turbine (Patent 767 078, filed 23 February 1934): (1) combustion chamber; (2) turbine inlet nozzles; (3) turbine wheel; (4) stator blade ring; (5) compressor; (6|) pre-heater; (7) exhaust nozzle; (8) reduction gears; (9) airscrew; (10) turbine cooling-air intake and baffles; (11) generator; (12) hot gas sector of turbine; (13) cooling air sector of turbine; (14) gearbox outline; (15) airscrew circle.

remaining 64 per cent, the turbine blades passed through or were immersed in a cooling air stream drawn from the atmosphere through an intake and baffles.

109-007 (ZTL 6001) ducted fan, contra-rotating turbojet

By the end of 1940 the layout of the DB turbojet, designated ZTL 6000, was established in its initial form and development of components began. The engine was to be in Class II and to develop a thrust of 1,400kp (3,087lb) at 900km/h (559mph) at sea level. By about June 1942, however, and possibly much earlier, the engine design had been revised and a less ambitious thrust of 960kp (2,117lb) was aimed for. The designation of the engine was now ZTL 6001 within the company and 109-007 officially. The ZTL designation stood for *Zweikreisturbinen-Luftstrahltriebwerk* (two-circuit turbojet engine). The first of these two airflow circuits in the 109-007 passed through the main, central compressor, providing primary and secondary air for the combustion system. The second airflow passed through the secondary compressor or ducted fan, most of this air bypassing the combustion and turbine sections to mix with the hot gases in the exhaust nozzle section; part of the ducted fan airflow, however, entered the turbine section for cooling purposes.

Complication arose from the fact that the ducted fan was built round and not in front of the main compressor. Thus, a rotating drum carried on its outside the moving blades of the ducted fan and on its inside what would have normally been the stator blades of the main compressor. This drum was driven at about half turbine speed via gearing from the turbine, while an inner drum, carrying the inner blades of the main compressor, was driven in the opposite direction at full speed from the turbine shaft. To obtain the necessary strength, compactness and lightness of the mechanical components was undoubtedly the major problem facing Leist and his team.

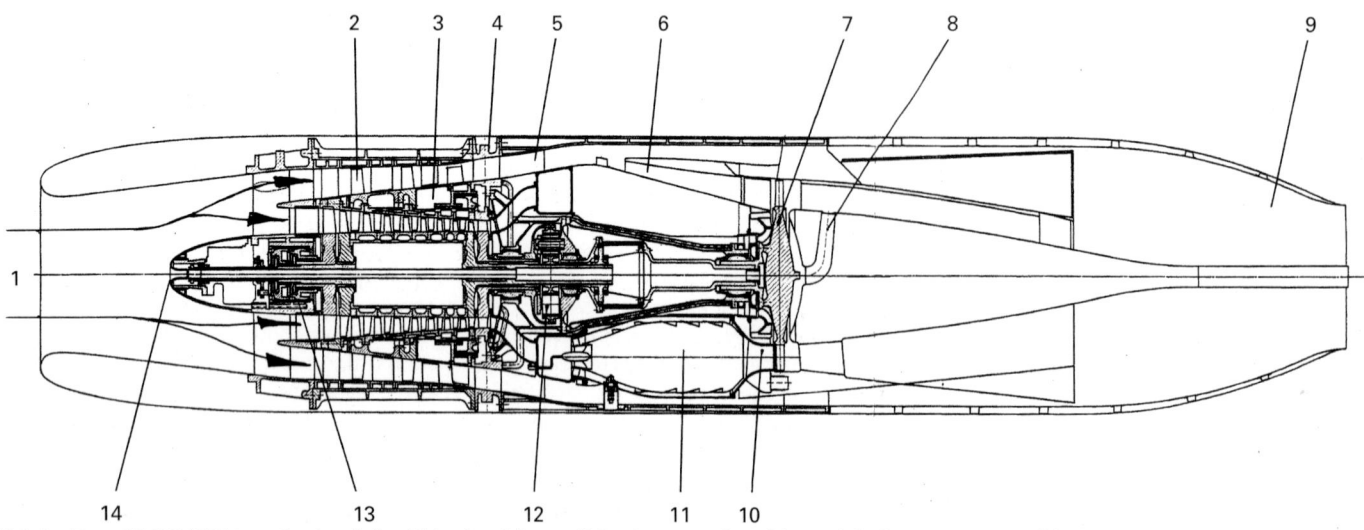

Daimler-Benz 109-007 (ZTL) two-circuit turbojet: (1) intake airflow, splitting into outer ducted fan and the inner compressor; (2) 3-stage ducted fan drum; (3) 17-stage contra-rotating compressor (8 stages on inner drum and 9 stages on outer revolving drum); (4) main support point; (5) angled support vanes; (6) turbine cooling air scoop; (7) single-stage, partially impinged turbine (upper third in cooling air, lower two-thirds in hot gases up to 1,100°C); (8) turbine cooling air outlet; (9) exhaust nozzle for cold air and hot gases; (10) turbine inlet nozzles; (11) combustion chamber; (12) drive gears for fan drum; (13) housing with front bearings for fan and compressor drums; (14) inlet for cooling air.

Daimler-Benz 109-007 (ZTL) two-circuit turbojet, with bellmouth intake for static testing.

This team was assigned the minimum of personnel: these included Dr Spieser and Dr Laukhuff, responsible for component construction, and Dr Kamps and Dr Stiefel, responsible for component experimentation. Further assistance was sought from outside the company. Discussions were held with the AVA, for example, concerning compressor and fan blade design, while the firm of Voith was consulted on the mechanical aspects of components. Not surprisingly, progress on the 109-007 was extremely slow and by autumn 1943 only one example of the engine was running.

Beginning on 1 April 1943, this prototype was first used for mechanical testing of components and to tune the lubricating oil circuits. For this purpose, the engine was run without combustion and without its exhaust nozzle. During the twenty-one hours' total running time, its rotational speed did not exceed 5,000rpm, well under half the design speed. Other important work during these first runs concerned the collecting of data on the compressor at various speeds. To run the engine without combustion, an 8,000hp electric motor was employed.

The first runs of the 109-007 engine with combustion began on 27 May 1943. As a precaution, watersprays were used to cool the turbine wheel and inlet nozzles. Subsequent development reached the stage where the water cooling could be dispensed with. Most, if not all, the engine runs were made without the exhaust nozzle attached, but a static thrust corresponding to a thrust of 600kp (1,323lb) at 900km/h (559mph) at 7,000m (23,000ft) altitude is said to have been reached.

By the end of 1943 Daimler-Benz could see the good results being achieved by other companies. Abandoning their conservatism, the team was anxious to catch up in the field of turbojets. Unfortunately, however, it was also plain that the 109-007 was still at an early stage in its development. Its complexity indicated that it would not be ready for production for some years, by when the war was expected to be over, one way or the other.

About May 1944, by when an order from the RLM had halted further development, the 109-007 had reached the following form. The central or main compressor was of the seventeen-stage axial type with all blades revolving. Eight stages of compressor blading were carried on an inner drum revolving at full turbine speed, while nine stages of compressor blading were carried on the inside of a contra-rotating outer drum revolving at about half turbine speed. The compressor supplied about 29 per cent of the total engine air mass flow at a compression ratio of about 8:1. Its efficiency was calculated as 80 per cent and the temperature of the air at the final stage was about 320°C.

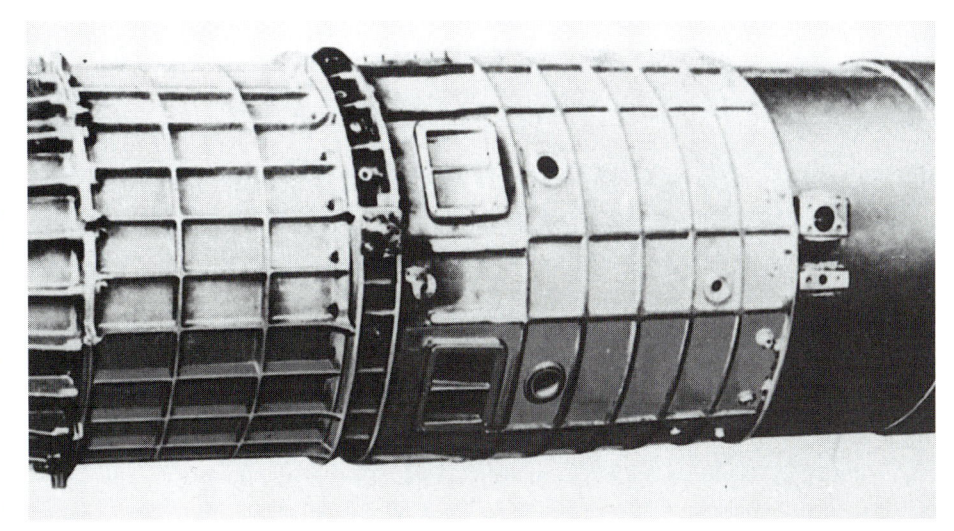

The hollow compressor drum was largely of aluminium alloy and was of almost constant outside diameter. It was constructed by mounting internally channelled rings (carrying compressor blades) onto a steel drum. Bolted to each end of the steel drum were forged flanges that extended into shafts for the front and rear bearings. The rear shaft was connected via a large flexible coupling to the turbine shaft and also carried the first driving gear for the ducted fan drum. The flexible coupling avoided excessive strain (caused by misalignment, distortion and acceleration) being placed on the gear unit.

The ducted fan was of the three-stage axial type, supplied about 71 per cent of the total engine air mass flow and had a calculated efficiency of 84 per cent. Three stages of rotor blades were attached to the outside of the contra-rotating drum and these were preceded by three stages of wide-chord stator blades attached to the inside of the external casing. In longitudinal section, the wall of the drum was wedge-shaped (giving progressively shorter fan and compressor blades towards the rear) and was of complicated construction to facilitate compressor and fan assembly.

The contra-rotating drum was supported at the front by the first stage of the inner compressor blades, which were attached to a disc in front of the compressor drum. This disc had a hollow shaft that fitted over the compressor shaft and led forward into a bearing. A similar arrangement of compressor blades, disc, hollow shaft and bearings supported the contra-rotating drum at its rear end, but here the hollow shaft had a bell housing with internal teeth to engage planetary gears. The planetary gears, stub shafts for which were secured inside the engine's cone-shaped central support, meshed also with a central gear on the compressor drive shaft in front of the coupling with the turbine. By this means the contra-rotating drum was driven.

Four tubular, interconnected combustion chambers were used, although the engine design made provision for a fifth chamber in case it was later found possible to reduce the amount of cooling air to the turbine. The combustion system aimed at an efficiency of 95 per cent, with a maximum combustion temperature of 1,300°C, a pressure loss of 3 per cent and a chamber pressure of 10 atmospheres. Curved diffuser ducts led back and outwards from the inner compressor outlet to a welded, annular plenum chamber that supplied air to the combustion chambers. Each of these consisted of an inner and outer tube. Primary air for combustion entered the inner or flame tube through a ring containing angled swirl vanes. Fuel pipes led from the centre of these swirl vanes to a nozzle that sprayed fuel against the swirling air, the pipe and nozzle assembly also acting as a flame-holder in a similar manner to the Junkers system. Secondary air passed around vanes, set at an opposed angle to the primary air vanes, and entered the annular space between the flame tube and the outer tube; it entered the flame tube via small scoops and fingers at a swirling angle to enhance mixing with the combustion gases.

The turbine was of the single-stage, axial type with an efficiency calculated as 64 per cent at full speed. Along 33 per cent of the circumference, the blades moved through a cooling air stream, while for the remainder of the circumference the blades worked in a gas temperature of between 1,070°C and 1,100°C. The cooling air was diverted from part of the ducted fan airflow and entered a scoop at the beginning of the inner shroud surrounding the turbine. Inlet guide vanes at the entrance to this scoop directed the angle of the cooling air and were of broader chord than the inlet guide vanes that preceded the turbine blades around the rest of the circumference.

A forged steel turbine wheel was used and was bolted on its forward face to a flanged, hollow shaft. This shaft was mounted in a bearing and was internally splined to fit onto a shaft flexibly coupled to the compressor shaft. The turbine blades and inlet nozzles were hollow and were made from nickel steel. In the design of these, assistance from Heinkel-Hirth was obtained.

From the turbine section to the end of the engine, the structure was of welded sheet-metal construction. From the rear diameter of the turbine wheel, a fixed cone extended rearwards to finish in a tubular section near the end of the engine. Surrounding this cone was a shorter, inner shroud that directed the exhaust gases and cooling air from the turbine into the exhaust nozzle. The main portion of the air from the ducted fan flowed through the annular space inside the engine wall and joined the hot gases just past the inner shroud, the two streams passing out through the exhaust nozzle. At one time it was proposed to use a variable-area exhaust nozzle of the ring type originally experimented with by BMW, but the final intention was to use a fixed-area exhaust nozzle. Although not tested with its exhaust nozzle, it seems likely that the 109-007 would have been relatively quiet due to the inner, high-speed exhaust jet being surrounded by the slower, ducted fan air stream.

The engine was started on petrol and then switched over to J2 fuel. Included in the control system was a barometric fuel control, overspeed governor and throttle valve. For lubrication, a dry sump system with pressure feed to the main bearings was used. Very few auxiliaries were fitted to the prototype, test-stand engine. The 109-007 was designed for a static thrust of 1,275kp (2,811lb) at a compressor speed of 12,600rpm and a ducted fan speed of 6,200rpm. Its weight was 1,300kg (2,887lb) and its lowest expected specific fuel consumption 1.05. The air mass flow through the compressor was 8.2kg/sec (18.08lb/sec) and through the ducted fan 19.9kg/sec (43.88lb/sec). The overall diameter of the engine was 1.625m (5ft 4in) and its length 4.725m (15ft 6in).

109-016 turbojet

With the cancellation of the 109-007, Daimler-Benz was ordered to work on a turboprop version of the Heinkel-Hirth 109-011, designated 109-021. At the same time, Karl Leist left the company and went to work at the Technisches Hochschule Braunschweig. Prior to this, other turbojet engines were projected or planned by DB. These were the P.100 of 13,000kp (28,665lb) thrust, the P.101 of 2,000kp (4,410lb) thrust and the P.102 of 6,500kp (14,333lb) thrust, all figures being for take-off thrusts. All were axial engines using, as far as possible, the same features and the so-called Mehrflutig type of construction, which, it is believed, consisted of sheet metal strengthened by multiple channel-section rings and stringers welded in place. All had nine-stage compressors giving a pressure ratio of 6.7:1.

The P.100 was to be an enormous engine for the time, far outside the Class IV size, and the biggest contemplated in Germany or anywhere else. Its purpose was to power a large *Schnellbomber* (fast bomber) of 70,000kg (154,350lb) loaded weight, including a bomb load of 30,000kg (66,150lb) and a fuel load of 18,800kg (41,454lb). This bomber had been worked out jointly by Daimler-Benz and Focke-Wulf by 11 December 1944. On 19 January 1945, Fritz Nallinger gave a discourse, apparently before high-ranking officials, in which he stated:

> The greatest problems foreseeable concern development of the 13,000kp take-off thrust TL unit intended for the Schnellbomber. This TL unit must therefore have ten times the thrust of the most powerful unit currently on the test stand. To solve such a task in a necessarily short period of time is an extraordinary one. It must, however, be realized that by coordinating experiences, this task, both on the airframe and engine side, can be rapidly solved such that both projects can be put to operational use in an acceptable time frame. The decision, nevertheless, must be taken immediately and, in addition, every possible support must be provided for this

development and thus enable it to be carried out in the shortest possible period of time. The prospects that are presented by these projects for the conduct of the war are such that, in my view, this proposal must not be laid aside but, instead, be tackled immediately.

He then demonstrated the illustrations and data for the projects.

For anyone to think that it was practical to begin such a large engine at that point in the war is extraordinary, but, late in 1944, DB's P.100 project received RLM backing and was officially designated the 109-016. The engine's weight was calculated as a massive 6,200kg (13,950lb), without auxiliaries, its diameter as 2.0m (6ft 7in), excluding underslung accessories, and its length as 6.70m (21ft 9in). The air mass flow was to be 400kg/sec (882lb/sec). After hundreds of sheets of calculations and some drawings were done, a mock-up of the axial compressor was made but the 109-016 proceeded no further.

The P.101 was intended to be interchangeable with the Heinkel-Hirth 109-011 but was to be more powerful and smaller. Its compressor was to deliver an air mass flow of 62kg/sec (136.7lb/sec). One feature of this engine, possibly a first for a German turbojet, was that of having reverse thrust, but how this was to be done is not known. Other data for the P.101 includes a weight of 1,050kg (2,315lb).

The P.102 fell into a class above IV but between the P.100 and P.102. Its data included an air mass flow of 200kg/sec (441lb/sec) and a weight, without auxiliaries, of 3,070kg (6,769lb). It was anticipated that two of these engines could be used to power the 70-tonne Schnellbomber giving, for example, a total thrust of 11,340kp (25,000lb) at 1,000km/h (621mph) at an altitude of 11,000m (36,000ft).

When the order came for DB to stop all work on its own turbojet development, in order to concentrate on the Heinkel-Hirth 109-021 turboprop, the new development appears to have been the responsibility of Dipl.-Ing. Hertzog under the supervision of Oberingenieur Freytag. The work was carried out largely at DB's dispersal at Backnang (Württemberg). Even before May 1944 DB had looked into the possibilities of such a turboprop development and had held discussions with the Arado company with a view to developing the 109-021 as the powerplant for a long-range reconnaissance version of the twin-engined Arado Ar 234 jet bomber. Another project study made by DB was for a curious transport aircraft that was to carry a fast jet bomber near to the bombing zone before releasing it, thus extending the range of the jet bomber. A completely different application, envisaged by Focke-Wulf, was for a fighter-bomber powered by a single 109-021.

Smaller than the Junkers 109-022 turboprop, the DB 109-021 was to consist basically of the projected Heinkel-Hirth 109-021 turboprop with an extra, third turbine stage added so that 2.50m (8ft 2½in) diameter contra-rotating, variable-pitch, three-bladed airscrews could be driven through reduction gearing. At least 50 per cent of the engine's power was to be diverted to airscrew thrust. The reduction gearing was to be of the planetary type, giving an airscrew speed ratio of 1:5.82 of the engine speed. A ground electrical source for starting the engines was by now favoured by the RLM. By the end of the war DB had not built a single 109-021 turboprop, but, in any case, the 109-011 turbojet upon which it was based was still underdeveloped. Power rating figures vary, reflecting the embryonic nature of the project, but a sea level, static figure of 2,400eshp with a turbine speed of 10,500rpm and an airscrew speed of 1,800rpm seems about average. An uncowled weight of 1,266kg (2,792lb) was expected together with a diameter of 0.910m (2ft 11⅞in) and a length of 3.696m (12ft 1½in).

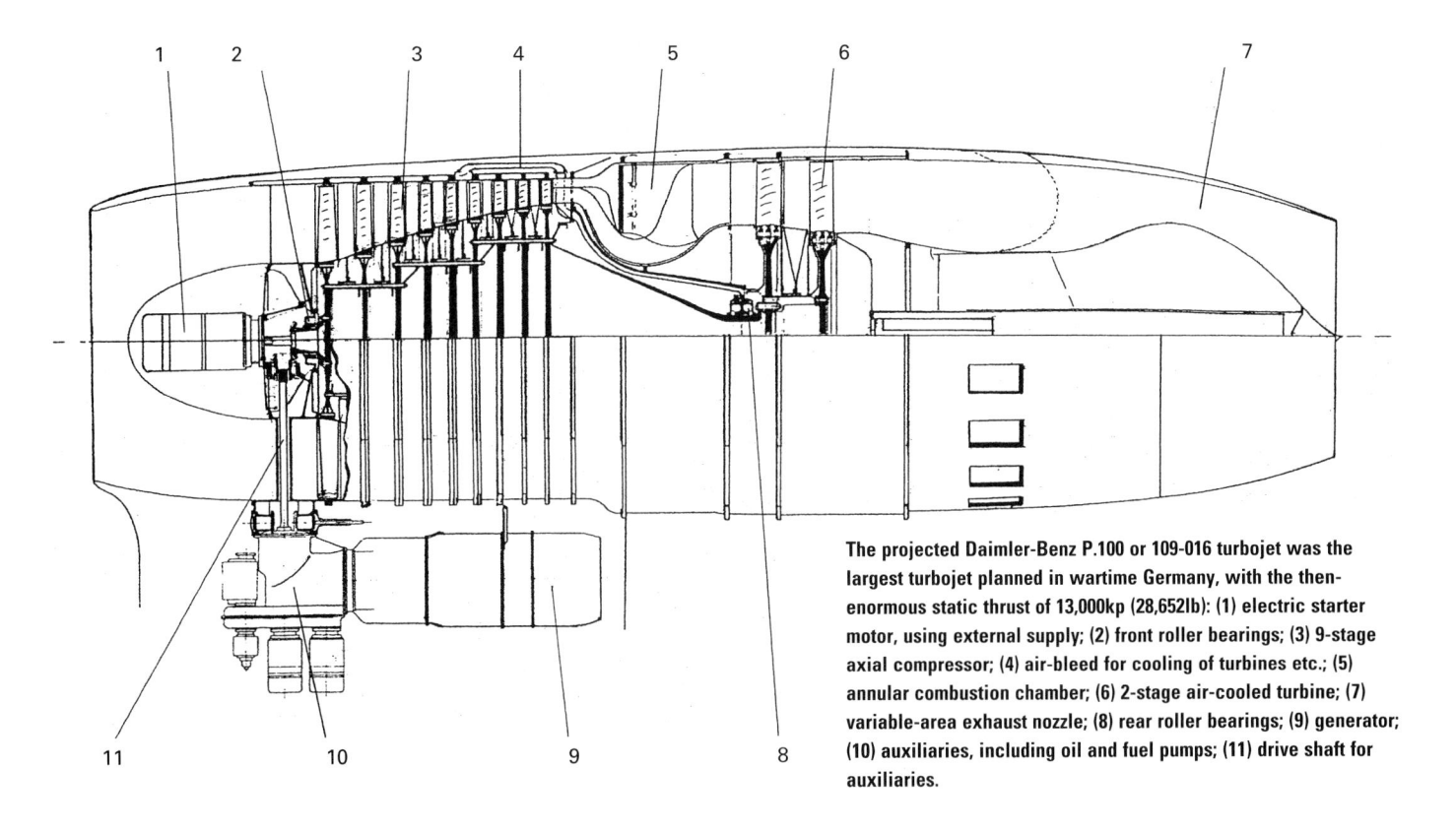

The projected Daimler-Benz P.100 or 109-016 turbojet was the largest turbojet planned in wartime Germany, with the then-enormous static thrust of 13,000kp (28,652lb): (1) electric starter motor, using external supply; (2) front roller bearings; (3) 9-stage axial compressor; (4) air-bleed for cooling of turbines etc.; (5) annular combustion chamber; (6) 2-stage air-cooled turbine; (7) variable-area exhaust nozzle; (8) rear roller bearings; (9) generator; (10) auxiliaries, including oil and fuel pumps; (11) drive shaft for auxiliaries.

CHAPTER TWENTY-NINE

Focke-Wulf Flugzeugbau GmbH

The famous Focke-Wulf company was formed in 1924 by Heinrich Focke, Georg Wulf and Werner Neumann. During World War Two its Chief Designer, Kurt Tank, headed the design of such famous machines as the Fw 190 fighter and Fw 200 Condor long-range aircraft. In February 1936 Otto Pabst joined the company at Bad Eilsen and by January 1940 he had risen to an appointment as a departmental chief entrusted with the formation of a new department for the study of gas dynamics. After completing work on a new Mach 1.0 wind tunnel, Pabst undertook the development of a turbine engine.

Initial investigations into jet propulsion were made by Hans Multhopp in the Project Office at Bad Eilsen during 1941. For comparative purposes, the Fw 190A fighter was used; at that time it was powered by a BMW 801 D twin-row, radial engine and had a maximum speed of 660km/h (410mph) at 7,000m (23,000ft). Herbert Wolff of the Propeller Department was sceptical about the jet engine because of its anticipated lower efficiency. In any event, investigations at the AVA Göttingen with wind tunnel models and various airscrew schemes indicated that the Fw 190 was capable of speeds up to 820km/h (509mph).

A turbojet for the Fw 190 fighter

By 1942 Pabst had come up with an ingenious scheme to fit a turbojet to the Fw 190 with, by order of Kurt Tank, minimal alteration to its airframe. The turbojet was to be fitted in place of the BMW radial engine and was to exhaust via an annular nozzle around the forward fuselage. Accordingly, the turbojet needed to be quite short and so had two centrifugal compressors in series, driven by a single-stage turbine. An annular combustion chamber was also very short and employed *Stauscheibenbrenner* or disc-shaped burners. Each disc-shaped burner was a simple, combined, injector and flame holder. Fuel entered a circular chamber in the burner and was ejected downstream from a slot around its circumference. The incoming air, flowing around the circular chamber, formed a turbulent cone in which the fuel could readily mix, and combustion could proceed in these lower-velocity zones. The burners were spaced one to two diameters apart so that their turbulent cones all mixed together within a short distance. The resulting short combustion flame length was governed by the burner diameter and the velocity of the fuel. Initial experiments were with gaseous fuel but the normal gasoline was to be used for the Fw 190 TL.

The annular thrust nozzle of Pabst's turbojet was interrupted in the region of the fighter's cockpit windscreen, but later the annular exhaust nozzle was replaced by a toroidal duct that exhausted into two slightly canted exhaust ducts. Behind the firewall the aircraft was virtually unchanged, although, due to the higher fuel consumption of the turbojet, the fuselage nose was lengthened to house an extra fuel tank. This increased the total fuel carried from 640ltr (141gal) to 1,400ltr (309gal) and maintained the normal endurance of 1.2 hours. One source gives the expected static thrust from this turbojet as 600kp (1,323lb). With the turbojet, the Fw 190 TL was expected to achieve a maximum speed of 760km/h (472mph) at sea level and 830km/h (515mph) at 9,000m (29,500ft).

By the end of 1942 Focke-Wulf's work on its turbojet had been cancelled by the RLM, since other companies were further ahead. Pabst continued work in the gas dynamics department and early in 1944 began developing a very short ramjet that used the burners already described for the turbojet. That summer he gained his doctorate in engineering. He was appointed Chief Engineer at Bad Eilsen in January 1945.

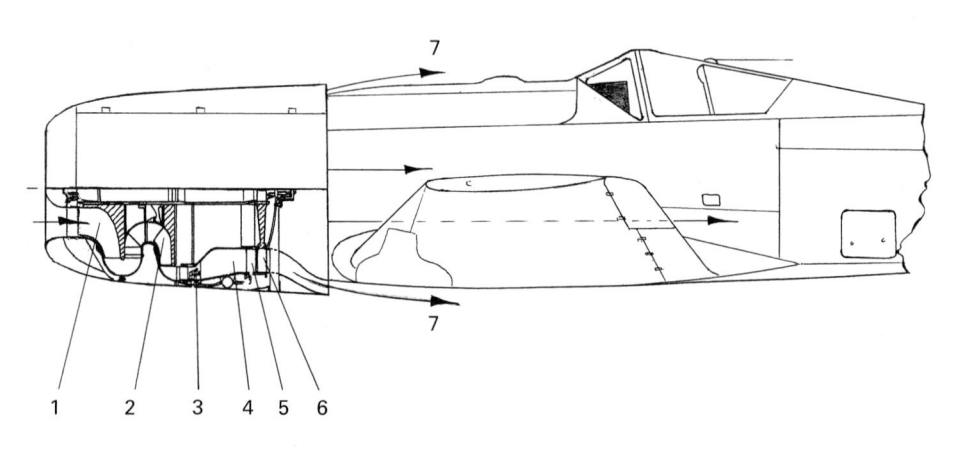

The projected Focke-Wulf Fw 190 TL standard fighter adaption for turbojet propulsion. Partial section (1) first centrifugal compressor; (2) second centrifugal compressor; (3) Pabst disc-type fuel burners; (4) annular combustion chamber; (5) turbine inlet nozzles; (6) single-stage turbine; (7) jet efflux from annular exhaust nozzle (later modified). Author

CHAPTER THIRTY

Dr-Ing. h.c. F. Porsche KG

Dr-Ing. Ferdinand Porsche established the independent design office bearing his name at Stuttgart-Zuffenhausen in 1930. Porsche, most famous for his designs of racing cars and the Volkswagen (people's car), had behind him a long and successful career in engineering. Born in 1875, he went to the Technisches Hochschule in Vienna in 1894. Following work at Bela Egger (later Brown Boveri) and other companies, Porsche designed his first air-cooled aircraft engine in 1912. His time was then spent mainly designing cars for such firms as Steyr in Austria and Daimler until he formed his own design bureau.

Initially, Porsche KG had only about thirteen staff, but by the end of the war in 1945 this had risen to about 40 staff and 120 workers, dispersed from Stuttgart to Rheinau and to Gmünd in Austria. From 1940 design work and a certain amount of research centred on armoured fighting vehicles (AFVs) and Porsche's name is associated particularly with tanks. Soon the Porsche design bureau became heavily involved in AFV design, all the more so when, in late autumn 1941, Soviet tanks were found to be of a higher quality than previously suspected and therefore very troublesome. Thus it was not until some three years later, when Porsche was asked to design an expendable turbojet for the Fi 103 (V1) flying bomb, that the company entered the turbojet field.

Origin of the expendable turbojet

Ferdinand Porsche had indeed already given some thought to the gas turbine and the turbojet, as his notebook shows. One page, dated 9 November 1930, has a sketch of an exhaust gas turbine with water injection for reducing the gas temperature. This features a combustion chamber and a two-stage turbine. Another sketch shows a turbojet with a radial-flow compressor. Further on in his notebook, Porsche made some calculations for E-Stelle Rechlin during the first months of the war. These give figures for thrust, power and fuel consumption for a turbojet at a flight speed of between 900 and 1,000km/h (559–621mph). Porsche's notes clearly show that, from early on, he was familiar with turbojet theory.

In the second half of 1944, as Allied forces moved across France, the Germans gradually lost the sites from which to launch the Fi 103 flying bombs against London. Forced back to the central Netherlands, the flying bomb units could only attack targets such as Antwerp, due to the limited range of their missiles. To use the flying bomb over ranges greater than the normal 240km (149 miles), it was necessary to resort to the unsatisfactory expedients of either air launching from He 111 bombers or increasing fuel load at the expense of warhead weight. Another solution was to improve the fuel consumption of the Argus 109-014 pulsejet or replace it with a new, more efficient power unit. In this last respect, the Technisches Amt saw a case for a small turbojet unit that would not only increase the range of the Fi 103 through better fuel consumption, but would also improve on other shortcomings of the pulsejet missile. Such shortcomings included the need for a special launching catapult, low speed and low operational ceiling, all of which contributed to the loss of a large percentage of missiles. Obviously, a prime characteristic of the missile turbojet had to be maximum simplicity and economy of manufacture, since the engine was expendable, but it could never approach the pulsejet for simplicity.

Porsche's Type 300 turbojet

Probably around October 1944, the official order was given for design studies to be made of expendable turbojets, but only the projects of BMW (P.3307) and Porsche are known. In any event, the Porsche design, known as its Type 300 L and officially designated 109-005, appear to be the only serious efforts in the field. Whereas the Argus 109-014 pulsejet had a static thrust of 350kp (770lb), the 109-005 turbojet was designed for a static thrust of 400kp (882lb).

In October 1944 Generaloberst Jodl of the Wehrmachtführungsstab reported on the status of development of Porsche's turbojet to the Chef TLR Generalmajor Diesing, who was responsible for the further development and manufacture of the Fi 103 flying bomb. Jodl's report stated:

> The project work on the further development of the FZG 76 (Fi 103), with focus on the attainment of increased range and higher speed, is not yet concluded. Investigations at present encompass determination of the optima between speed and range. It appears not improbable that speeds close to 800km/h and ranges of 600km can be attained. The prerequisite for these is the development of a new turbojet whose projection at the present time is being worked upon by Prof. Porsche. Definitive figures, however, can first be advised when details and evaluation of this powerplant project become available. Even with the employment of increased manpower than is currently available, its realization will take at least 8 months of effort. At the same time, measures must be taken to increase the impact accuracy, as the linear extension of the present impact distribution would lead to unsatisfactory values.

Mueller's redesign: the 109-005 turbojet

Although the turbojet was designed by Porsche and his team, the project was given to Dr Max Adolf Mueller in the last months of the war in order that he could bring his greater gas turbine experience to bear and incorporate all the latest design practice known to him. As related elsewhere, Mueller's experience included work on turbojets for Junkers and Heinkel-Hirth. In February 1945 he also replaced Dr Alfred Müller on an AFV gas turbine project under way at Porsche KG. Mueller did more

245

than fine-tune Porsche's turbojet design, however, and actually made a complete redesign so that the engine became known only as the 109-005. The most obvious changes that he made were the introduction of an axial inducer before the compressor and the deletion of the variable-area exhaust nozzle. The latter was not necessary because the missile was only operating through a narrow range of flight velocities and altitudes. Mueller's redesign increased the static thrust to 500kp (1,103lb) with a higher rotational speed, lower fuel consumption and considerably less weight, but greater overall size. Mueller also appears to have completely redesigned the axial compressor and reduced the pressure ratio from Porsche's 4.2:1 to 2.8:1. With increased rotational speed and increased efficiency, Mueller's compressor delivered about the same air mass flow but with easier starting characteristics. The following describes the 109-005 turbojet as it stood at the end of the war, but the design work was not finalized and no parts were constructed. Some test work was performed, however, for example on the combustion system.

Its basic layout consisted of an eight-stage axial compressor, a cannular combustion chamber, a single-stage turbine and a fixed-area exhaust nozzle. The whole engine was to be manufactured in 130 to 140 man-hours, or a little more than one-fifth of the time required to make a BMW or Junkers production engine. The engine was to be of the simplest possible construction and make the maximum use of *Sparstoffe* (inexpensive materials).

The intake duct diverged towards the rear to form a diffuser to reduce the velocity of incoming air before the compressor. Well forward of the compressor, the intake duct housed an axial flow inducer that was geared to run at slightly less than one third of the compressor speed. Eight stages were used for the axial compressor and it was designed for an efficiency of 78 per cent and a compression ratio of 2.8:1 under static conditions with 50 per cent reaction blading. A Mach number of 0.78 constant through all stages was chosen, since, although a higher Mach number would have increased efficiency, the lower value permitted the use of blades pressed from sheet metal or cast with less accuracy. Both rotor and stator blades had a thickness ratio of 12 per cent, their profiles being a modified NACA section with increased curvature towards the trailing edge. The compressor blades were also twisted in order to maintain a constant angle of attack along their length, while for the rotor blades the tip velocity was the same for all stages, partly to ease manufacture.

There were eight individual combustion chambers, enclosed in an annular shell. Each chamber consisted of a fuel nozzle and an annular pilot baffle inside a primary air cylinder, the end of which projected inside the front of a larger secondary cylinder. Combustion began in the vortex of the annular baffle and was completed in the primary cylinder. The resulting hot gases joined the secondary air entering through the annular space between the two cylinders. Variable-area, spring-loaded fuel nozzles were planned, supplied with J2 fuel by a gear-type pump. On single chamber tests, the combustion efficiency was recorded as 94 to 95 per cent.

The air-cooled turbine blades were to be made from chrome-nickel steel sheet and were similar to the blades for the BMW 109-003. However, instead of using a purely mechanical method such as pins to attach the blades to the turbine wheel, the blades were to be welded on after their V-shaped feet had been fitted into cutouts in the wheel rim. The weld, it was hoped, would enable stresses to be spread over a much wider area. Also, the welding of the blades allowed the use of an economical wheel made from two sheet-metal discs to be used. From 800°C in the combustion chambers, the gas temperature dropped to 660°C leaving the turbine inlet nozzles and then 610°C when leaving the turbine blades. Unlike general German practice in cooling hollow turbine blades, cooling air was not to be tapped off the compressor but scooped from the atmosphere since the suction action inside the revolving blades was considered quite adequate, especially in view of the short life required in missile use.

The static thrust of 500kp (1,102lb) of the 109-005 was to be produced at 14,500rpm. Its weight was 200kg (441lb) and its specific fuel consumption at speed and altitude was expected to be 1.7. At static thrust, its fuel consumption was expected to be 1.38, which compares with the 2.88 of the Argus 109-014 pulsejet. For the fully cowled turbojet, the dimensions were a diameter of 0.650m (2ft 1⅝in) and a length of 2.850m (9ft 4¼in). Using the 109-005 turbojet, the Fi 103 flying bomb was expected to fly at about the same speed of 650km/h (404mph), though at the slightly higher ceiling of 4,000m (13,100ft), whereas the range was expected to almost treble to 700km (435 miles) using, it is believed, the normal fuel load.

CHAPTER THIRTY-ONE

Dispersal and Revival

The effort made by German industry in getting turbojets into military service during the Second World War was remarkable in view of the shortages of special metal alloys and the general difficulties that worsened as the war went on. The German programme, having been kick-started by the partnership of the physicist Hans von Ohain and the aircraft industrialist Ernst Heinkel, was ably directed by Helmut Schelp and his associates in the RLM. The pioneers von Ohain and Heinkel did not, however, receive their just rewards. Von Ohain became tied to overseeing largely dead-end engine developments when he would have preferred a return to immersing himself in pure research physics. Heinkel, on the other hand, after expending much of his industrial resources on turbojet development, found that he was forced to attempt development of the unpromising 109-011 turbojet while the official orders for jet aircraft eluded him.

The Allies showed some interest in Heinkel-Hirth turbojets after the war. A few examples of the 109-011 and 109-006 were built for flight-testing in the USA, but construction work at Zuffenhausen was soon closed down when the Soviets got wind of it and strongly objected. Part of Heinkel's failure to go into production was that its early efforts were dissipated over too many projects and it was soon overtaken by Junkers and BMW. This mistake was not made at Junkers: having ignored the early, promising work of Herbert Wagner, Anselm Franz concentrated all Junkers' efforts onto getting one design of a simple turbojet into production and service. In this, Junkers was successfully following the official RLM programme and its 109-004 B engines powered Germany's first jet fighters and light bombers into Luftwaffe service.

Success also attended BMW's work, this company having been instructed to develop a more advanced and more economical turbojet to follow on from Junkers' engine. After a slow start, BMW was on the verge of overtaking Junkers technically and its 109-003 A and E turbojets went into production just before the war ended. Daimler-Benz was, along with Junkers and BMW, in the top three of Germany's piston aero-engine manufacturers, but it failed to make much progress in turbojet development due to a late start: only one prototype of its complicated 109-007 engine was produced.

More powerful turbojets were under development at Heinkel-Hirth, Junkers and BMW but time ran out before these got very far. On the turboprop front, only designs were made. After the war, the main interest in German turbojet work was in the Soviet Union and France (see under those sections in Volume 2). In Great Britain and the USA, turbojet and turboprop development had progressed beyond the German work and continued at an ever-increasing tempo. Nevertheless, most aspects of the German work were, of course, free to be scrutinized. German turbojets available in England were so numerous that many were used as targets in experiments. In a typical test, a turbojet was run while fired upon, usually with a machine gun: a spectacular explosion was the usual result, often during a demonstration before high-ranking personnel.

Post-war developments

Germany was in a chaotic state at the end of war in May 1945, but soon its aircraft and engine specialists found work to do once more. Their expertise was too valuable to waste. Germany was divided by its conquerors into the East, controlled by the Soviets, and the West, controlled by the Allies: the Americans, British and French. Berlin, within East Germany, was similarly divided into Eastern and Western zones. The enforced transportation of many German aircraft and engine specialists to work in the Soviet Union is described in Volume 2.

In 1950 a small number of these specialists and their families were sent back to Germany, their work in the Soviet Union completed. A larger number followed them in November 1953. The remainder of the German engine specialists went to Savyolov (100km north of Moscow) and joined up with the German airframe specialists, headed by Brunolf Baade and Prof. Günther Bock. Ferdinand Brandner headed the turbojet group and Bock headed the airframe group, overall leadership being assumed by Baade. These specialists were, as we shall see, soon moved to East Germany to work on various projects.

Heinkel work in Spain and the INI-11 turbojet

In the meantime, the Americans were working on the Marshall Plan to bring West Germany (and Japan) back into economic viability. By 1955 Allied restrictions on German aero-engine work, and much else, had been lifted. However, the irrepressible entrepreneur Prof. Ernst Heinkel could not wait for all restrictions to be lifted. Heinkel had already manufactured a few turbojet engines for the Allies to test. By 1951 the opportunity had arisen for engineers from Heinkel, Dornier and Messerschmitt to work in Spain at the Instituto Nacional de Industria (INI) in Madrid: six engine specialists from Heinkel were headed by Fritz Schäfer, who, it will be recalled, had piloted Heinkel's He 280 V1 on its maiden flight on 30 March 1941.

The task of the Heinkel engineers was to work with their Spanish colleagues of the Hispano-Suiza company to develop the Heinkel-Hirth 109-011 turbojet, its goal being a static thrust of 1,500kp (3,308lb). Although Hispano-Suiza had, as early as 1946, obtained a licence to build the Rolls-Royce Nene turbojet (and, later, the Tay and Verdon engines), the desire of the Spanish government was to have an indigenous turbojet industry. Designated the INI-11, the 109-011 was redesigned to take into account shortages of manu-

An INI-11 turbojet on the test stand at INTA, Spain, 1954.

facturing facilities and special materials, and to allow for some components and accessories to be purchased from abroad. No data or manufacturing documents were available on the 109-011 since most had been removed by the Allies, but a few drawings were available. The Germans' own recall of tests on the 109-011 was useful in the redesign.

Changes in the 109-011 design to produce the INI-11 included an enlarged air intake and a reduction of the front bearing support struts from five to only two vertical ones. The shaft drives for the auxiliaries and oil pumps passed through these streamlined struts. No changes were made to the axial inducer, diagonal or three-stage axial compressors. However, the annular combustion chamber needed redesign since this had originally suffered burning of the secondary air mixing fingers. The new combustion chamber was therefore made longer and, instead of using mixing fingers, holes allowed secondary air to pass from around the combustion chamber to the inside. The combustion chamber had a one-piece cast casing around it, this being easier to manufacture than the previous sheet metal one. It was equipped, as Heinkel-Hirth had planned, with duplex fuel injectors from the L'Orange company.

Owing to manufacturing limitations in Spain, it was not possible to use the original air-cooled, two-stage turbine. Instead, a single-stage turbine with blades of heat-resisting metal was employed. These blades were supplied, pre-formed, from Leistritz in Germany. Cooling air, tapped from the compressor, was only used for the turbine wheel and blade roots. It was hoped to use the, by then, almost universal fir-tree roots for the turbine rotor blades, but again there were manufacturing difficulties. Instead, the turbine blades were made with cylindrical ends that were clamped between two turbine wheels. As in the earlier Heinkel-Hirth design, the turbine blades were able to rock slightly to avoid root stresses. Finally, the original variable-area exhaust nozzle was replaced with a fixed one. With an overall diameter of 0.762m (2ft 6in)

Turbine blade cylindrical feet (roots) of an INI-11 turbojet.

and a length of 3.05m (10ft), the INI-11 came out smaller than the 109-011 but, at about 1,000kg (2,205lb), was slightly heavier. The air mass flow of 31.75kg/sec (70lb/sec) and pressure ratio of 5:1 were greater, while the specific fuel consumption of about 1.0 was lower.

During late 1955 three INI-11 prototypes

An HeS 053 turbojet in Egypt, 1953.

were built and tests began at the Instituto Nacional de Tecnica Aeronautica (INTA), where special test facilities were constructed for turbojets. All was not well, however, and one of these engines was destroyed after about three hours running at a thrust of 1,300kp (2,866lb). The cause of the failure was found to be the fracture of an axial compressor blade. A Spanish He 111 was to be used as a flying test-bed, but in 1956 the INI-11 was cancelled. This came about, it is said, because in that year Spain gained access to more modern technology due to the Agreement on Military Bases with the USA. This seems curious when the country was already building excellent Rolls-Royce turbojets. From 1956 the Empresa Nacional de Motores de Aviacion SA (ENMA) obtained licences to manufacture small gas turbines and the acquisition of a knowledge base was thus begun. A final projection of the INI-11 was seen in the INI-12, which was to have a seven-stage axial compressor (no diagonal stage) and a static thrust of 2,400kp (5,292lb). It seems that this turbojet was intended for the projected Hispano-Suiza delta-winged jet fighter, which was also cancelled.

Heinkel work for Egypt and the HeS 053 turbojet

While the Heinkel team was at work in Spain, Ernst Heinkel concluded a contract with the Egyptian government in November 1953. This was to develop a military aircraft complete with its turbojet: the Heinkel He 011 Mach 1.2 delta-winged fighter. By the end of 1953 work began on a new turbojet for Egypt, designated the HeS 053. Because the Allied ban on such work in Germany had not then been lifted, Heinkel set up a company in Vaduz in Lichtenstein, this company being known as EKOBA-Anstalt. Later the work was transferred to the old Stuttgart-Zuffenhausen works, where about forty-five individuals worked on the HeS 053 and another group, led by Karl Schwärzler, worked on the design of the airframe. During these early postwar years the Heinkel company, like that of Messerschmitt, made the so-called small bubble cars to bring in some revenue.

The turbojet team was headed by Christian Schmidt (from SNECMA in France), who had, in the Heinkel-Hirth days, dealt with the construction of experimental engines. Later the team was joined by Dr-Ing. Fritz Gosslau, who had been Director of the Argus company that developed the V1 flying bomb's pulsejet engine during the war. In addition, as part of the contract conditions, about twelve Egyptians were added to the team in order to gain experience.

The HeS 053 was designed as a powerful turbojet, its planned thrust being 6,500kp (14,333lb), or 9,000kp (19,845lb) with afterburning. This was to be one of the world's most powerful engines at that time. It featured an eleven-stage axial flow compressor, the first stage being adjustable, and had an air mass flow of 100kg/sec (220.5lb/sec) with a pressure ratio of 7.4:1. The combustion system was curious since it had nine tubular combustion chambers that led into an annular section (the reverse layout was used in the East German Pirna 014 engine). There were nine duplex fuel injectors. The turbine was a two-stage type, the first stage being air-cooled. A variable-area exhaust nozzle was provided by a pair of two-position eyelid-type doors at the end and a hydro-mechanical system provided speed control.

German companies that previously had been wartime suppliers to Heinkel-Hirth provided components, including WMF (sheet metal, including combustion chamber parts), Leistritz (turbine blades) and L'Orange (duplex fuel injectors). Work on components, such as the combustion chamber and speed governor, was undertaken at Zuffenhausen, albeit secretly until the Allied ban was lifted. Finally, under German supervision, turbojet test stands and facilities were erected in Heluan, near Cairo.

By May 1956 three prototypes of the HeS 053 turbojet had been built and readied for testing but without their afterburners. Heinkel's plans were once again thwarted, however, since the jet aircraft and engine programme was cancelled by Nasser's government that spring. All the engines, equipment, drawings and reports at Zuffenhausen were turned over to the Egyptians and no known testing of the HeS 053 took place. Meanwhile,

A Messerschmitt HA 300 fighter taxiing past an Antonov An-12 with an E 300 under its wing for flight tests. Both are of the Egyptian Air Force.

Heinkel's coffers. Thus, although many of the turbojet engineers at Zuffenhausen departed with the end of the Egyptian work, there were still about fifteen left and Ernst Heinkel asked Fritz Gosslau to build up a new team in autumn 1956. A number of specialists in the new team had previously worked for BMW and, later, for SNECMA in France. Others came from a small group working under the Junkers (JFM) name in Essen. This group was led by Dr-Ing. August Lichte, formerly of Jumo, but he did not follow the engineers over to Heinkel. In 1957 the new, small Junkers enterprise was purchased from the Federal government jointly by Messerschmitt at Augsburg and by Heinkel at Zuffenhausen. In order to keep the Junkers name from dying out, the as the hostilities of the Suez crisis drew near, the Egyptian Air Force acquired Soviet jet aircraft, including the outstanding MiG-15 fighter and the Il-28 bomber.

At 6,000rpm the compressor of the HeS 053 gave an air mass flow of 100kg/sec (220.5lb/sec) for a pressure ratio of 7.4:1. Since the compressor had no blow-off valves, it seems likely that starting would have been difficult. The weight of the engine was 1,565kg (3,451lb), its diameter 1.105m (3ft 7½in) and its length 4.025m (13ft 2½in). Specific fuel consumption was estimated at 0.930.

Heinkel's third turbojet team and the Junkers group

Although the Egyptian venture was not completed, it did bring some money into

A view from the exhaust end of an E 300 turbojet on a test stand at Heluan in Egypt.

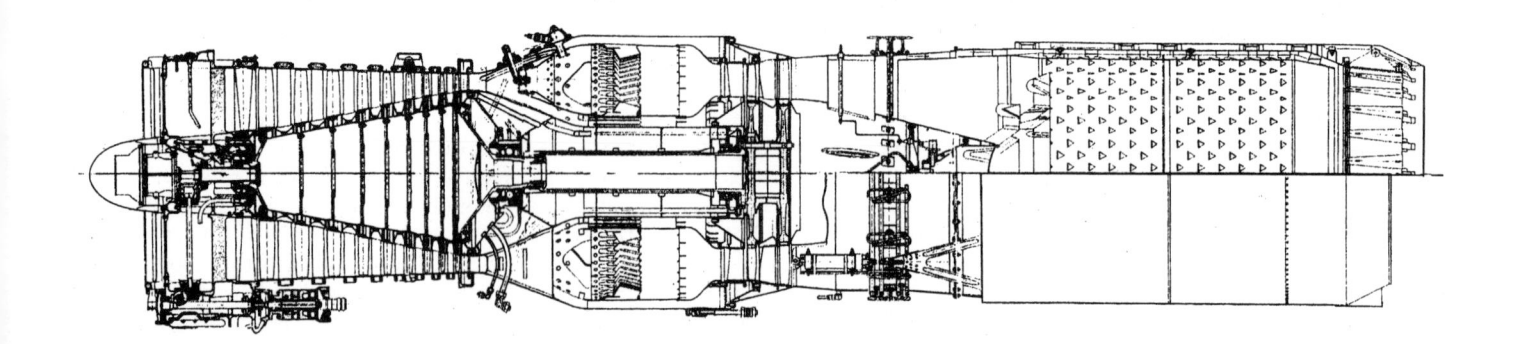

The E 300 turbojet showing its 9-stage axial compressor, 2-stage turbine, afterburner and petal-type variable-area exhaust nozzle.

designation of engines at Zuffenhausen was changed from 'He' to 'Ju'. Thus, by 1958 the HeS 053 had become the He 003 and then the Ju 003.

Heinkel's new engine team took on the task of investigating new engines suitable for a jet fighter required by the Federal government. The first engine projected was the He 004, which featured a six-stage axial compressor, with a pressure ratio of 4.0:1, and a single-stage turbine. Simplicity was its keynote. There followed the He 005 with a pressure ratio of 4.5:1 and a single-stage turbine, and the He 006 with a pressure ratio of 5.0:1 and a two-stage turbine. Pressure ratios up to 9.0:1 were studied and the projects concluded with the Ju 015 turbojet, which had a nine-stage axial compressor and a pressure ratio of 6.5:1. Also included in the studies were low bypass, afterburning turbojets.

During 1960 the Junkers group, now located in Munich, comprised more than 120 individuals and work began on the construction of manufacturing and test facilities. Unfortunately the company was now in financial difficulties and its engineers were being hired by Ferdinand Brandner for another jet aircraft project in Egypt.

Brandner and another jet project in Egypt

The incomplete test stand and facilities still remained at Heluan in Egypt. Prof. Willi Messerschmitt was developing the HA 200 jet trainer and the small HA 300 Mach 2.2 interceptor at Hispano Aviacion in Spain, and a contract was made to help the Egyptians to develop these aircraft and build them in Heluan. Originally it was intended to power the interceptor with the British Orpheus turbojet, but development of this engine was dropped by the Bristol company.

Early in 1960 Brandner undertook the task of developing up to production status an engine for the HA 300 and a smaller one for the HA 200 trainer. Brandner's recruits came from various sources and included some of his old colleagues from the Soviet Union days as well as people from France, Germany, Austria and Switzerland. Ferdinand Brandner (1903–1986) obtained his Diploma Engineer qualification in 1925 at the Technische Hochschule in Vienna. His early interest and experience was in automobile design and then piston engines, especially diesels. In 1936 he joined the Junkers-Motorenwerke in Dessau, where the chief aim at that time was a dramatical increase in aero-engine power. It was Brandner who designed the Jumo 222 engine for the Ju 288 bomber, which by the war's end was giving 3,000hp and was Junker's most powerful piston engine. However, it was never fully developed. His contact with turbojets came after the war when he was deported to the Soviet Union and worked under the great Nikolai Kuznetsov. By 1947 Brandner was leading the German engine collective in Upravlencheskiy, where the world's most powerful turboprop, the NK-12, was developed (see in the Soviet section in Volume 2). As already mentioned, Brandner left the Soviet Union for Germany after meeting up with Brunolf Baade and others at Savyolov.

At the beginning of 1961 Brandner's group of specialists, with their families, had mostly moved to Cairo. Brandner's deputy, as development head, was Bruno Schuldt. The small E 200 turbojet was developed for the HA 200 jet trainer. It had a radial compressor and was closely based on the French Turboméca Marboré to give a static thrust of 400kp (882lb). The number of E 200 engines built totalled 170.

More challenging was the larger E 300 turbojet for the HA 300 fighter. This axial engine eventually gave 3,260kp (7,188lb) without reheat and 4,380kp (9,658lb) with reheat; a 150-hour run was successfully carried out in May 1964. Flight tests were carried out using an Antonov An 12 transport aircraft, the number two turboprop being replaced with an E 300 turbojet. The HA 300 fighter was apparently taxied and was to have made its maiden flight in July 1969 but, just one month before, the Egyptian government terminated all work at the by-now extensive Heluan facility. This also brought to an end the possibility of E 300s powering India's twin-jet supersonic military aircraft, which was being developed at Hindustan Aeronautics under Prof. Kurt Tank's design leadership.

Rebirth of the East German aircraft industry

While this Egyptian saga was being played out, momentous events were occurring in East Germany. In 1953, following the death of its dictator, Joseph Stalin, the Soviet Union ceased plundering the part of

Dipl.-Ing. Brunolf Baade was largely responsible for the post-war revival of the East German aeronautical industry. He had worked in aeronautics before and during the war and was an inspirational leader.

Germany under its occupation and turned to improving its economy. A large part of this plan was to be a new and revitalized East German indigenous aircraft industry. This idea had been championed by Dipl.-Ing. Brunolf Baade (1904–1969), the leader of the German airframe specialists working in the Soviet Union. In December 1953 Baade received permission from Moscow for his team to begin work on the design of a jet airliner. It was designated the Model 152 and was based on their previous EF 150 swept-wing twin-jet bomber built in the Soviet Union.

Brunolf Baade had completed his studies by 1926 and had attended the Technische Hochschule in Munich and assisted Prof. Wunibald Kamm at the DVL in Berlin. He then worked at BMW, in the USA and, by 1936, at Junkers in Dessau. As Chief Designer he was engaged in the development of the Junkers Ju 88, Ju 188 and Ju 288 bombers, and also the experimental Ju 287 jet bomber. Baade was a well-liked, inspirational leader who could quickly see to the heart of a problem and set his team on the right path.

By June 1954, when the last Germans and Austrians left the Soviet Union, the

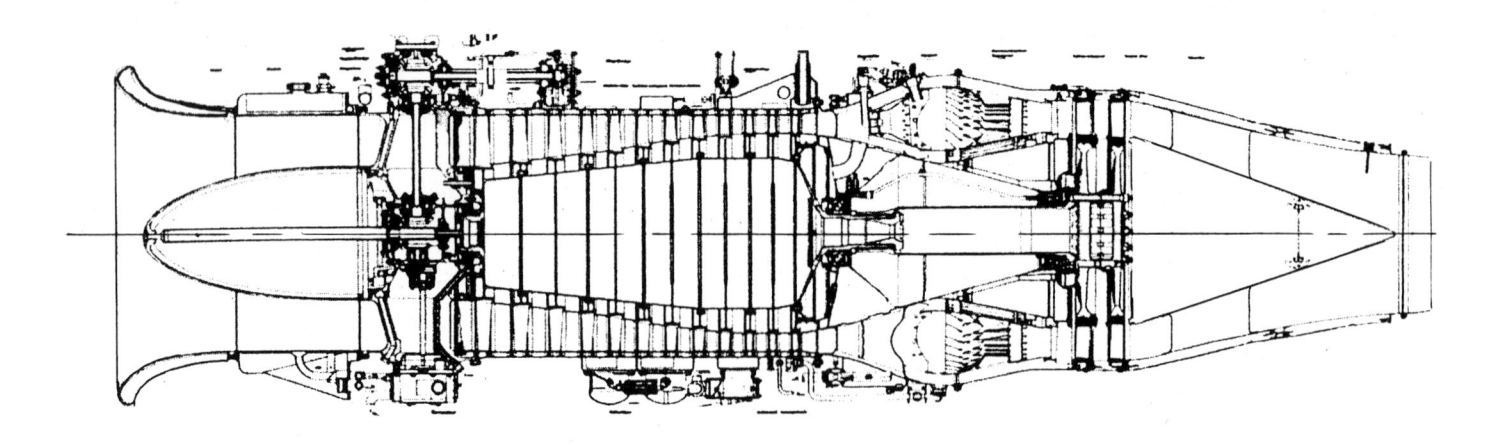

ABOVE: The Pirna 014 axial turbojet, showing its 12-stage axial compressor, combination combustion chamber system, 2-stage turbine and fixed-area exhaust nozzle.

ABOVE: The combination combustion chamber system of the Pirna 014 turbojet. This curious design comprised an annular combustion chamber that led into twelve conical sections (Brennerköpfen) before the turbine inlet nozzles. Control of secondary air to the combustion chamber was made by rotation of the slotted rings over the fixed slots.

Views of the Pirna 014 axial turbojet, one on the test stand.

plans for the Model 152 airliner were complete. Baade established the headquarters for the new aircraft industry at a former Luftwaffe airfield in Klotzsche, on the outskirts of Dresden. Modern offices, test and production facilities for airframes and turbojets were set up. The project was given the highest priority in the East German economy and Baade himself became a member of the East German Central Committee. From this position, he ordered the recruitment and training of many engineers and technicians, since the 500 who had come back from the Soviet Union were nowhere near enough.

Parts were also to be made all over East Germany: eventually about 25,000 workers were involved in Baade's project. At Pirna, about 25km southeast of Dresden, a bureau was set up to design turbojets that were to be made in a factory just south of Berlin. In a reversal of their activities in the years just after the war, the Soviets sent trainloads of machinery, equipment and materiel to the factories in East Germany. Also, to kick-start the fledgling industry, they sent the means to assemble Ilyushin Il-14 short-range airliners. Meanwhile, Fritz Freytag, a former designer at Junkers, was putting the finishing touches to the design of the Model 152 airliner, now known as the Baade BB 152, and organizing the construction of prototypes.

BB 152 airliner and the Pirna 014 turbojet

The BB 152 had a shoulder-mounted, swept-back wing with pronounced anhedral and upper surface fences. At the wing tips were small nacelles to house outrigger wheels when retracted. The main undercarriage was an unusual tandem type, having a four-wheel bogie beneath the rear of the fuselage and a two-wheel bogie beneath the front. Beneath each wing was a forward-swept pylon carrying a twin pod (*Zwillingsdiffusor*) for two turbojets. The tail empennage was conventional and there was accommodation for up to seventy-two passengers.

At Pirna (Werkes 802), work on developing the turbojets for the BB 152 began under works Director and Chief Constructor Dr-Ing. Rudolf Scheinost. The engine was designated the Pirna 014 and the company was known as VEB Entwicklungsbau Pirna.

Initially designated Projekt 014, the axial turbojet was aimed at a static thrust of 3,150kp (6,946lb), a specific fuel consumption of 0.85 and a weight of 1,100kg (2,426lb). Its dimensions were to be a length of 3.520m (11ft 6½in) and a diameter of 0.981m (3ft 2⅜in).

The 014 turbojet had a twelve-stage axial compressor, a two-stage turbine and a fixed-area exhaust nozzle. The 'combination' combustion chamber was somewhat curious and consisted of an annular chamber that led into twelve conical sections (*Brennerköpfen*) before the turbine inlets. Air was tapped off at the compressor's seventh stage for turbine cooling and at the ninth stage for cabin pressurization. To ease the task, one-twelfth of the combustion chamber was used in the test phase.

On 12 October (or December) 1956, the prototype 014 V-01 engine was on the test stand but the development that followed was very slow and no 014s were ready for the BB 152/1-V1 (DM-ZYA prototype). On 25 February 1958 the second turbine wheel of the 014 V-01 engine cracked, flew apart and totally destroyed the exhaust duct assembly. Setbacks like this meant that the maiden flight of the BB 152/1-V1 on

A Pirna 014 turbojet mounted beneath an Ilyushin Il-28R (DM-ZI) to conduct flight tests.

Two views of the BB 152 V1 (DM-ZYA) jet airliner prototype just before its first flight on 4 December 1958. Because the Pirna 014 turbojets were not ready, early flights had to be made using four Soviet-built RD-9b turbojets. The revived East German aircraft industry relied almost totally upon the success of this programme.

BB 152 airliner		
Span	27.0m	(88ft 7in)
Length	31.4m	(103ft 0¼in)
Estimated wing area	120sq m	(1,292sq ft)
Height	9.0m	(29ft 0¼ in)
Empty weight	28,920kg	(63,769lb)
Loaded weight	48,000kg	(105,840lb)
Maximum speed	920km/h	(571mph)
Cruising speed	800km/h	(497mph)
Range with 48 passengers	2,500km	(1,553 miles)

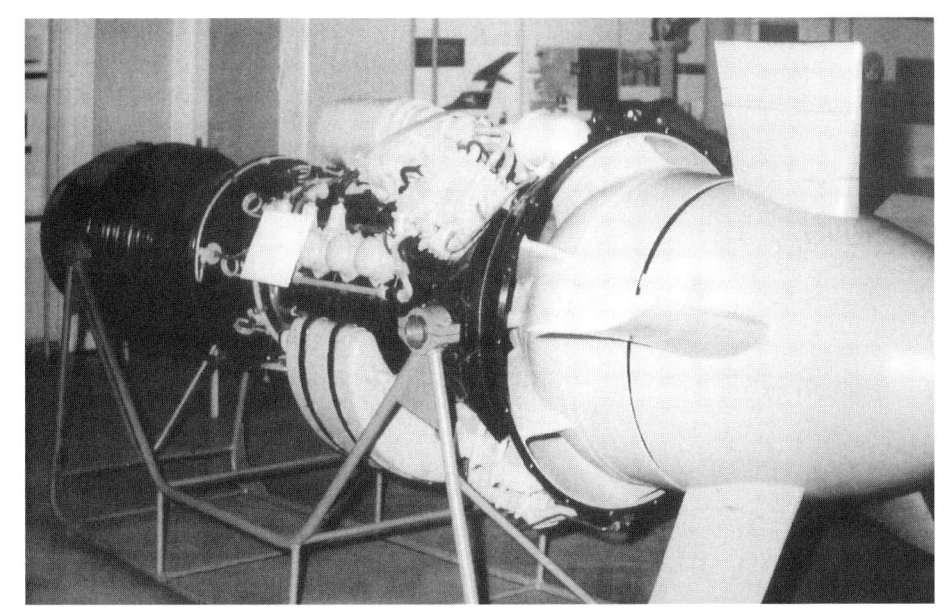

Mock-up of the Pirna 018 (PTL) turboprop.

4 December 1958 had to be made using four Soviet-built RD-9b turbojets of 2,600kp (5,733lb) thrust, this being less than the power designed for. However, Willi Lehmann, the pilot, flew the aircraft for 25 minutes without incident. Soon, 014 engine flight tests were being performed using an Ilyushin Il-28R (DM-ZZI). The Il-28R (DM-ZZK) later joined the programme to test the 014 A-0 pre-production engines. The turbojet under test was installed in a nacelle beneath the fuselage of the twin-jet Il-28R. After intensive flight tests, the V-10, V-12, V-21 and V-23 prototypes of the 014 A-0 were fitted to the BB 152/11-V4 (DM-ZYB), which then made its first flight with these engines on 26 August 1960.

The East German airline Deutsche Lufthansa (later Interflug) was to be the first customer for the BB 152 airliner. Sadly, all was not well with this aircraft. On 4 March 1959 the Soviet Premier, Nikita Kruschev, and an enormous crowd waited in vain at the Leipzig trade fair to watch the BB 152 make a low-level fly-past: the airliner failed to appear because it had earlier crashed near Klotsche, killing the pilot Lehmann and the other three crew members. It was later thought that the crash happened when the flexible fuel tanks in the wings were torn apart by pressure differences. Normally, tests of a new fuel system would be carried out using a tilt table, but this had not been done. It was the end of the dreams and hopes of so many in the East German aviation industry. Owing to the protracted development of the BB 152 and its Pirna 014 engines (the first BB 152 for airline service could not have been delivered until 1962), the airliner was easily overtaken by Western aircraft such as the Comet, Caravelle and 707.

More than 2 billion Marks from the East German economy had already subsidized the venture; although more help from the Soviet Union was sought, it was not given. On 5 April 1962, on orders from Moscow, the aircraft industry was virtually terminated: only repair work on MiG aircraft and helicopters was allowed to continue. The twenty-six BB 152 airframes under construction were cut up for scrap, as presumably were most of some sixty-six 014 turbojets built. Only three examples of the more refined 014 C turbojet were built, this engine giving a static thrust of 3,380kp (7,453lb). A marine use was made of a few 014s in a scheme designated 0511, for which an 014 was used as a gas generator to supply two propellers driven by their own turbines. The 0511 was fitted to the Type 'Hai' anti-submarine boat.

Other Pirna engines

With the demise of the BB 152, there did not seem to be a market for the 014 turbojet, but the Pirna company went on to develop various small gas turbines (up to 394hp) to drive electrical generators, water pumps and as APUs for fighters,

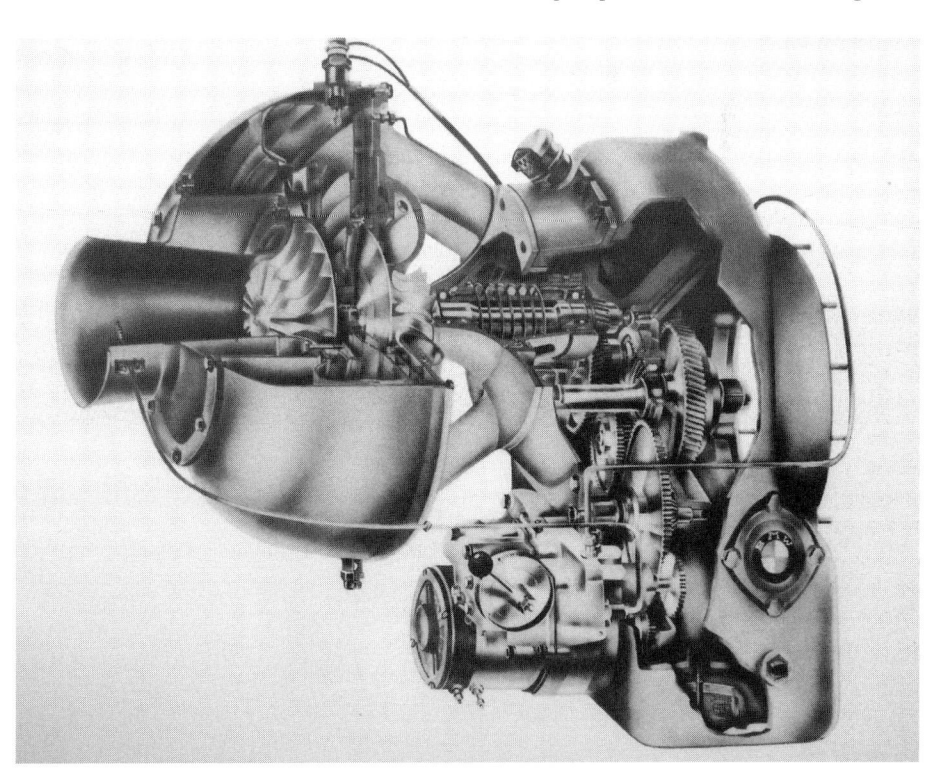

The BMW 6002 small gas turbine was first run in November 1958 and featured a novel disc-type combustion chamber.

helicopters and airliners. There were also a number of projects for larger engines, which will be briefly mentioned. Following on from the 014 was the 015 project for a bypass version of 4,500kp (9,923lb) static thrust. A specific fuel consumption of 0.6 to 0.7 and a weight of 1,800kg (3,969lb) were anticipated. The 016 was a 3,500kp (7,718lb) static thrust axial turbojet, of which only one was built, and the 017 was a small gas turbine.

In 1955 the 018 turboprop engine was planned. Although much smaller, it was based on the Germans' experience in developing the huge Kuznetsov NK-12 in the Soviet Union. The 4,932eshp 018 had a twelve-stage axial compressor driven by a two-stage turbine. There was the usual Pirna annular combustion chamber that led into nine *Brennerköpfen*. A four-bladed airscrew was driven from the front of the compressor shaft via reduction gearing, the gearing being surrounded by the annular air intake. The 018/V-1 prototype was first run on the test stand on 28 November 1958. Its compressor's air mass flow was 31.31kg/sec (69.04lb/sec) and its specific fuel consumption 0.345. Excluding the airscrew, its weight was 1,300kg (2,866lb) and its length was 3.68m (12ft 0¾in). The purpose of the 018 turboprop was to power a successor to the twin-engined Ilyushin Il-14P. It was to be designated the 153, but only a mock-up of the 153 A was constructed. Furthermore, only three prototypes of the 018 engine were built. Other projected uses for the 018 included a four-engined airliner.

The projected Pirna 019, a small 600kp (1,323lb) thrust turbojet, was followed by the most advanced of Pirna's turbojets, the 020 bypass engine. This was a continuation of the earlier Projekt 015. Its low-pressure system consisted of a six-stage axial compressor, extended out to a three-stage ducted fan, all driven by a two-stage turbine. In between, and coaxially, was an eight-stage high-pressure axial compressor, driven by a single-stage turbine. There was an annular combustion chamber leading into eight *Brennerköpfen*. The total pressure ratio of the compressors was 12:1 with an air mass flow of 96kg/sec (211.68lb/sec). For a weight of 950kg (2,095lb) and a diameter of 1.120m (3ft 8in), the 020 was expected to give a static thrust of 3,700kp (8,159lb). However, by 1962, this project only reached the stage of testing the three-stage fan and eight-stage HP compressor. A mock-up of the 155 airliner was built, showing an 020 engine fitted above each wing. With the end of the 020 project, only work on small gas turbines continued at Pirna.

BMW reborn

Small gas turbines also became the business of BMW and Daimler-Benz in the Federal Republic, following the lifting of the Allied restrictions by 1955. On 22 January 1954 the BMW Studiengesellschaft für Triebwerkbau GmbH was established in Munich. Its Director was Helmut Sachse, who had been the technical head of BMW-Flugmotorenbau from 1937. The new study group was very small at first and included former BMW and Junkers personnel, most of whom had returned from the Soviet Union in 1954. Experienced former BMW personnel included Dr Hermann Hagen and Karl Prestel. In March 1956 work began on the BMW 6002 gas turbine, which lay in the 50 to 74hp bracket and weighed 65kg (143lb). It was partly financed by the BMVg (German Ministry of Defence). The BMW 6002 featured a single-stage radial compressor with a 3:1 pressure ratio, a compact, flat disc-type combustion chamber and a single-stage radial turbine. Power take-off was via reduction gearing. The novel combustion chamber was fed with fuel through the hollow rotor shaft. Rotational atomization in the combustion chamber was achieved by use of a centrifugal injection ring.

The first run of the BMW 6002 in November 1958 was followed by a 50-hour run in April 1959, the tests being made at Allach, near Munich. Most 6002 gas turbines were used to run electricity generators, water pumps and similar devices, but soon the BMW 8025 (originally designated BMW 8011) was derived from the 6002 as a small turbojet. On 25 September 1960 a BMW 8025 was used to power the Hütter-Allgaier H-30 TS turbine-glider (D-KABA) from Göttingen. This was the first time in postwar West Germany that a German aircraft had been powered by a German engine.

In 1957 Ferdinand Brandner replaced Helmut Sachse. He, in turn, was replaced in 1958 by Helmut Graf von Zborowski, the former Director of BMW's rocket developments. Intensive development of small gas turbines followed, their employment mainly being to power helicopters and provide auxiliary power for aircraft.

Daimler-Benz reborn

In 1955 Daimler-Benz A.G. once again began the development of aero-engines as a separate business to that of vehicle production. Prof. Fritz Nallinger and his deputy Dr Hans Scherenberg were responsible for development. The technical leadership was the responsibility of Dr-Ing. Bruno Eckert, whose team included Christian Schmitz and Emil Waldmann, the latter of whom had been a key member in Helmut Schelp's department in the old Technisches Amt. The personnel of the new organization, many of whom had formerly worked for BMW, were recruited from those who had returned from France and the Soviet Union or had been members of the Heinkel team that was dissolved in 1956.

Following comprehensive design studies during 1956, the first project was the DB 720/PTL 6 turboprop engine with a useful power of 1,000 to 1,600hp. In addition to being an aircraft powerplant, a modified version of this engine for locomotive and marine applications was also considered. The DB 720/PTL 6 had a four-stage axial compressor followed by a radial compressor, giving a total compression ratio of 5.5:1. An annular combustion chamber was used, with twelve injection and two starting fuel injectors. Restricted by the metals available at that time, a temperature of 800°C at the turbine inlets was used. A two-stage turbine was used to drive the compressors at 21,000rpm and a separate or free turbine was used to provide the power take-off. Without starter, the DB 720/PTL 6 weighed 220kg (485lb) and all auxiliaries, including the speed governor, were developed by DB.

On test stands at Stuttgart-Untertürkheim, component testing began in 1958. Next, the gas-producing section of the engine was tested and then the power turbine was added for testing of the complete unit. Most problems centred on the material strength of the combustion chamber. An official 50-hour acceptance run was made in autumn 1962. The engine eventually reached 1,400hp with high reliability and long life, and was tested as a turboshaft engine in a Bell UH-1D helicopter. Daimler-Benz went on to develop other small turboprops and also two-circuit turbojets beyond the time-scale of this history.

In the meantime, in West Germany, the Luftwaffe was reborn in September 1956.

APPENDIX ONE

Report on the Whittle System of Aircraft Propulsion

(Theoretical Stage) October 1935
M.L. Bramson, ACGI FRAeS

TERMS OF REFERENCE

The purpose of this Report is to record the result of an independent step-by-step check of the theories, calculations and design proposals originated by Flt. Lt. Whittle, and having for their object the achievement of practical stratospheric transport. No investigation of the patent situation has been attempted.

MATERIAL SUBMITTED BY FLT. LT. WHITTLE

The inventions and discoveries of Flt. Lt. Whittle have not yet reached the experimental stage, and so the material available for investigation is necessarily confined to a reasoned statement of the principles involved, coupled with justifying aerodynamic and thermodynamic calculations and design proposals.

DESCRIPTION

The Problem

The desirability of stratospheric flight arises from the rapid decrease of air-resistance (drag) experienced at high altitudes, due in turn to the low air densities obtaining there. For example, the density at an altitude of 69,000 feet is only one-sixteenth of normal atmospheric density at sea level. The principal difficulty to be overcome for this purpose is the maintenance of power notwithstanding the rarification of the atmosphere available for combustion. This difficulty has, in a moderate measure, been overcome by supercharging aero-engines of orthodox type, but it can be shown that supercharging to the extent which would be necessary for maintenance of adequate power in the stratosphere would not be feasible on account of the power which would be consumed by such a supercharger even if the size and weight were not prohibitive.

Furthermore, even assuming that means were found to maintain adequate power at the altitude mentioned, the orthodox means, to wit the aircraft propeller, of applying that power effectively to the atmosphere for propulsive purposes would need to be of impracticable dimensions. Those, briefly, are the reasons underlying the search for some alternative mode of propulsion which shall both maintain and apply propulsive power with adequate efficiency at those altitudes where such power will produce the greatest speed, economy and range.

It has long been recognized that for this purpose some form of jet propulsion would be necessary. Many suggestions have been made to that end, mostly based on the use of some explosive as propellant, but none has been practical. They have failed to provide a solution in the main, either because they involved carrying in the aircraft not only the fuel but also the oxygen required for combustion, or because though theoretically capable of functioning in the stratosphere the means proposed were incapable of raising the aircraft to the stratosphere.

Solution Proposed by Flt. Lt. Whittle

The system of propulsion proposed by Flt. Lt. Whittle falls and will be treated under three headings: Aerodynamic principles, Thermodynamic principles, and Engineering. The general scheme is as follows:

An aeroplane of the 'cleanest' possible aerodynamic form is provided with a circular or annular forward orifice facing the air stream. (NOTE. In this report, all dimensions, temperatures, pressures and other figures given will, unless otherwise stated, relate to the particular case of an aircraft and reaction engine designed to operate at an altitude of 69,000 ft and at a speed of 500 mph.) This orifice communicates directly with what may be termed 'the engine room' in which will exist a pressure exceeding that of the surrounding atmosphere by the pressure head corresponding to the kinetic energy of the air stream meeting the said orifice. The air thus partly compressed is drawn into a centrifugal air compressor which delivers into a heat-insulated combustion chamber. Oil fuel is admitted into this combustion chamber, where it burns, thus relating the temperature of the air which is allowed to expand at constant pressure in the combustion chamber. The mixture of air and products of combustion then flows to the intake nozzle or nozzles of an impulse turbine so designed that the temperature and pressure drop is only sufficient to enable the turbine to drive the centrifugal compressor to which it is directly coupled. The energy contained in the air stream issuing from the turbine exhaust is thereupon converted into kinetic energy in a propulsion nozzle which delivers a high velocity jet of air and combustion gases through an orifice in the tail of the aeroplane. The total forward thrust imparted to the aeroplane is equal to the force required to accelerate the mass of air flowing through the machine in unit time from rest to the velocity (absolute) of the propulsion jet.

Aerodynamic Principles

When an aeroplane is in steady horizontal flight the force of propulsion must be equal to the total air resistance or drag. For a very clean streamline aeroplane without excrescences such as undercarriages and the like, the drag may, under the most economical flight conditions, amount to one-eighteenth of the total weight of the aeroplane. Such an aeroplane might be designed to have its most economical speed at ground level at, say, 125 mph. Under corresponding flight conditions the speed is, for practical purposes, inversely proportional to the square root of the air density. Therefore, at 69,000 ft, for example, where the relative density is 1/16, the corresponding speed of such an aircraft is 125 x √16 = 500 mph. At this speed drag will still be the same fraction (1/16) of the total weight of the aeroplane and if, therefore, a thrust of equal amount can be maintained, the speed of 500 mph will be achieved. Any excess of thrust available between ground level and the operating altitude can be used for climbing in overcoming the component of gravity which lies along an inclined flight path.

Thermodynamic Principles

The Whittle Reaction Engine is based upon a heat engine cycle of the 'combustion at constant pressure'

variety. The inventor has prepared pressure-volume diagrams and entropy diagrams for the particular case of his reaction engine which forms the subject of the critical discussion below. One set of diagrams represent conditions at 500 mph at the operating height of 6,000 ft, and the other set represents starting conditions at no airspeed and at sea level. These diagrams have been attached as Appendix No. 1 to this Report.
(NOTE: Appendices I and III are not included in this paper.)

The thermodynamic cycle is as follows: Air at 220°C absolute and 0.702 lb/sq in absolute pressure (atmospheric conditions at 69,000 feet) is compressed adiabatically (due to the forward speed of the aircraft as mentioned above) to 245° absolute and 1.04 lb/sq in. It is thereupon further compressed adiabatically in the centrifugal compressor to 6.76 lb/sq in which gives a theoretical temperature increase to 420°C absolute. All losses in the compressor (assumed efficiency 80 per cent) with the exception of bearing and radiation losses, which are negligible, will be transformed into heat contained in the air so compressed. This produces a further temperature rise from 420° to 464° absolute. Heat is then added at constant pressure (fuel oil is introduced and burnt in the combustion chamber) raising the temperature to 1,092° absolute. Expansion now takes place in two stages: the first stage takes place in the turbine nozzles through which there is a pressure drop from 6.776 lb/sq in to 2.3 lb/sq in, the corresponding theoretical temperature drop being from 1,092° to 800° absolute. During this expansion the gases accelerate from 300 ft/s up to a velocity of 2,500 ft/s. They thereupon impinge on and pass through the turbine blades to which, assuming a 75 per cent turbine efficiency, they give up 75 per cent of their kinetic energy. The losses, amounting to 25 per cent of the said kinetic energy, are essentially fluid friction losses and are therefore transformed into heat, which raises the temperature of the turbine exhaust from 800° absolute to 875° absolute.

Thus in the case of an aeroplane weighing 2,000 lb and requiring a thrust of 1/18 of that weight, i.e. 111 lb, 111/49.3 = 2.25 lb/s of air will be the necessary capacity of the reaction engine to provide the thrust required.

The interesting case of non-level conditions at no forward speed of the aircraft is dealt with in the inventor's second Pressure-Volume Diagram in Appendix 1. The main differences are:

The initial temperature, which is 288°C abs instead of 220°C abs.

The smaller amount of heat added per lb of air so not to exceed the same maximum temperature of the cycle and the same blade temperature as that adopted for the high altitude conditions.

The greatly increased throughput in lb/s of the reaction engine due to the increased density of the atmosphere.

The fact that, due to the aircraft being stationary, there is only one compression stage.

The net result of these changes in conditions is, it will be seen, that the thrust per lb of air per second is slightly greater, namely 52.3 lb, but the thrust due to the 26 lb/s throughout is 36 × 53.2 = 1,915 lb, which is nearly equal to the weight of the aircraft. (Should these figures actually be obtained, it is clear that both acceleration and climb will be very rapid.)

Efficiencies
The overall efficiency of a reaction engine of this type is the thrust horsepower divided by the input of heat energy in unit time. (This corresponds to the thermal efficiency of an aero-engine multiplied by the propeller efficiency.)

In the particular case referred to, thermal efficiency (i.e. the kinetic energy given to the working fluid divided by the heat energy input) would be 48 per cent, giving an overall efficiency 17.13 per cent. For the sea-level conditions, and assuming a flying speed of 125 mph, the thermal efficiency would be 22.9 per cent and the overall efficiency would be 4.5 per cent.

Engineering

The Power Unit
The Whittle Reaction Engine consists of a single-stage turbo-compressor directly coupled to and driven by a gas turbine of the pure impulse type. Taking the case of a unit capable of a throughput of 2.25 lb of air per second at 69,000 ft, the impeller diameter would be 19 in and its speed would be 17,850 rpm, giving a linear tip speed of 1,470 ft/s. (The overall diameter of the compressor would be 43 in.)

The compressor has a double inlet and its designed capacity would be 470 cu ft/s giving an inlet velocity of 400 ft/s. The turbine may, alternatively, consist of a double row velocity compounded impulse wheel or of two single row impulse wheels working in parallel. The latter arrangement is probably preferable, as it permits direct coupling between the compressor and the turbine. (The two row turbine wheels would have to be geared down in relation to the compressor: this complication might, however, be balanced by the advantage of lower peripheral speed of the turbine wheels.) For efficiency, the linear speed of the single row turbine blades should be one half of that of the gases issuing from the turbine nozzle (*see* Appendix III) which is 2,500 ft/s. The turbine blade speed should therefore be 1,250 ft/s and the effective diameter of the turbine wheels 16.15 inches.

The turbine exhaust gases pass straight to the propulsion nozzle where, as already mentioned, the speed of the gases is accelerated to 2,320 ft/s. The volume per lb of gas has at this point expanded to 591 cu ft.lb giving a total of 591 × 2.25 = 1,330 cu ft/s in the particular case considered. This gives a propulsion nozzle outlet diameter of 10.25 in.

The Aeroplane
The aeroplane consists of a fuselage of correct streamline form, the forward portion of which is a sealed air reservoir capable of withstanding an internal pressure of 15 lb/sq in and containing the pilot, passengers and controls. An annular opening facing the air stream is formed between the circumference of this sealed portion and the monocoque shell of the rest of the fuselage. The total cross-sectional area of this annular opening need only be about 100 sq in (in the particular case considered) which, assuming 4 ft 6 in to be the diameter of the sealed portion, gives a width of the annular opening of only 0.6 in. (In actual practice, the width of this opening would be made greater to make certain of getting the full necessary flow.)

A cantilever monoplane wing of 52 sq ft area would be fitted, giving a wing loading of 19.3 lb/sq ft. As there is no propeller, there is no need for large ground clearances during the landing and take-off, and the retractable undercarriage can be short. Furthermore, as the machine is designed and intended for flight at its most economical angle of incidence, the wings can be set at a larger angle of incidence in relation to the fuselage than is the present practice. Thus, even from the point of view of getting correct wing incidence for take-off and landing, a high undercarriage is not necessary. An auxiliary compressor would be fitted, drawing air from the pressure side of the main compressor and delivering at normal atmosphere pressure into the sealed cabin.

The inventor claims that, by fitting a correctly designed sleeve of venturi shape over the propulsion nozzle, an increased propulsion efficiency can be obtained, but any such possible advantage has been ignored in his calculations.

CRITICAL DISCUSSION
The following critical discussion is based entirely upon the particular case to which reference has already been made and which has been worked out by the inventor. The calculations have been checked. The data are as follows:

Given

Aeroplane

Weight	2,000 lb
Wing loading	19.3 lb/sq ft
Most economical speed (speed of minimum drag) at ground level	125 mph
Minimum drag	111 lb
Corresponding speed at 69,000 ft altitude (where effective efficiency = 1/16 is 500 mph)	733 ft/s

Engine Thermal Cycle

Assumed compressor efficiency	80%
Assumed turbine efficiency	75%

Total theoretical temperature rise due to compression	200°C
Maximum turbine blade temperature	527°C = 800°C abs
Actual temperature rise due to 1st stage of compression (pilot head)	25°C
Therefore, rise of temperature due to 2nd stage of compression (compressor)	175°C
Actual temperature rise in compressor 175/0.8 =	219°C
Initial air temperature	220°C abs
Temperature after pilot head	245°C abs
Effective heat drop in turbine	219 units

(*Note 1.* The unit of heat chosen in these calculations is the pound calorie multiplied by the specific heat of air at constant pressure or, in other words, the quantity of heat required to raise one pound of air one degree centigrade at constant pressure.)

Theoretical heat drop in turbine	219/0.75 = 292 units

Deduced

Final compression temperature 245°C + 219°C =	464°C abs
Maximum temperature of cycle 800°C + 292°C =	1,092°C abs
Overall temperature ratio of compression 420/220 =	1.91
Temperature ratio of first stage (turbine) of expansion 1,092/800 =	1.366
Therefore temperature ratio of final expansion 1.91/1.366 =	1.40
Temperature before final expansion 1,092°C − 219°C =	873°C abs
Temperature at end of final expansion 873/1.40 =	623°C
Therefore useful heat drop is 873°C − 623°C	250 units

To obtain the effective output of the engine, the heat equivalent of the pitot compression must be deducted.

Therefore effective output of engine = 250 − 25	225 units
Heat addition = 1,092°C − 464°C	628 units
Therefore thermal efficiency = effective heat drop/heat addition = 225/628	35.8%
The jet velocity resulting from a useful heat drop of 250 units is 146.7 × √250	2,320 ft/s

(*Note:* The constant 146.7 is derived from the mechanical equivalent of the pound calorie.)

Hence, net change of gas velocity produced = 2,320 − 773	1,587 ft/s
Therefore propulsive thrust per pound of gas per second = 1,587/32.2 =	49.3 lb
Therefore weight of air throughput required per second = 111/49.3	2.25 lb
Propulsive efficiency	48%
Overall efficiency	17.2%
Jet horsepower	308 hp
Thrust horsepower	148 hp

AERODYNAMICS

Flt. Lt. Whittle shows, by the application of Professor Melville Jones' formulae for induced power and profile drag, that a well-streamlined aeroplane with the proposed wing loading of 19.3 lb/sq ft may, under conditions of *minimum drag,* be expected to have a lift/drag ratio of 21. He adopts however, the figure of 18, which more nearly corresponds to flight conditions giving *maximum range.*

There is no serious doubt that an aircraft of the type contemplated could be made to approach the ideal streamline aeroplane as closely as, for example, the modern glider, whose best lift/drag ratio has been known to attain the figure of 23 and over. Flt. Lt. Whittle's figure for drag, and for speed at which such drag will be experienced at 69,000 ft, may, therefore, be accepted unreservedly.

THERMODYNAMICS

A Consideration of Basic Assumptions

(a) *Compressor Efficiency.* The compressor efficiency assumed, of 80 per cent, is unusually high. There are published test results (ARC R&M 1336) showing adiabatic temperature efficiencies for a single-phase centrifugal compressor up to 73 per cent. Dr A. Rateau, probably the greatest authority on exhaust driven turbo-compressors, stated, in an article in the *Revue Générale des Sciences* of the 15 January 1930 and reproduced in the *Génie Civil* of the 15 February 1930 as follows: (The subject of the article is the supercharging of Diesel Engines by means of exhaust driven turbo-compressors) . . . 'On the other hand in designing the compressor exactly for the required throughput of air, efficiencies of 82 per cent for the compressor and 78 per cent for the turbine or, in other words, an overall efficiency of 64 per cent can be counted on . . .'

The inventor supplies the following interesting information which he has obtained from the compressor experts of the British Thomson-Houston Co. Ltd. This company is said to have obtained compressor efficiencies of 76 per cent and 82 per cent on actual test. Moreover, they have, on the basis of their experience, established a non-dimensional figure of merit for centrifugal compressors, according to which Flt. Lt. Whittle's proposed compressor should compare favourably with their best existing examples.

The main features distinguishing the inventor's compressor from normal practice is the high-pressure ratio obtained in a single stage and the high volumetric output. The former is almost a natural function of the peripheral impeller speed and should be realized. The latter is obtained mainly by the double intake arrangement and without adopting excessive intake velocities of 400 ft/s. (A blower built by British Thomson-Houston for Messrs Charles Nelson and Company, had a maximum intake velocity well in excess of this figure.)

A very large mass flow obtained through the double intake tends towards increased efficiency since the fluid friction losses cannot increase proportionately.

In view of these considerations, I regard an 80 per cent efficiency as a probability, but by no means a certainty. I do feel confident, however, that with skilful design a compressor efficiency between 70 per cent and 75 per cent will be obtained.

(b) *Turbine Efficiency.* Referring again to the quotation from Dr Rateau's paper, it will be seen that 73 per cent is given as obtainable in practice. Having regard to the fact that the blade speed adopted approaches one half the gas speed at the turbine nozzle (which is a condition for maximum efficiency), it is probable that 75 per cent will be achieved.

(c) *Temperature Rise due to Compressor.* For practical purposes adiabatic compression may be assumed with negligible error. The theoretical temperature rise required is 175°C. For a compressor having a sufficient number of vanes to make the peripheral component of the gas speed discharging from the impeller equal to the peripheral speed of the impeller, the relationship between impeller speed and temperature rise is given by $U^2 = 32.2 \times 333 \times$ the temperature rise, from which can be derived the impeller speed required, U = 1,380 ft/s. The inventor's figure is 1,470 ft/s which is, therefore, in excess of the speed theoretically required. The actual temperature rise in the compressor 175/0.8 = 219°C, is a measure of the actual power required per pound of throughput. Therefore, actual power required equals (2.25 × 333 × 219)/559 = 300 hp. For adiabatic compression, the pressure ratio equals the temperature ratio$^{3.5}$. Therefore, as the overall temperature ratio of compression is 420/220 = 1.91, the compression ratio will be $(1.91)^{3.5}$ = 9.6. (This ratio, of course, includes the pitot compression.) I see no reason why these temperatures and compression ratios should not be obtained in practice.

(d) *Turbine Blade Temperature.* A turbine blade temperature of 800°C absolute (527°C) has been taken as the basis of the turbine design. Since the effective power of the turbine must be equal to the power absorbed by the compressor, it follows that the effective heat drop in the turbine must be equal to the actual temperature rise in the compressor which is 219°C. We therefore have actual heat drop in turbine nozzles 219/0.75 = 292 units. So that the blade temperature shall be 800°C abs, the temperature at the beginning of the first expansion stage through the turbine must be 800 + 292 = 1,092°C abs. This, therefore, limits the heat addition per pound to 1,092 – 464 = 628 units. Herein lies the justification for the assumption, or rather the stipulation, that the blade temperature shall not exceed 800°C abs.

(e) *Thrust*. All the further thermodynamic deductions enumerated previously follow directly from the fact that, in a heat cycle of the type adopted, the overall compression ratio is equal to the overall expansion ratio. One or two points should be noted in this connection. Great emphasis is rightly laid by the inventor on the fact that almost the entire losses incurred in the turbine are transformed into heat in the gas stream, whereas in all existing applications of a gas turbine such heat would be entirely lost. This is not the case in the reaction engine, since part of such heat is recovered in the form of additional kinetic energy in the jet. The inventor has made no allowance for fluid friction losses in the propulsion nozzle. These should, however, be very small and should be provided for by a slight increase in the throughput capacity of the unit.

The performance of the reaction engine for ground level conditions has been obtained on the same basic assumptions. The calculations for horizontal flight conditions are given in Appendix II. It will be seen that, although the thermal efficiency is low at ground level, the thrust is exceedingly large in relation to the total weight of the aircraft.

ENGINEERING

The Reaction Engine

The peripheral speed adopted by the inventor for the compressor rotor is 1,470 ft/s. This is considerably in excess of existing practice. (I am informed by Flt. Lt. Whittle that the British Thomson-Houston Centrifugal compressor Design Department are aware of certain cases of speeds of 1,250 ft/s.) The speed now proposed involves an increase of stress, all other things being equal, of about 38 per cent. Provided one of the modern high tensile steels is used, I believe that, with careful design, it will be possible to make an impeller capable of standing up to the peripheral speed proposed. Great care must be taken to avoid the risk of vibration of the impeller blade tips. Much depends upon the skill and care employed in the detail design, and one must attach great importance to the employment for this purpose of all available expert advice. Subject to the foregoing, the engineering features of the compressor should be trouble-free.

Combustion Chamber and Burners

These do not call for special comment and any minor problems arising should yield to ordinary skilful design.

Turbine

As already mentioned, the turbine consists of two single-row impulse wheels working in parallel and, in fact, having a common shaft. It will probably be proved desirable to machine the two wheels and their shaft from one solid forging. The turbine blade speed of 1,250 ft/s gives rise to a stress at the blade root of about 12.6 tons per sq in, for a blade length of 1.33 in. At the very reasonable blade temperature of 527° adopted, this stress would give a 'creep rate' of 2×10^{-7} inch per inch per hour if the steel used is Kayser Ellison 953, which creep must be allowed for in the design.

The turbine wheel rim will necessarily be considerably hotter than the rest of the wheel and will suffer tangential compression stresses in addition to the stresses caused by the centrifugal tension of the blade. The whole design of the turbine discs and blade root and their method of attachment is a very delicate and important matter, and should be submitted to experts. I do not expect trouble due to heat transmission from the rim to the turbine discs. If a small amount of cool air is allowed to enter the turbine casing near the shaft, it will flow outward while being 'sheared' at an exceedingly high rate in the small clearance space between the turbine wheel and the casing. This will effectively prevent excessive temperatures from reaching the shaft and/or the bearings.

It is appropriate here to point out that, should detailed consideration of the engineering design problems lead to unexpected difficulties, the alternative of adopting a two row 2-stage impulse wheel is available and would lead to considerably reduced peripheral speed. It would, however, be at the expense of simplicity, as such an arrangement would necessitate gearing between the turbine and the impeller shaft. In my view, it is worthwhile going to considerable lengths to avoid such gearing. (Likewise, in the case of the compressor, a 2-stage compression could be resorted to at the expense, however, of increase, but not prohibitive, weight and bulk.)

The design problems and difficulties to be overcome, in their probable order of importance, may be summarized as follows:

1. To make provisions for the combined heat and centrifugal stresses at the turbine blade roots.
2. The design and manufacture of a compressor rotor capable of withstanding the centrifugal and bending loads on the vanes.
3. To guard against turbine blade and compressor blade vibration.
4. Design of main shaft to avoid torsional vibration periods, and to resist gyroscopic couples.

I do not regard any of these problems as insurmountable, but I do consider it possible that they may not all be satisfactorily overcome in the first reaction engine produced.

Weight

The inventor estimates the weight of the complete reaction engine unit as 500 lb. Unless or until designs are available, it is impossible to form a reliable opinion on this estimate. This much may, however, be said. The working elements of the engine operate at extremely high velocities, a fact which tends towards decreased size and weight for a given power. The same tendency results from the fact that the working elements are purely rotary, as distinct from reciprocating and rotary. Furthermore, the engine has no equivalent to cooling fins, water jackets or radiators, nor has it any airscrew. For these reasons, it can be safely assumed that even the first complete engine will not so far exceed weight estimates as to render flight tests impossible or inconclusive.

The Stratospheric Aeroplane

Structurally, the proposed stratospheric aeroplane presents no new problem, with the one exception of providing a hermetically sealed cabin with safe and satisfactory doors permitting the crew's getting in and out. The problem of flying controls and engine controls to be operated from inside a sealed cabin without having numerous sources of air leakage, can be simplified by operating all controls hydraulically. By tapping the pressure side of the compressor, an ample supply of air for breathing is available. It will be noticed that the compression pressure is near 1/2 atmosphere which is ample for most people. An auxiliary booster can, of course, be fitted.

DEGREE OF PERMISSIBLE ERROR IN FUNDAMENTAL ASSUMPTIONS

An investigation has been made with the object of ascertaining the lowest compressor and turbine efficiencies at which stratospheric flight would be possible. The result of this investigation shows that, even if both the compressors and the turbine efficiencies were each only 60 per cent, then without raising the turbine blade temperature, a thermal efficiency of 14.2 per cent would still be obtained, and an overall (thrust) efficiency of 9.1 per cent. The principal disadvantages experienced in the event of such low efficiencies being obtained would be, first, that the throughput capacity of the unit would have to be approximately doubled and, secondly, that at ground level an adequate thrust would only be obtainable by permitting an increase of the turbine blade temperature.

SUMMARY

The stratospheric aeroplane and the Whittle Reaction Engine have been described in principle and with particular reference to the case of an aeroplane of 2,000 lb all-up-weight capable of a speed of 500 mph at 69,000 ft altitude.

The inventor's calculations for the aforesaid particular case have been checked and the results are discussed critically from the aerodynamic, thermodynamic and engineering points of view. It is shown that, although the inventor in some respects goes beyond existing experience, he does not appear to go beyond the temperatures, stresses and speeds that are possible with modern technique and materials.

The author's figures are based on a compressor efficiency of 80 per cent and a turbine efficiency of 75 per cent, the attainment of which is considered probable. Nevertheless, it is considered that, if these efficiencies are both 60 per cent, stratospheric flight with the Whittle Reaction Engine would still be possible.

CONCLUSIONS

1. Flt. Lt. Whittle's theoretical calculations and deductions therefrom are substantially correct.
2. His fundamental discovery is that the gas turbine, although very inefficient as a prime mover when power is required in the form of shaft horsepower, can be adequately efficient as an auxiliary to the production of a power jet.
3. Should the discovery be successfully put into place, the points of superiority over existing aeroplanes would be:
(a) Economical speeds of 500 mph and over.
(b) Probable ranges of 5,000 miles and over.
(c) The use of non-volatile fuel.
(d) Freedom from noises and vibration.
4. The proposed development, though necessarily speculative as regards time and money required, is so important that it should, if possible, be undertaken.

RECOMMENDATIONS

The 'Brief Outline of Development Procedure' appended to this Report (Appendix III) has, by request, been prepared by the inventor.

I recommend the adoption of the procedure therein proposed, with the proviso that all designs should be submitted to an independent authority on turbine and compressor design before actual construction is undertaken.

M. L. BRAMSON
8 October 1935

APPENDIX II

Flight at Sea Level

Air temperature	288°C abs
Aircraft speed 125 mph	182 ft/s
Assumed compressor efficiency	80%
Assumed turbine efficiency	78%
Theoretical temperature rise in compressor	175° C
Actual temperature rise in compressor 175/0.8 =	219°C
Temperature rise due to pitot compression	1.5°C
Total temperature rise due to compression 219 + 1.5 =	220.5°C
Actual compression temperature 220.5 + 288 =	508.5°C
Temperature ratio (175 + 1.5 + 288)/288 =	1.612
Temperature drop in turbine 219/0.75 =	292°C
Blade temperature (stipulated)	800°C abs
Maximum temperature cycle 800°C + 292°C =	1,092°C abs
Heat addition, 1,092 − 508.5	583.5 units
Temperature ratio in turbine 1,092/800 =	1.365
Temperature ratio of final expansion 1.612/1.365 =	1.181
Exhaust temperature of turbine 1,092 − 219 =	873°C abs
Final temperature 973/1.181°C =	739°C abs
Final temperature drop 873 − 739 =	134°C
Therefore jet speed = 146.7 × √134	1,695 ft/s
Thrust (of air) = (1,695 − 182)/32.2	47.0 lb/lb
Capacity of compressor = 16 × 2.25	36 lb/s
Total thrust = 36 × 47	1,690 lb
Thermal efficiency = 134/583.5	22.9%
Propulsive efficiency	19.4%
Overall efficiency	4.45%
Jet horse-power	2,880 hp
Thrust horse-power	561 hp

Background Comments by the Reporting Engineer 33 years later

'This *must* be done!'

In 1935 my practice as a consulting engineer was conducted from an office in Bush House, London, and Flt. Lt. Frank Whittle appeared there one day. He wanted financing for the development of a system of jet propulsion of aircraft which he had invented. About two years earlier he had submitted it to the British Air Ministry, who had turned it down.

His material consisted solely of thermodynamic and aerodynamic calculations and diagrams; there were no engineering designs.

He was a bright, confident, young officer-pilot in the Royal Air Force, who seemed to know what he was talking about. This impression was somewhat qualified by the eyebrow-raising improbability of this basic thesis that aeroplanes could be made to fly without propellers. Moreover, my own retention of thermodynamic theory was rusty, which made me distrust my own judgement. Nevertheless, or perhaps for this reason, as well as because of my general favourable impression of Whittle, I decided to study his theories and proposals thoroughly. This took two weeks.

At the end of that period I got quite excited. First because of the insight, clarity and accuracy of his presentation and calculations; secondly because my scepticism of any project based on internal combustion turbines (which had hitherto resisted all practical development efforts) disappeared when I realized that here, for the first time, was an application where maximum energy was needed in the turbine exhaust, instead of in the shaft. This was, of course, the reverse of all past objectives for such turbines. And, thirdly, because of the dramatic advance in aviation technology implicit in Whittle's theories.

I suddenly felt 'This *must* be done!' (which meant 'financed'). A survey of my clients produced a London firm of investment bankers, O.T. Falk & Co., whom I approached. They sent along to see me a man who turned out to be an astonishing professional hybrid, to wit: a financial expert and a theoretical physicist! His name was Lancelot L. Whyte and, thanks to his comprehension and sense of perspective, a favourable decision was made. There was, however, one proviso: an independent engineer's report must be produced and must be conclusive.

When this decision was conveyed to Whittle, he said, in effect: 'Very well, but Bramson must be that engineer, for it is premature and potentially dangerous [he meant nationally] to spread detailed knowledge of this discovery any farther.'

So, instead of my initial role of capital-raising intermediary (entitled to some minor participation) I had, for the good of the cause, to become an ethically independent (i.e. non-participating) reporting engineer. It pleases me to remember that I did so enthusiastically.

My report was indeed conclusive. And three years later the first Whittle jet engine was tested in Rugby at the British Thomson-Houston Turbine Factory, and the first jet-powered aircraft flew in 1941.

Re-reading the report after all these years, I find there are only minor points, of emphasis rather than of substance, that would need amendment. But any temptation to feel smug about that is immediately squelched by one's immense admiration for the originator of one of the most striking and consequential technological revolutions of our time.

M.L. BRAMSON
12 November 1968

Comment (1968)

We consider the publication of the Report to be of importance for four reasons:

(1) It was on the basis of Mr Mogens Louis Bramson's favourable judgement, formed against much adverse expert opinion, that the development of the jet engine was originally financed and organised. The Report, which was directed to the original Whittle proposals and the analyses which he had made, was commissioned by one of the signatories (L.L.W.) on behalf of Falk and Partners (O.T. Falk and Sir Maurice Bonham Carter). Together with them, Whittle's friends, Messrs Williams and Tinling, adopted the Bramson Report and so Power Jets Limited came into being.

(2) The Report, as will be seen, is a model of clear and consistent writing and, as such, deserves to be studied by all technical people whose duties involve reporting. It is remarkable in content and exemplary in style.

(3) Mr Bramson has never, in our opinion, received the credit due to him as one of the constructive early proponents: he remained as an actively participating Consultant to the project for several years.

(4) It has been insufficiently recognized that the Whittle project essentially depended on the marriage of a jet engine with a new subgenus of airframe, and that the boldness and completeness of the total concept (as appreciated by Bramson) went beyond the mere proposal of using a gas turbine to produce a propulsive jet. The third signatory (W.E.P.J.) spent many hours with Frank Whittle (whose Patent Agent he was) discussing the inventive features thus involved, such as boundary-layer control, cabin-pressurisation, by-pass-engine feasibility, and the then revolutionary idea of a squat undercarriage.

(5) It is with pleasure that the three signatories below, all intimately connected with the earliest phase of jet development, have submitted the Report for publication.

FRANK WHITTLE (Hon. Fellow)
LANCELOT LAW WHYTE
W.E.P. JOHNSON (Fellow)

APPENDIX TWO

Engine Data Tables

These tables are not exhaustive but, in 286 entries (incl. Volume 2), they give representative examples of the turbojet, turboprop and turboshaft engines, more details of which are to be found in the text. Regarding the data, it is not always possible to be exact since an engine may be in a prototype or developmental stage, or else exact data simply is not available. Fuel consumption for turboprops is given as lb/eshp/hr, turboprop weights are without the airscrew and thrust is dry. Space precludes exact definitions.

All figures are in metric.

ABBREVIATIONS

Headings
Comp.	compressor
Comb.	combustion chamber/s type
sfc	specific fuel consumption
eshp	equivalent shaft horsepower
L	length
D	diameter
Wt	weight

Tables
A	axial compressor
Ann	annular combustion chamber
BP	bypass turbojet
C	can-type or cannular combustion chamber/s
C	centrifugal compressor (single- or double-sided)
D	diagonal-flow compressor
D	duct
DF	ducted fan turbojet
F	fan
GP	gas producer
P	turboprop
T	turbojet
TF	turbofan
TS	turboshaft

Engine designation	Type	Comp.	Stages	Comb.	Turbine stages	Thrust or eshp	sfc	L	D	Wt.	Date	Remarks
Great Britain												
Power Jets												
WU	T	C	1	D	1						1937	First Whittle engine
WU (2nd)	T	C	1	D	1	218					1938	U1 rebuilt
WU (3rd)	T	C	1	C	1						1938	U2 rebuilt
W.1	T	C	1	C	1	562					1941	Flight engine for E.28/39
W.1X	T	C	1	C	1						1940	Early flight hops
W.1A	T	C	1	C	1	390						
W.2	T	C	1	C	1							Failed. Too ambitious
W.2/4	T	C	1	C	1						1941	Basis for W.2B. Wrecked
W.2B	T	C	1	C	1	692					1942	
W.2B/23	T	C	1	C	1	454					1942	Rover-built. Developed by R-R
W.2/500	T	C	1	C	1	1,127					1942	
W.2/700	T	C	1	C	1	1,134					1943	
L.R.1	P	C	1	C	2	770			1.143	1,395	1943	Project
L.R.1 (BP)	BP	DF+DC	2+1	C		1,250				1,256	1943	Project
Metropolitan-Vickers												
F.2 Series 1	T	A	9	Ann	2	816	1.19	3.25	0.915		1941	Flight tested
F.2/2	T	A	9	Ann	2	1,089	1.14				1942	
F.2/3	T	A	9	C	2	1,225					1943	
F.2/4 Beryl	T	A	10	Ann	1	1,588	1.05	4.04	0.963	794	1945	
F.3	DF	A	2+10	Ann	4+4	1,814	0.65	3.56	1.17	998	1943	First practical turbofan
F.5	DF	A	10+2	Ann	4+4	2,136				953	1945	Rear fan
F.9 Sapphire	T	A	13	Ann	2	3,175					1948	Taken over by AS
de Havilland												
H.1	T	C	1	C	1	1,360		2.55	1.27		1942	Development engine
Goblin II	T	C	1	C	1	1,360	1.233	2.55	1.27	703	1945	Reliable. In production
Goblin IV	T	C	1	C	1	1,667	1.15	2.55	1.27	735	1948	
DGT/40	T	C	1	C	1	1,996			1.448		1944	Forerunner of Ghost

ENGINE DATA TABLES

Engine designation	Type	Comp.	Stages	Comb.	Turbine stages	Thrust or eshp	sfc	L	D	Wt.	Date	Remarks
DGT/50 Ghost	T	C	1	C	1	2,268	1.02	3.07	1.35	986	1945	
H.3	P	C	2	C	3	500					1946	Run. Not developed
H.4 Gyron	T	A	7	Ann	2	9,080					1953	
Gyron Junior	T	A	7	Ann	2	3,320					1957	0.45 scale Gyron
Rolls-Royce												
W.R.1	T	C	1	C	1	900			1.37		1942	Based on W.2B
W.2B/23	T	C	1	C	1	770	1.12		1.09		1942	Named Welland I
B.37	T	C	1	C	1	900	1.08	2.13	1.05	417	1943	Named Derwent I
Derwent II	T	C	1	C	1	998						
Derwent III	T	C	1	C	1	1,098						For boundary layer exp.
Derwent IV	T	C	1	C	1	1,111					1945	
Derwent V	T	C	1	C	1	1,588	1.4	2.11	1.09	567	1945	New design
R.B.40	T	C	1	C	1	1,900			1.04	998	1944	Forerunner of Nene
R.B.41 Nene	T	C	1	C	1	2,268		2.458	1.258	708	1944	
Nene 3	T	C	1	C	1	2,310	1.06	2.46	1.26		1948	
RB.44 Tay	T	C	1	C	1	2,835	1.06				1948	Last R-R centrifugal turbojet
Griffith C.R.1	GP	A	3	Ann	3							Gas producer and ducted fan versions
A.J.65	T	A		C	1	2,268						Named Avon
R.A.1 Avon	T	A		C	1	2,948			1.05			
R.A.2 Avon	T	A	12	C	1	2,560	1	3.17	1.05	1,157	1948	
R.A.3 Avon	T	A	12	C	2	2,948			1.07	1,020	1948	
R.A.7 Avon	T	A	12	C	2	3,400	0.92		1.07	1,116	1950	First of 100 Series
R.A.9 Avon	T	A	12	C	2	2,948	0.9				1952	Mk 502 for Comet 2X
R.A.25 Avon	T	A	12	C	2	3,220	0.91	2.6	1.07		1953	Mk 504/505 for Comet 2
R.A.21 Avon	T	A	12	C	2	3,629					1950	Military Mks 113 & 115
R.A.14 Avon	T	A	15	C	2	4,223	0.84	3.39	1.05	1,297	1951	First of 200 Series
R.A.26 Avon	T	A	15	C	2	4,535	0.86	2.88	1.05		1954	Mk 521
R.B.39 Clyde	P	A+C	9+1	C	2	4,000	0.71	3.08	1.19		1945	Compound, 2-shaft engine
Dart 1	P	C	2	C	2	890	0.87	2.41	0.98	385	1946	Dart 501
R.Da.6 Dart	P	C	2	C	2	1,600	0.71	2.46	0.97			Dart 510
R.Da.7 Dart	P	C	2	C	3	1,800					1954	Dart 520 with 3-stage turbine
R.B.109 Tyne	P	A+A	6+9	C	3+1	4,500					1956	2-shaft engine
RCo.2 Conway	BP	A+A	4+8	C	2+2	4,536					1952	2-shaft engine. One built
RCo.5 Conway	BP	A+A	6+9	C	2+1	5,897					1953	Development engines
RCo.10	BP	A+A	7+9	C	2+1	7,780					1957	Civilian Conway
RCo.12	BP	A+A	7+9	C	2+1		0.87			2,043	1958	Civilian Conway
RB.108	T	A+A	8	Ann	2	912		1.06	0.508	129	1955	First lift engine
Armstrong Siddeley												
ASX	T	A	14	C	2	1,157	1	4.24	1.07	862	1943	With RAE Sarah comp.
ASP	P	A	14	C	2	3,600		3.5	1.39	1,565	1943	Turboprop version of ASX
Python 1	P	A	14	C	2	4,110	0.6	3.12	1.38	1,429	1948	
Mamba 1	P	A	10	C	2	1,135		1.75	0.76		1945	
Mamba 2	P	A	10	C	2	1,475	0.67	2.22	0.737	350	1946	
ASA.1 Adder	T	A	10	C	2	476	1.2	1.86	0.71	249	1948	Created from Mamba 2
Viper 1	T	A	11	Ann	1	744				165	1950	Short-life engine
Viper 2	T	A	11	Ann	1	744	1.09	1.66	0.63		1952	Mk 101 long-life version
Double Mamba	P	A	2x10	C	2x2	2,950	0.67	2.51	1.34	907	1947	ASMD.1
Double Mamba	P	A	2x10	Ann	2x2	3,190	0.71	2.6			1951	ASMD.3
Sa.1 Sapphire	T	A	13	Ann	2	3,175					1948	Research engine
Sa.3 Sapphire	T	A	13	Ann	2	3,443	0.91	3.4	0.95		1951	
Sa.6 Sapphire	T	A	13	Ann	2	3,720	0.9	3.4	0.95		1952	First production version
Sa4/7 Sapphire	T	A	13	Ann	2	4,763					1952	Redesigned 200 Series
Sa.7 Sapphire	T	A	13	Ann	2	4,990	0.89	3.353	0.89	1,395	1955	Mk 202. In production

Engine designation	Type	Comp.	Stages	Comb.	Turbine stages	Thrust or eshp	sfc	L	D	Wt.	Date	Remarks
Bristol												
Theseus I	P	A+C	8+1	C	3			2.68	1.24		1945	With heat exchanger
Theseus II	P	A+C	8+1	C	3	2,450	0.77	2.098	1.372	1,000	1947	Without heat exchanger
Proteus	P	A+C	12+1	C	2+1						1947	Prototype
Proteus II	P	A+C	12+2	C	2+1	2,500		2.88	0.978	1,724	1950	600 Series
Proteus III	P	A+C	12+1	C	2+2	4,250		2.88	0.978	1,315	1952	700 Series. Redesign
Cpled. Proteus	P	A+C	2(12+1)	C	2(2+2)	7,000	0.8	4.245	2.042	3,675		
BE.25 Orion	P	A+A	7+5	C	3+1	5,150	0.39	2.85	1.06	1,452	1956	Two-spool engine
BE10 Olympus	T	A+A	6+8	C	1+1	4,540			1.02		1950	
Olympus 200	T	A+A	5+7	C	1+1	7,260	0.77	3.16	1.02		1954	
Olympus 300	T	A+A	6+7	C	1+1	9,075						
BE26 Orpheus	T	A	7	C	1	1,360		2.82	0.81		1954	
Orpheus 803	T	A	7	C	1	2,268						
Pegasus I	T	A	2+7		1	4,085					1959	First vectored thrust engine
Pegasus 2	T	A	2+7		1	5,220					1960	
D. Napier												
NNa.1 Naiad	P	A	12	C	2	1,590	0.73	2.59	0.71	497	1947	One of first axial turboprops
Cpled Naiad	P	2xA	2x12	C	2x2	3,150	0.72	2.62	1.44	1,157		NaC.1
Nymph	P	A		C		535						Design stage only
NE.1 Eland	P	A	10	C	3	3,000	0.62	3.1	0.92		1952	
NE.4 Eland	P	A	10	C	3	4,200		3.1	0.92		1955	
NOr.1 Oryx	GP	A+A	12+4	C	2	754	0.68	1.93	0.49	225	1952	Gas generator. For helicopters
NGa.1 Gazelle	TS	A	11	C	2+1	1,260					1955	
Germany												
Ernst Heinkel												
HeS 2	T	C	1	Ann	1	136		0.9	0.905		1937	Demonstration engine
HeS 3A	T	C	1	Ann	1						1938	Test flown
HeS 3B	T	A+C	1+1	Ann	1	450	2.16	1.48	0.93	360	1939	Used in He 178
HeS 6	T	A+C	1+1	Ann	1	550	1.6			420	1939	Test flown only
HeS 8	T	A+C	1+1	Ann	1	600		1.6	0.775	380	1941	109-001. Abandoned
HeS 30	T	A	5	C	1	860		2.72	0.62	390	1942	109-006. Abandoned
HeS 40	T	A	5	C	1	940?					1942	Constant-volume engine
Heinkel-Hirth												
HeS 11	T	A+D+A	1+1+3	Ann	2	1,115					1941	Prot. V1. Static engine
109-011 V6	T	A+D+A	1+1+3	Ann	2	1,000					1944	Some flight tests
109-011A-0	T	A+D+A	1+1+3	Ann	2	1,300?	1.31?	3.45	0.864	950	1945	Pre-production. Underdeveloped
Junkers												
109-004A	T	A	8	C	1	430					1941	Flight tested
109-004A-0	T	A	8	C	1	840	1.4	3.8	0.96	850	1942	Pilot production
109-004B-1	T	A	8	C	1	900	1.38	3.86	0.8	720	1943	In production
109-004H	T	A	11	C	2	1,800	1.2	4	0.86	1,200	1945	Projected
109-012	T	A	11	C	2	2,780		4.862	1.063	2,000	1945	Projected
109-022	P	A	11	C	3	5,000		5.64	1.063	2,600	1945	Projected
BMW												
P.3304	T	A+A	4+5	Ann	3+4						1940	109-002. Contra-rotating
P.3302 V1	T	A	6	Ann	1	150					1940	Prototype
P.3302 V11	T	A	6	Ann	1	600	1.6		0.69	650	1942	Flight tests
109-003A-0	T	A	6	Ann	1	800	1.35	3.565	0.69	570	1943	Pre-production
109-003A-1	T	A	7	Ann	1	800	1.35	3.565	0.69	570	1944	In production

Engine designation	Type	Comp.	Stages	Comb.	Turbine stages	Thrust or eshp	sfc	L	D	Wt.	Date	Remarks
109-003C	T	A	7	Ann	1	900	1.27	3.415	0.69	610	1945	Prototype only
109-018	T	A	12	Ann	3	3,400	1.1	4.95	1.25	2,500	1945	Parts made
109-028	P	A	12	Ann	4	6,570	1.2	6	1.25	3,500	1945	Project
P.3306	T	A	7	Ann	1	1,700	1.2	3.2	0.85	900	1945	Project
P.3307	T	A	6?	Ann	1	500	1.5	2.85	0.69	650	1945	Project. Expendable engine
Daimler-Benz												
109-007	DF	F+A	3+17	C	1	1,275	1.1	4.725	1.625	1,300	1943	Contra-rotating. Abandoned
109-016	T	A	9	Ann	2	13,000		6.7	2	6,200	1944	Projected
109-021	P	A+D+A	1+1+2	Ann	3	2,400		3.696	0.91	1,266	1944	Project, based on 109-011
Porsche												
109-005	T	A	8	C	1	500	1.7	2.85	0.65	200	1944	Expendable. For missiles

German post-war

Engine designation	Type	Comp.	Stages	Comb.	Turbine stages	Thrust or eshp	sfc	L	D	Wt.	Date	Remarks
Heinkel INI-11	T	A+D+A	1+1+3	Ann	1	1,300	1	3.05	0.762	1,000	1955	Built in Spain. Abandoned
Heinkel 053	T	A	11	C+Ann	2	6,500	0.93	4.025	1.105	1,565	1956	Built in Egypt. Abandoned
E300	T	A				3,260					1964	Built in Egypt
Pirna 014	T	A	12	Ann+C	2	3,150	0.85	3.52	0.981	1,100	1956	Built in East Germany
Pirna 018	P	A	12	Ann+C	2	4,900	0.345	3.68		1,300	1958	3 prototypes
Pirna 020	BP	F+A+A	3+6+8	Ann+C	2	3,700			1.12	950	1962	Project

Abbreviations and Acronyms

A & AEE	Aircraft and Armaments Experimental Establishment (Boscombe Down)
ARC	Aeronautical Research Committee
AS	Armstrong Siddeley
AVA	Aerodynamische Versuchanstalt (Aerodynamic Research Institute)
BEA	British European Airways
BMW	Bayerische Motorenwerke (Bavarian Engine Works)
BOAC	British Overseas Airways Corporation
Bramo	Brandenburgische Motorenwerke (Brandenburg Engine Works)
BTH	British Thompson-Houston Co. Ltd.
CFS	Central Flying School
CSU	constant speed unit
DFS	Deutsches Forschungsinstitut für Segelflug (German Gliding Research Institute)
DGLR	Deutsche Gesellschaft für Luft- und Raumfahrt (German Society for Air and Space)
DSR	Director of Scientific Research
DVL	Deutsche Versuchsanstalt für Luftfahrt (German Experimental Institute for Aviation)
EGT	exhaust gas temperature
E-Stelle	Erprobungsstelle (Testing or Trial Station/Proving Ground)
FAA	Fleet Air Arm
G (after registration on aircraft)	denotes requiring an armed guard
GEC	General Electric Company
GT	gas turbine
HMSO	His (or Her) Majesty's Stationery Office
HP	high pressure
Jumo	Junkers Motorenbau (Junkers engine manufacturer)
kp	kiloponds (kilograms thrust)
LFA	Luftfahrtforschungsanstalt Hermann Göring (Hermann Goering Aeronautical Experimental Establishment)
LP	low pressure
MAP	Ministry of Aircraft Production
ML	Motorenluftstrahl (piston-driven ducted fan)
MLS	ML with afterburner
MoS	Ministry of Supply
MTL	ducted fan with combined piston engine and turbine drive
MV	Metropolitan-Vickers Electric Co. Ltd
NPL	National Physical Laboratory
OR	Operational Requirement (official aircraft specification)
P (on aircraft)	denotes 'experimental' or 'prototype'
PTL	Propellerturbinenluftstrahl (turboprop)
RAE	Royal Aircraft Establishment
RAF	Royal Air Force
RLM	Reichsluftfahrtministerium (German Aviation Ministry)
R-R	Rolls-Royce
TBO	time between overhauls
Technisches Amt	Technical Office (of the RLM)
TL	Turbinenluftstrahl (turbojet)
TL-Antrieb	propulsion by turbojet
TLR	TL combined with rocket (Rakete) booster
TLS	TL with afterburner
TL-Triebwerk or TL-Gerät	turbojet engine
Verkehrsministerium	Ministry of Transport
ZTL	twin-circuit or bypass turbojet

Bibliography and Further Reading

Many of the following titles have been consulted during the creation of this book, and others have been listed to provide further relevant reading. In addition, many articles from periodicals and journals, plus other papers and reports too numerous to mention, were read. Various issues of *The Aeroplane, Aeroplane Spotter, Air International, Air Pictorial, Aerofan, Engineer, Flight, Flug Revue, Flying Review, Flypast, Interavia, Jet & Prop, Luftfahrt International, RAeS Journal, Wings, Airpower* and *Aero Magazine* were consulted.

GREAT BRITAIN

Banks, F.R.: *I Kept no Diary* (Airlife, 1978)
Barnes, C.H.: *Bristol Aircraft Since 1910* (Putnam, 1964)
Beamont, Roland: *The Years Flew Past* (Airlife, 2001)
Birtles, Philip: *Vampire, Venom and Sea Vixen* (Ian Allan, 1986)
Caygill, Peter: *Jet Jockeys* (Airlife, 2002)
Doust, Michael J.: *Wyvern* (Ad Hoc Publications, 2006)
Duke, Neville: *Test Pilot* (Allan Wingate, 1953)
Hardy, M.J.: *Gloster Meteor* (Winchmore, 1958)
Hooker, Sir Stanley: *Not Much of an Engineer* (Airlife, 1984)
James, Derek N.: *Gloster Aircraft Since 1917* (Putnam, 1971)
Judge, Arthur W.: *Gas Turbines for Aircraft* (Chapman & Hall, 1958)
Kershaw, Tim: *Jet Pioneers, Gloster and the Birth of the Jet Age* (Sutton Publishing, 2004)
Lithgow, Mike: *Mach One* (Allan Wingate, 1954)
Middleton, Don: *Test Pilots* (William Collins, 1985)
Pugh, Peter: *The Magic of a Name: The Rolls-Royce Story*, 3 vols (Icon Books, 2002)
Vicary, Adrian: *British Jet Aircraft* (Patrick Stephens, 1982)
Whittle, Sir Frank: *Jet: the Story of a Pioneer* (Frederick Muller, 1953)
Wilson, C.H., and W.J. Reader: *Men and Machines: D. Napier & Son, 1808–1958* (Weidenfeld and Nicolson, 1958)
Wood, Derek: *Project Cancelled* (Jane's, 1986)

GERMANY

Boyne, Walter, and Donald Lopez: *The Jet Age* (NASM Smithsonian Institution Press, 1979)
Ebert, H., J.B. Kaiser and K. Peters: *Willy Messerschmitt: Pioneer der Luftfahrt und des Leichtbaues, Die deutsche Luftfahrt*, xvii (Bernard & Graefe Verlag, 1992)
Gersdorff, Kyrill, and Kurt Grasmann: *Die deutsche Luftfahrt, Flugmotoren und Strahltriebwerke* (Bernard & Graefe Verlag, 1981)
Hellmold, Wilhelm: *Die V-1: Eine Dokumentation* (Bechtle Verlag, 1988)
Kay, Antony L.: *German Jet Engine and Gas Turbine Development 1930–1945* (Crowood, 2002)
Kay, Antony L., and J.R. Smith: *German Aircraft of the Second World War* (Putnam, revised 2002)
Mewes, Klaus-Hermann: *Pirna 014: Flugtriebwerke der DDR* (Aviatic Verlag, 1997)
Müller, Rheinhold: *Junkers Triebwerke* (Aviatic Verlag, 2006)
Nowarra, Heinz: *Focke-Wulf Fw 190–Ta 152, Entwicklung, Technik, Einsatz* (Motorbuch Verlag, 1987)
Scalia, Joseph Mark: *Germany's Last Mission to Japan* (Naval Institute Press, 2000)
Schubert, Helmut: *Deutsche Triebwerke: Flugmotoren und Strahltriebwerke von 1934 bis 1999* (Aviatic Verlag, 1999)
Smith, R. Richard, and Eddie Creek: *Jet Planes of the Third Reich* (Monogram, 1982)
Sobolev, D.A., and D.B. Khazanov: *The German Imprint on the History of Russian Aviation* (Rusavia, 2001)
Wagner, Wolfgang: *The History of German Aviation. The First Jet Aircraft* (Schiffer, 1998)

GENERAL

Brown, Capt. Eric: *Wings of the Weird and Wonderful*, 2 vols (Airlife, 1985)
——: *Wings On My Sleeve* (Airlife, 1961)
Collier, Basil: *A History of Air Power* (Weidenfeld and Nicolson, 1974)
Godsey Jr, F.W.: *Gas Turbines for Aircraft* (McGraw Hill, 1949)
Green, William, and Roy Cross: *Jet Aircraft of the World* (Macdonald, 1955)
Gunston, Bill: *Faster Than Sound* (Patrick Stephens, 1992)
——: *Fighters of the Fifties* (Patrick Stephens, 1981)
——: *World Encyclopaedia of Aero Engines* (Guild Publishing, 1986)
Judge, Arthur, W.: *Gas Turbines for Aircraft* (Chapman & Hall, 1958)
Kármán, Theodore von: *The Wind and Beyond* (Little Brown & Co., 1967)
Lambermont, Paul, and Anthony Pirie: *Helicopters and Autogyros of the World* (Cassell, 1958)
Scaife, W. Garrett: *From Galaxies to Turbines* (Institute of Physics Publishing, 2000)
Smith, G. Geoffrey: *Gas Turbine and Jet Propulsion* (Associated Iliffe, 1942/1946/1955)
Spick, Mike: *The Ace Factor* (Airlife, 1988)
Swanborough, F.G.: *Turbine-Engined Airliners of the World* (Temple Press Books, 1962)
Vogelsang, C. Walther: *Die 2. Etappe* ('Astra' Josef Penyigey-Szabó Verlag, 1955)

Index

For convenience, this Index has been divided into **General, Aircraft and Engine** sections.

General

247 Squadron RAF 69, 73
39 (1PRU) Squadron RAF 95
504 Squadron RAF 63
60 Squadron RAF 76
616 Squadron RAF 62, 63, 66
77 Squadron, RAAF
813 Squadron 121
890 Sqn, Royal Navy 74

A&AEE, Boscombe Down 64, 104
Ackeret, Jacob 14
Adenstadt, Dr. Heinrich 189
AEG (company) 189, 193
Aeronautical Research Committee (ARC) 12, 13, 45, 50
Air Ministry 20, 22, 28, 34, 45, 54
Air Ministry Laboratory 12
Algerian War 75
Alice compressor 15
Amman, Dr. Rolf 200
Anne compressor 14, 15, 118
Antz, Hans 175
Armstrong Siddeley (company) 19, 54, 118–136
Armstrong Whitworth (company) 82
Asiatic Petroleum 26
AVA, Göttingen 176, 189, 196, 198, 201, 204, 220, 222, 228, 230, 241

Baade, Brunolf 247, 251, 253
Bancroft, Dennis (designer) 38
Barry, Sqn Ldr Dennis 66
Bates, Dr. 196
Baumann, Dr. Karl 46, 49
Baur, Karl (pilot) 232
Bayne, B. (engineer) 159
Beamont, R.P. (pilot) 93
Beck, Dr (engineer) 189
Beck, Ernst 176
Bedford, Bill (pilot) 155
Bennett, Dr. J.A.J. (designer) 163
Bentele, Dr. Max 182, 192, 221
Berlin Air Museum 174
Betty compressor 15
Betz, Albert 168, 182, 187
Biefang, Dr. 200
Bishop, R.E. (designer) 68, 77
Bispink (pilot) 208
BMW (company) 198–208, 230–238, 256
Bock, Prof Günter 175, 247
Böhm, H. 82
Bone, Dr. Geoffrey 28
Booth, J.S. (pilot) 124
Brabazon Committee 108
Brabazon, Lord 77, 140
Bramo (company) 198
Bramson, M.L. 21, 257
Brandner, Ferdinand 247, 251, 256
Brisken, Walter 176

Bristol (company) 137–157
Bristol Siddeley organisation 83
British European Airways (BEA) 108, 111, 112, 163
British Overseas Airways Corporation (BOAC) 77, 113, 145
British Thomson-Houston (BTH) 21, 22, 28, 34, 54, 55
Brodie, J.L.P. 50
Broeker, Christian 167
Brooke-Smith, Tom (pilot) 80, 117
Brown Boveri company 14, 230
Brown, Capt. Eric (pilot) 7, 33 39, 70
Brown, D.L. 37
Brownsover Hall 28, 38, 137
Bruckmann, Dr.-Ing Bruno 200, 238
Brückner-Kanis (company) 231
Bryce, G.R. (pilot) 111
Bubble cars, German 249

Cameron (designer) 139
Camm, Sir Sydney (designer) 99, 154
Campini 41
Carter, D 14
Carter, George (designer) 29, 34, 51, 56, 61
Chadwick, Roy (designer) 150
Chaplin, H.E. (designer) 104
Chatterton, Ernest (engineer) 159
Churchill, Winston, PM. 33
Clark, Dr. 142
Colingham, R.W. 21
Constant, Hayne 13, 45, 46, 89
Courant, Prof Frederick 168
Cranwell, RAF College 20, 32
CRD Flight 62
Cunningham, John (pilot) 73, 77

Dabruck, H. 230
Daimler-Benz (company) 178, 239–243, 256
Daunt, Michael (pilot) 33, 48, 51, 56, 61
Davie, Sqn Ldr D. (pilot) 33
de Havilland (company) 50–53, 77–84, 152, 168
de Havilland, Geoffrey 50
de Havilland, Geoffrey Jr. 70, 71, 73
Dean, T.D. 'Dixie' (pilot) 62
Decher, Dr. Siegfried 189
Denkmeir, Dr. 209
Derry, John (pilot) 73
Donaldson, Capt E.M. (pilot) 58, 60, 64
Duke, Neville (pilot) 99, 103
DVL, Berlin 175, 200, 209, 221, 251

Eckert, Bruno 256
Edwards, George (designer) 108
Eisenlohr, Wolfram 176, 178
EKOBA (company) 249
Encke, W. 168, 182, 187, 196, 201, 222, 228
English Electric (company) 70, 93, 104, 158
ENMA (company) 249
Evans, F.G. (designer) 137

Fabrique National (company) 60, 104
Fairey (company) 163
Falk, Roly (pilot) 150
Falklands War 166
Fattler, Dr. 207, 209
Fawn, Geoffrey 91
Fedden, Roy 137
Feilden, Dr. Robert 28, 29
Fell, Jimmie (engineer) 140
Fernibough, W.F. 11
Fickert, Karl (engineer) 234
Firth-Vickers (company) 22, 27
FKFS 185, 221
Focke-Wulf (company) 244
Forsyth, Capt. A.G. (designer) 163
Fortescue, P. (designer) 137
Foster, Squ Ldr B.H.D. 98
Fottinger, Prof Hermann 170
Fozzard, J.W. (designer) 155
Franz, Anselm 187–189, 192
Freda compressor 19, 46
French Navy 73
Freytag (engineer) 243
Freytag, Fritz (designer) 253
Friedrich, Rudolph 182, 187–189, 199, 231
Fritz, Dr. 221

Galland, Gen Lt Adolf (pilot) 191, 215, 218
Genders, Sqn Ldr 73
Gloster Aircraft Company 28, 32, 50
Goering, Hermann 174, 175
Gogerly, Fl Off Bruce (pilot) 75
Gosling, Flt Lt C. 63
Gosslau, Fritz 250
Göttingen University 168
Graalmann, Martin (engineer) 187
Graves, Mike (pilot) 120
Greenstead, Brian (pilot) 66
Greenwood, Eric (pilot) 105
Grierson, John (pilot) 33
Griffith, Dr. A.A. 11–13, 20, 89,113, 115, 118, 150
Gündermann, Wilhelm (engineer) 169, 170
Günter, Siegfried (designer) 169, 216
Günter, Walter (designer) 169

H. Wiggin (company) 27, 35
Hadfield Steel (company) 27
Hagen, Dr Hermann 189, 200, 204, 256
Hahn, Max (engineer) 168–171
Hale, Sgt George (pilot) 75
Halford, Maj Frank Bernard 50, 77, 158
Harris, Dr. 41
Hartenstein, Werner 182
Haworth, Lionel (designer) 106, 111
Hazelden, H.G. (pilot) 136
Heinkel (company) 66
Heinkel, Prof Ernst 168–174, 247, 249
Heinkel-Hirth (company) 220–226
Helmbold, Heinrich (designer) 169
Helmschrott, Josef (engineer) 213

269

INDEX

Henschel (company) 187
Herriott, John 55, 57
Hertel, Prof. Heinrich 82, 173
Hertzog (engineer) 243
High Duty Alloys company 19, 49, 58
Hille, Fritz 200
Hindustan Aeronautics (company) 251
Hirth-Motorenwerke 178, 182
Hispano Aviacion (company) 251
Hitler, Adolf 167, 174, 177, 192, 212, 217
Hives, Ernest 9, 54, 89, 91, 104, 147
HMS *Albion* 103, 121
HMS *Antrim* 166
HMS *Ark Royal* 103
HMS *Eagle* 121, 122
HMS *Illustrious* 87, 128
HMS *Ocean* 70
HMS *Plym* 98
HMS *Victorious* 82
Hoffmann, Ludwig (pilot) 219
Holzwarth, Hans 137, 167, 226
Hooker, Sir Stanley 44, 54, 55, 85, 89, 91, 142, 148, 151, 152, 154
Howell, A.R. 'Taffy' 14

Imperial College 13, 53, 54
India/Pakistan conflict 75, 154
INI (company) 247

J.M.Voith (company) 221
Jägdverband 44 (JV44) 215, 218
Janssen, Ubbo (pilot) 207
Jessop (company) 27
Johnson, Claude 54
Johnson, W.E.P. 39
Joseph Lucas (company) 35, 53, 55–58, 81, 118
Junkers (company) 82, 182, 186–197, 227–229
Junkers, Klaus 186
Junkers, Prof Hugo 186
Jüttner, SS Obergruppenführer Max 217

Kaiser-Wilhelm-Institut, Göttingen 82, 168
Kamm, Prof Wunibald I.M. 185, 251
Kammler, SS Gen Hans 167, 217, 218
Kamps, Dr. 241
Kappus, Peter G. (pilot) 209, 210, 231
Kármán (von), Theodore 168
Keller, Curt 187
Kennedy, H.V. (pilot) 39
Knemeyer, Siegfried 177, 225
Kommando Nowotny (Luftwaffe) 62, 213
Koppenburg, Klaus 186
Korean War 75, 87, 94, 99, 101, 104, 121, 130, 135, 142
Kraft, Prof. 189
Krauss, Joseph 230
Krupp (company) 197
Kruppe, Dipl.-Ing 201
Kruschev, Nikita 255
Küchemann, Dr. 204
Kuznetsov, Nikolai 251

L'Orange (company) 248, 249
Laidlaw, Drew & Co. 21
Laukhuff, Dr. 241
Lawrence, Peter (pilot) 133
Leduc, René 40
Lees, Wg Cdr G. 39
Lehmann, Willi (pilot) 255
Leist, Dr.-Ing Karl 239, 240, 242
LFA 204, 230
Lichte, Dr.-Ing August 250
Lickley, R.L. (designer) 104
Lindsey, W.H. (designer) 118
Lithgow, Mike (pilot) 87, 101
Loehner, Kurt 199

Lombard, Adrian A. 55, 85, 89, 106, 113
Lubbock, I. 26
Lucht, Wilhelm 174, 178
Lusser, Robert (designer) 178, 181
Lynch, Bernard 66

Mader, Otto 187, 189
Mansell, Harvey (designer) 140
Mansell, Stanley (designer) 137, 149
Marchant, Charles (designer) 140, 149
Marshall Plan 247
Martin Baker (company) 66, 88
Martin, Dicky (pilot) 133
Martin, James 66
Mauch, Hans A. (engineer) 175, 178, 187, 198, 239
McDowall, Wg Cdr (pilot) 63
McFarlane, Lt Bruce (pilot) 121
McKenna, Frank 56
Messerschmitt, Prof Willy 169, 175, 251
Metropolitan-Vickers (company) 14, 18, 19, 41, 45–49
Milch, Erhard 167, 174, 177, 182, 221
Miles, F.G. 38
Miles, George 38, 39
Ministry of Aircraft Production (MAP) 26, 28, 34, 37, 39, 49, 55, 56, 68
Ministry of Supply (MoS) 77, 80, 98, 99, 103, 108, 113, 116, 133, 136, 145, 152, 153, 155, 159, 163, 165
Morgan (pilot) 101
Moult, E.S. 50, 68
Mueller, Dr. Max Adolf 178, 182, 186–189, 199, 220, 221, 245, 246
Müller, Dr Alfred 237, 245

Nagel, Dr. 201
Nallinger, Fritz 239, 242, 256
Napier (company) 158–166
Napier, David (engineer) 158
National Gas Turbine Establishment (NGTE) 37, 44
National Physical Laboratory (NPL) 12, 38
NATO 155
Nimonic alloys 27
Nuclear bombs (British) 98, 99

O.T.Falk & Partners 21
Oestrich, Dr.-Ing Hermann 198, 207, 208
Ohain (von), Hans-Joachim Pabst (physicist) 167–171, 220, 221, 226, 247
Oswatitsch, Dr. Klaus 82, 104
Owner, Frank M. 137, 149, 152

Pabst, Otto 244
Page, F.W. (designer) 104
Papper (Pepler?), Wilhelm 187
Parsons, Charles A 11, 14, 15
Pegg, A.J. 'Bill' 145, 146
Penn, A.J. (engineer) 159
Penrose, Harald (pilot) 120
Peter, Gotthold (pilot) 216
Petter, W.E.W. (designer) 93, 94, 104, 124, 150
Phillips, Mary (Whittle's secretary) 39
Pike, Geoffrey (pilot) 70
Plumb, Bob (engineer) 142, 147
Pohl, Prof Robert W. 168, 170
Polaris submarine 151
Porsche (company) 245, 246
Porsche, Ferdinand 175, 238, 245
Power Jets, company 9, 13, 19, 21 26, 28, 34–37, 44, 54, 55
Prandtl, Ludwig 12, 168, 176
Prestel, Karl 256
Pye, Dr.D.R. 19, 28

Quill, Jeffrey (pilot)

Rechlin, E-Stelle 174, 180, 209, 218
Regner, Hans (designer) 173
Reitsch, Hanna (aviatrix) 198
Reuter, Hermann 230
Ricardo, Harry R. (engineer) 159
RLM 167, 171, 172, 174, 175, 178, 184, 187, 189, 201, 209, 220, 228, 241, 256
RNAS Ford 70, 121
Rohrbach, August 167
Rolls, Charles S. 54
Rolls-Royce (company) 54–60, 85–104, 105–117
Roskopf, Hans (designer) 200, 204
Rover (car company) 34, 35, 54
Royal Aircraft Establishment (RAE) 12–14, 28, 33, 37, 38, 46, 49, 62, 70, 83, 104, 118
Royal Flying Corps 50
Royal Navy 75, 76, 79, 82
Royce, Henry 54
Russell, Dr. A.E. (designer) 145
Ruth compressor 19

Sachse, Helmut 256
Sammons, Herbert (designer) 158, 159
Samuelson, F 21
Sandys, Duncan, MP 154
Santa Fe submarine 166
Sarah compressor 19, 118
Saur, Karl Otto 215
Sawert, Dr. 201
Saxton, Bill (designer) 118, 122, 123, 124
Sayer, P.E.G. (pilot) 30, 31, 32
Schaaf, Dr. Wilhelm 200
Schäfer, Fritz (pilot) 180, 247
Scheinost, Dr.-Ing Rudolf 253
Schelp, Helmut 167, 171, 184, 198, 220, 239, 256
Schenk (pilot) 181
Scherenberg, Prof Fritz 256
Schmidt, A.F. 221
Schmidt, Christian 249
Schmidt, Paul 175
Schmitz, Christian 256
Schneider, Hermann 230
School of Gas Turbine Technology 37
Schrader, Wg Cdr 63
Schuldt, Bruno 251
Schwärzler, Karl (designer) 216
Science Museum, London 33
Scott, J.E.D. (pilot) 66
Schwärzler, Karl 249
Seitz, Rudolf (designer) 212
Selle (pilot) 207
Shannon, Dr. I.S. 46
Shepherd, Capt R.T. (pilot) 85, 93
Sinai Campaign 75
Slade, Gp Capt (Pilot) 127
Smith, Dr. D.M. 46
SNECMA (company) 250
Soestmeyer, Dr Christoph 201
Special alloys 27
Speer, Albert 200, 217
Spieser, Dr. 241
Stabernack, Hans 187
Staege (pilot) 209
Stafford, R.S. (designer) 136
Stalin, Joseph 251
Stein, Dr. Wolfgang 189, 228
Stern, Dr. 12
Stiefel, Dr. 141
Stoeckicht, Wilhelm 235
Stoffregen, Dr. 200
Suez Crisis 75, 76, 99, 121
Summers, J. 'Mutt' (pilot) 98, 108

TAL 201
Tamblin, W.A. (designer) 73
Tank, Kurt (designer) 244, 251

270

Technisches Amt 167, 175, 198, 220, 228, 230, 232, 235, 237, 238, 239
Tennant, E.A. (pilot) 153
Thomas, A. (designer) 118
Tinling, J.C.B. 21, 39
Tizzard, Sir Henry 132, 50
Todt Organisation 222
Triple Crown (world speed records) 104
Twiss, Peter (pilot) 104, 128
Tyson, Geoffrey (pilot) 48

Udet, Ernst 167, 174–177
USAAF 8th Air Force 63

VDM (company) 230, 235
VEB (company) 253
Vickers (company) 56

W.G.Armstrong Whitworth (company) 67
Wade, Sqn Ldr T.S. 87
Wagner, Herbert 167, 178, 186, 187, 189
Waldmann, Emil 176, 256
Waldmann, Karl 230
Walker, D.N. 39
Walter, Helmut 175
Warsitz, Erich (pilot) 171, 173, 174
Waterton, W.A. (pilot) 133
Watson, Dr. 142
Watson, Dr.E.A. 55, 56
Weinrich, Helmut (engineer) 175, 198
Wendel, Fritz (pilot) 202
Wertheim, Joseph 167
Westland (company) 163, 166
Whitehead (designer) 152
Whitehead, Lt Cdr D. (pilot) 82
Whittle, Sir Frank 11, 18, 20, 34, 38, 44, 49, 50, 54, 58, 64, 120, 137, 167, 170, 171, 262
Wilde, Geoffrey 57, 91
Wilkes, Spencer 55
William Prym (company) 197
Williams, R. Dudley 21, 39
Wilson, H.J. (pilot) 62
Wimpenny, John (designer) 73
WMF (company) 199, 221
Wolff, Harald 221, 226
Wunderwaffen 217

Zborowski (von), Helmut Graf 200, 232, 256
Zobl, Dr H. 201

Aircraft

A-32 Lansen fighter 99
Alizé airliner 111
Ambassador airliner 111
Antonov An-12 test bed 251
Apollo airliner 108, 123
Ar 234 bomber 191, 192, 196, 205–209, 211–215, 219, 243
Argosy transport 111
Ashton test bed 88, 93, 115, 153
Athena trainer 123
Atlantic ASR 112
Attacker fighter 86
AW.52 experimental 88

B-17 Flying Fortress bomber 56
B-29 Superfortress bomber 67, 98
B-36 Peacemaker bomber 67
B-47 Stratojet bomber 67
Balliol trainer 106, 108, 123
Balzac V/STOL aircraft 117
BB 152 airliner 251, 253–255
Beaufighter 137
Belfast transport 112
Bell UH-1 helicopter test bed 256

Bell X-1 research 40
Belvedere helicopter 166
Bf 110 test bed 189, 204
Blenheim 137
Blue Steel missile 151
Boeing 707 airliner 79, 113–115
Boxkite 137
BP.111A experimental 88
BP.120 experimental 88
Brabazon airliner 144
Bristol Type 188 research aircraft 82, 83
Britannia airliner 111, 145–148
Buccaneer bomber 81
Buckingham bomber 137

CA-27 Sabre fighter 99
Canberra bomber 67, 93–95
Canberra test bed 95, 96, 130, 134
Caravelle airliner 103
CF-100 fighter 97
Comet airliner 67, 73, 77–80, 96, 103
Concorde supersonic airliner 104, 124
Convair 340 airliner 162
Convair 660 airliner 111

Defiant fighter 66
DH.100 Vampire fighter 50, 51, 68–73
DH.108 Swallow research 71, 73
DH.110 fighter 73, 97
DH.112 Sea Venom fighter 71, 74
DH.4 50
DH.9 50
Douglas DC-8 airliner 79, 113–115

E.28/39 experimental 28, 29, 31–35, 40, 173
EF 150 bomber 251

F 27 airliner 111
F.2B fighter 137
F.D.2 research aircraft 102–104
F-80 Shooting Star fighter 75
F-84 Thunderjet fighter bomber 67, 75
F-86 Sabre fighter 67, 75, 99
F-100 Super Sabre fighter 104
F-104 fighter 38
F-111 bomber 82
F9/40 (Meteor forerunner) 36, 38, 47, 51, 53, 56, 57, 61, 93
Firestreak missile 124
Flying Bedstead 88, 90, 91, 115
Fw 190 fighter 244
Fw 44 Steiglitz 198

G.82 trainer 88
Gannet AS 122, 125–128
Gnat fighter 152, 155
Gyrodyne convertiplane 163

H.D. 45 airliner 104
H-30 TS turbine-glider 256
HA 200 trainer 251
HA 300 interceptor 251
Hastings test bed 130
He 011 fighter 249
He 111 test bed 174, 181, 249
He 118 test bed 171
He 162 fighter 66, 209, 210, 215–218
He 176 rocket aircraft 173, 174
He 177 Greif bomber 237
He 178 experimental 171–176
He 219 test bed 210
He 219A Uhu 66
He 280 fighter 178, 180, 181
He 343 bomber 225
Herald airliner 111
Hermes airliner 139

Hornet fighter 50
HP.88 research aircraft 88
HP.115 experimental 124
Hs 125 executive jet 124
Hs 293 guided missile 187
Hs 748 airliner 111
Hunter fighter 97, 99, 100, 130, 136

Il-14 airliner 253
Il-28 bomber 250
Il-28 test bed 253

J-35 Draken fighter 104, 124
Javelin fighter 67, 130–133
Ju 287 experimental 218, 219, 251
Ju 288 bomber 251
Ju 52 test bed 209
Ju 88 test bed 193, 210, 222

Lancaster test bed 46, 48, 106, 118, 120, 123
Lancastrian test bed 68, 77, 85, 86, 93, 95, 129
Lightning interceptor 104
Lincoln bomber 67
Lincoln test bed 118, 140, 141

M.3B Falcon 39
M.52 supersonic project 38, 43, 103, 104
Marathon airliner 123
Me 163 interceptor 63
Me 262 fighter 62, 63, 177, 189, 191, 192, 202, 203, 207, 209–215, 219
Me 264 bomber 237
Messerschmitt P.1101 fighter 225
Meteor fighter 34–36, 48, 56–58, 60–67, 88, 89, 211, 212
Meteor test bed 93
Midge fighter 124, 153
MiG-15 fighter 75, 250
Mosquito 50, 67, 68, 70
Mystére IV fighter 75, 99

P.1 interceptor 103
P.1121 fighter 81
Pathfinder bomber 113
Percival helicopter 164
PIK-A experimental 124
Polaris missile 151
Pulqui fighter 88

Rotodyne convertiplane 163, 164

S.14 Mach Trainer II 88
S.O.4050 Vautour bomber 104
S-58 helicopter 165, 166
SA.4 Sperrin bomber 80, 97
SAAB 210 experimental 123
SC-1 V/STOL aircraft 104, 116
Scimitar fighter 103
Scout fighter 137
Sea Fury fighter 67
Sea Hawk fighter 86
Sea King helicopter 84
Sea Vixen fighter 101–103
Seamew AS 123
Shackleton patrol aircraft 124, 161
Skybolt missile 151
Spitfire fighter 120
SR.45 Princess flying boat 145
SR.53 interceptor 83, 124
SR.A/1 flying boat 48
Strikemaster 124
Super Mystére B-1 fighter 104
Swift fighter 97, 99–101

T-33 Silver Star trainer 88
Ta 183 fighter 225
Tempest fighter 158

271

INDEX

Tennant, E.A. (pilot) 124
Transall C-160 transport 112
Trent turboprop Meteor 67
TSR-2 bomber 82
Tudor test bed 88, 90
Typhoon fighter bomber 93, 158

V1 flying bomb 62, 63, 75, 158, 181, 245
V-1000 transport 113
V2 rocket 38
Valiant V-bomber 80, 96–99, 113
Vanguard airliner 111
Varsity test bed 162
VC-10 airliner 114, 115
VC-7 airliner 113
Venom fighter 71–73
Victor V-bomber 83, 113, 135, 136
Viking test bed 77, 87, 88
Viscount airliner 108–110, 123
Viscount test bed 89, 91
Vulcan test bed 113
Vulcan V-bomber 150, 152, 153

Wellington test bed 36, 37, 55, 56, 57, 106, 108
Wessex helicopter 165, 166
Westminster helicopter 163, 164
Wyvern fighter 120–122

XF-109 VTOL aircraft 104
XF9F-2 Panther fighter 88
XP-80 Shooting Star fighter 51, 70

Engines

Adder turbojet 123, 124
ASP turboprop 118, 119
ASX turbojet 118–120
Avon turbojet 79, 85–104, 128

B.23 Welland turbojet 55–57
B.37 Derwent I turbojet 57, 58, 153
B.H.P. piston engine 50
BE.17 expendable turbojet 152
Beryl turbojet 48
BMW 109–002 turbojet 176, 198, 199
BMW 109–003 turbojet 181, 184, 203, 209, 230–233
BMW 109–018 turbojet 233, 234
BMW 109–028 turboprop 235, 236
BMW 6002 gas turbine 255
BMW 801 piston engine 54
BMW F.9225 turbojet 199, 200
BMW P.3302 turbojet 200–203
BMW P.3304 turbojet 199
BMW P.3306 turbojet 225, 237, 238
BMW P.3307 turbojet 238
By-pass turbojets, Whittle 41

Centaurus piston engine 137
Clyde turboprop 105, 120, 121
Conway by-pass turbojet 112–115
CR-1 turbojet
Cyclone piston engine 166

D.11 Doris turbine engine 16, 17
Daimler-Benz 109–007 ZTL turbofan 49, 240
Daimler-Benz 109–016 turbojet 242, 243
Daimler-Benz P.100 turbojet 242
Daimler-Benz P.101 turbojet 243
Daimler-Benz P.102 turbojet 243
Dart turboprop 105–111, 123

Derwent I turbojet 57, 58, 153
Ducted fans, Whittle 41

E 200 turbojet 251
E 300 turbojet 250
Eagle piston engine 54, 120
Eland turboprop 161–163

F.1 turbojet 17, 19
F.1A turbojet 17
F.2 turbojet 19, 45, 46
F.2/1 turbojet 46, 47
F.2/2 turbojet 47
F.2/3 turbojet 47
F.2/4 Beryl turbojet 48
F.2/4A turbojet 49
F.3 turbofan 49
F.5 turbofan 49
F.9 turbojet (Sapphire forerunner) 48
Falcon piston engine 54

Gazelle turboshaft 165
Ghost turbojet 68, 78
Gnome (T-58) turboprop 83, 84, (coupled) 166
Goblin II turbojet 52, 53, 70
Goblin IV turbojet 53
Goblin Mk 35 turbojet 53
Goblin turbojet 51, 52
Griffith engines 18
Griffon piston engine 54
Gyron Junior turbojet 81
Gyron turbojet 79, 80

H.1 turbojet (later, Goblin) 50
Hawk piston engine 54
He 004 turbojet 251
He 006 turbojet 251
Hercules piston engine 137
HeS 011 (109–011) turbojet 184, 220–226
HeS 053 turbojet 249
HeS 10 (109–010) ducted fan 182
HeS 2 turbojet 170, 171
HeS 3 turbojet 171, 172, 174
HeS 30 (109–006) turbojet 178, 182–184
HeS 40 turbojet 184, 185
HeS 50 ducted fan motor engine 185
HeS 6 turbojet 174
HeS 60 ducted fan motor engine 185
HeS 8 (109–001) turbojet 174–181
HeS 9 turbojet 182
H-H 109–021 turboprop 226, 243
Hornet piston engine 198

INI-11 turbojet 247–249

J-42 turbojet 88, 89
J-65 turbojet 130
Janus turboprop 139
Ju 015 turbojet 251
Jumo 222 piston engine 251
Junkers 109–004 turbojet 176, 181, 184, 189–197, 209, 227, 228
Junkers 109–012 turbojet 227–229
Junkers 109–022 turboprop 229
Jupiter piston engine 137

Kestrel piston engine 54

L.R.1 turboprop and turbojet 36, 43, 44
Lion piston engine 158

Mamba turboprop 122, 125
Medway by-pass turbojet 115

Merlin piston engine 54–56

Naiad turboprop 161, (coupled) 160–162
Nene turbojet 73, 85–104
NK-12 turboprop 256
No.3 thrust augmenter 42, 43
Nomad compound engine 158–161
Nymph turboprop 162

Olympus turbojet 149, 150
Orion turboprop 148, 149
Orpheus turbojet 152–154, 251
Oryx gas generator 164, 165

Pegasus piston engine 137
Pegasus turbojet 154–157
Phoebus turbojet 140, 141
Pirna 014 turbojet 249, 252, 253
Pirna 015 turbojet 256
Pirna 018 turboprop 255, 256
Pirna 019 turbojet 256
Porsche Type 300 (109–005) turbojet 245
Proteus turboprop 112, 140, 144
Pulsejet (Argus 109–014) 181
Pulsejet (Schmidt) 175
Puma piston engine 50, 118
Python turboprop 118, 119–121

RAE dispersed compound engine 17
RB-108 lift engine 115–117
RD-9 turbojet 254, 255
RTO/RT1 turbojet 167, 186–188

Sabre piston engine 158
Sapphire turbojet 96, 99, 104, 127, 128, 133–136
Saturn turbojet 153
Scorpion rocket motor 166
Snarler rocket motor 87
Spectre rocket motor 83
Spey by-pass turbojet 115
Spey Junior by-pass turbojet 115
Sprite rocket motor 78
Super Sprite rocket motor 83

Tay by-pass turbojet 115
Tay turbojet 89, 91, 108
Theseus turboprop 137, 138
Trent turboprop 105, 106
Turbo-Wasp JT-6B turbojet 89
Tuttlingen engine 226
Tweed turbojet 112
Tyne turboprop 110–112

Verdon turbojet 89
Viper turbojet 123, 124
Vulture piston engine 55

W.1 turbojet 28, 35
W.1/3 turbojet 36
W.1A turbojet 33–35
W.1X turbojet 28
W.2 turbojet 34, 35, 54, 56
W.2/4 turbojet 34
W.2/500 turbojet 33, 35, 36, 43, 55
W.2/700 turbojet 36, 38, 43, 44
W.2B turbojet 34–36, 56
W.2B/23 turbojet 36, 37, 55
Walter rocket motor 168
Wasp piston engine 198
Welland turbojet 51
WR.1 turbojet 55
WU turbojet 22, 23